# Transport Phenomena and Kinetic Theory

## Applications to Gases, Semiconductors, Photons, and Biological Systems

Carlo Cercignani
Ester Gabetta
*Editors*

Birkhäuser
Boston • Basel • Berlin

Carlo Cercignani
Dipartimento di Matematica
Politecnico di Torino
32 Piazza Leonardo da Vinci
20133 Milano
Italy

Ester Gabetta
Dipartimento di Matematica "F. Castorati"
Università degli Studi di Pavia
1 via Ferrata
27100 Pavia
Italy

Mathematics Subject Classification: 82B31, 82B40, 82C22, 82C40, 82D75, 85A25, 92C60, 92D25

**Library of Congress Control Number:** 2006934735

ISBN-10: 0-8176-4489-X          e-ISBN-10: 0-8176-4554-3
ISBN-13: 978-0-8176-4489-5      e-ISBN-13: 978-0-8176-4554-0

Printed on acid-free paper.

©2007 Birkhäuser Boston          *Birkhäuser*

9 8 7 6 5 4 3 2 1

*www.birkhauser.com*                                                                    (TXQ/EB)

# Preface

This volume aims to provide an overview of some recent developments of mathematical kinetic theory focused on its application in modelling complex systems in various fields of applied sciences.

Mathematical kinetic theory is essentially based on the Boltzmann equation, which describes the evolution, possibly far from equilibrium, of a class of particles modelled as point masses. The equation defines the evolution in time and space of the distribution function over the possible microscopic states of the test particle, classically position and velocity. The test particle is subject to pair collisions with the field particles.

The interested reader can find in the book, *Theory and Application of the Boltzmann Equation*, by C. Cercignani, R. Illner, and M. Pulvirenti, Springer, Heidelberg, 1993, all necessary knowledge of the physics and mathematical topics related to this celebrated model of non-equilibrium statistical mechanics.

Another important model of mathematical kinetic theory is the Vlasov equation, where interactions between particles are not specifically collisions, but mean field actions of the field particles over the test particle. The model defines again an evolution equation for the one-particle distribution function over the microscopic state of the test particle.

The two models briefly mentioned above can be regarded as the fundamental models of mathematical kinetic theory and the essential background offered from the kinetic theory for classical particles towards the modelling of large systems of several particles undergoing non classical interactions.

Applied mathematicians have been constantly attracted by various challenging problems posed by this equation: traditionally, the initial value problem in unbounded domains, the initial-boundary value problem in closed domains or past obstacles, and the asymptotic theory in the limit when a hydrodynamic macroscopic description is possible.

In recent years, there has been a constantly growing interest in several developments (including technical modifications) of the above model to analyze the behavior of complex systems of interest in applied sciences.

One important field is the analysis of large systems of quantum particles with applications in modelling semiconductor devices and, more generally, in the nano-sciences. Additional fields of application, among others, refer to traffic flow modelling in roads and internet networks, multicellular systems in biology, and modelling of the dynamics of swarms and crowds.

This book follows the tradition of the Birkhäuser Series on *Modelling and Simulation in Science and Technology*. Indeed, it is the third of three edited books published following analogous aims.

This book is intended for both scientists and engineers operating in applied sciences with interest in modelling real systems using differential or operator equations and also university students who have a good knowledge of fundamental mathematics and differential calculus, at the level of master courses, and are interested in the application of mathematics in technology and applied sciences in general.

## Content

The various chapters of this book are authored by applied mathematicians active in research fields related to conceptual aspects and applications of mathematical kinetic theory.

The content ranges from theoretical aspects of kinetic theory to various applications. The book is organized into three parts.

The first part concerns some fundamental aspects of the Boltzmann equation; specifically, it refers to fundamental topics: existence of solutions and trend to equilibrium for the Boltzmann equation and velocity averaging methods for the Vlasov–Maxwell system.

The second part deals with modelling of semiconductor devices and related applications and computational topics. The various chapters attempt to cover four different relevant aspects: modelling, optimization, inverse problems, and computational solvers.

The third part covers a variety of miscellaneous applications in physics and natural sciences with the aim of offering a range of very different conceivable developments of mathematical kinetic theory for application.

We stress that the overall presentation attempts to cover not only modelling aspects and qualitative analysis of mathematical problems generated by applications of models, but also inverse problems which lead to a detailed assessment of models in connection with their applications, and to computational problems which lead to an effective link of models to the analysis or real world systems.

In detail, the contents are as follows.

## Part I: Analytic Aspects of the Boltzmann Equation

- Chapter 1, by Carlo Cercignani, provides a detailed survey on the existence theory for the initial value problem for the Boltzmann equation for classical particles. The analysis is mainly related to the equation for Maxwell molecules without any cutoff when the solution depends on just one space coordinate.

- Chapter 2, by Ester Gabetta, deals with a unified presentation of some recent results on the trend to equilibrium both for the solution of a Maxwellian pseudomolecules model of the Boltzmann equation and for the solution of the Kac equation. The methodological approach is innovative with respect to the traditional one as it pursues the optimal convergence rate related to suitable assumptions on the initial data.
- Chapter 3, by Francois Golse, deals with velocity averaging methods for the Vlasov–Maxwell system, namely a coupled system of hyperbolic-type equations, specifically a Vlasov equation for particles subject to a Maxwell-type field. Velocity averaging methods enlighten the qualitative properties of the solution, providing useful information for the application of computational methods.

## Part II: Modeling Applications, Inverse and Computational Problems in Quantum Kinetic Theory

- Chapter 4, by Luigi Barletti, Lucio Demeio, and Giovanni Frosali, deals with a review of multiband quantum transport models essentially divided into two classes: Schrödinger-based models and Wigner function-based (or density matrix-based) models. The former aim at the calculation of the wave function for the system or device under study. The latter involve statistics or transport theory concepts.
- Chapter 5, by Martin Burger, Michael Hinze, and Rene Pinnau, deals with optimization problems for dopant profiles in semiconductors with special attention to microelectronics devices. The analysis starts from modeling aspects and is developed up to computational aspects finalized to optimization objectives.
- Chapter 6, by A. Leitao. P.A. Markowich, and J.P. Zubelli, deals with modeling aspects and solution methods related to the solution of inverse problems for semiconductors. The main objective is the identification of the dopant profiles from data obtained by different models connected to the voltage-current map. As in the previous chapter the analysis is developed up to the exposition of computational techniques.
- Chapter 7, by M.J. Cáceres, J.A. Carrillo, I.M. Gamba, A. Majorana, and C.-W. Shu, deals with the development of the Boltzmann transport equation (BTE) for semiconductors in the semiclassical approximation. Various deterministic solvers are reported with detailed references to the existing literature and are critically analyzed, offering to the interested reader some useful overall information.

## Part III: Miscellaneous Applications in Physics and Natural Sciences

- Chapter 8, by Nicola Bellomo, Abdelghani Bellouquid, and Marcello Delitala, deals with the modeling of complex multicellular biological systems, focusing on immune competition and on the derivation of macroscopic equations describing the behavior of biological tissues. The mathematical approach looks at the development of methods suitable for describing the behavior of living matter.

- Chapter 9, by Guido Manzi and Rossana Marra, deals with the kinetic modeling of a fluid where two phases are present separated by a sharp layer which moves according to the dynamics of the system. The kinetic model refers to the Fokker–Planck equation, while the application looks at various interesting problems related to gas-fluid dynamics.

- Chapter 10, by Weizhu Bao, proposes a general review of various analytic and computational studies on the ground states and dynamics in rotating Bose–Einstein condensates. The first part of the chapter deals with modeling and analytic topics. Specifically, it refers to the Gross–Pitaevskii equation with an angular momentum rotation term and develops models related to semiclassical scalings and geometrical optics. The existence and nonexistence of the ground state is analyzed. The second part deals with the development of computational schemes based on the backward Euler method for time and second-order centered finite differences for spatial derivatives. Numerical simulations are compared with some particular analytic solutions.

- Chapter 11, by Aldo Belleni-Morante, deals with the modeling of photon transport in an interstellar cloud using the equations of mathematical kinetic theory. Specifically, two inverse problems are dealt with, finalized to a proper assessment of the model. The first problem refers to the identification of a time- and space-dependent source, the second one to the identification of the time-dependent cross section of the photons.

## Closing Thoughts

The idea of editing a book on the developments and modeling applications of mathematical kinetic theory arose during a workshop with selected participation held in Mantova, an artistically attractive town on the river Mincio in Italy. From the lively discussion of perspective ideas, the editors decided to work on this project, selecting (not only from among the participants) authors suitable for proposing chapters consistent with the aims of the project.

Each chapter provides the *state of the art* in the field, an overview of the existing literature and, in some cases, various research perspectives.

The editors are confident that the overall content provides applied mathematicians with an overview useful for research activity in this attractive and challenging field of applied mathematics.

*Carlo Cercignani and Ester Gabetta*

# Contents

# List of Contributors

**Carlo Cercignani**
Dipartimento di Matematica,
Politecnico di Milano
Piazza L. da Vinci 32
20133 Milano, Italy
carlo.cercignani@polimi.it

**Ester Gabetta**
Dipartimento di Matematica
Università degli Studi di Pavia
Via Ferrata, 1
27100 Pavia, Italy
ester.gabetta@unipv.it

**François Golse**
Université Paris 7 & Laboratoire
Jacques-Louis Lions
Boîte courrier 187
F75252 Paris cedex 05, France
golse@math.jussieu.fr

**Luigi Barletti**
Dipartimento di Matematica "U.
Dini"
Università degli Studi di Firenze
Viale Morgagni 67/A
50134 Firenze, Italy
barletti@math.unifi.it

**Lucio Demeio**
Dipartimento di Scienze Matematiche
Università Politecnica delle Marche
Via Brecce Bianche 1,
60131 Ancona, Italy
demeio@dipmat.unian.it

**Giovanni Frosali**
Dipartimento di Matematica
"G.Sansone"
Università di Firenze
Via S.Marta 3
50139 Firenze, Italy
giovanni.frosali@unifi.it

**Martin Burger**
Institut für Industriemathematik
Johannes Kepler Universität
Altenbergerstr. 69
A 4040 Linz, Austria
martin.burger@jku.at

**Michael Hinze**
Institut für Numerische Mathematik
Technische Universität Dresden
Willersbau, C318
D-01062 Dresden, Germany
hinze@math.tu-dresden.de

**Rene Pinnau**
Fachbereich Mathematik
Technische Universität Kaiser-
slautern
Erwin Schrödinger Str.
D-67663 Kaiserslautern, Germany
pinnau@mathematik.uni-kl.de

**A. Leitão**
Department of Mathematics
Federal University of St. Catarina
P.O. Box 476
88040-900 Florianopolis, Brazil
aleitao@mtm.ufsc.br

**P.A. Markowich**
Department of Mathematics,
University of Vienna
Boltzmanngasse 9
A-1090 Vienna, Austria
peter.markowich@univie.ac.at

**J.P. Zubelli**
IMPA
Estr. D. Castorina 110
22460-320 Rio de Janeiro, Brazil
zubelli@impa.br

**M.J. Cáceres**
Departamento de Matematica
Aplicada
Universidad de Granada
18071 Granada, Spain
caceresg@ugr.es

**J.A. Carrillo**
ICREA and Departament de
Matemàtiques
Universitat Autònoma de Barcelona
E-08193 Bellaterra, Spain
carrillo@mat.uab.es

**I.M. Gamba**
Department of Mathematics and
ICES
The University of Texas at Austin,
USA
gamba@math.utexas.edu

**A. Majorana**
Dipartimento di Matematica e
Informatica
Università di Catania
Catania, Italy
majorana@dmi.unict.it

**C.-W. Shu**
Division of Applied Mathematics
Brown University
Providence, RI 02912, USA
shu@dam.brown.edu

**Nicola Bellomo**
Department of Mathematics,
Politecnico
Torino, Italy
nicola.bellomo@polito.it

**Abdelghani Bellouquid**
University Cadi Ayyad
Ecole Nationale
des Sciences Appliquées
Safi, Maroc
bellouq2002@yahoo.fr

**Marcello Delitala**
Department of Mathematics,
Politecnico
Torino, Italy
marcello.delitala@polito.it

**Guido Manzi**
Max-Planck-Institut für Mathematik
in den Naturwissenschaften
04103 Leipzig, Germany
manzi@mis.mpg.de

**Rossana Marra**
Dipartimento di Fisica
Università di Roma Tor Vergata e
INFN
00133 Roma, Italy
marra@roma2.infn.it

**Weizhu Bao**
Department of Mathematics and
Center for Computational Science
and Engineering
National University of Singapore
117543, Singapore
bao@cz3.nus.edu.sg

**Aldo Belleni-Morante**
Dipartimento di Ingegneria Civile
Università degli Studi di Firenze
Via di S. Marta 3
50139 Firenze, Italy
abelleni@dicea.unifi.it

*Transport Phenomena and Kinetic Theory*

# Part I

# Analytic Aspects of the Boltzmann Equation

# 1

# Rigorous results for conservation equations and trend to equilibrium in space-inhomogeneous kinetic theory

Carlo Cercignani

Dipartimento di Matematica, Politecnico di Milano, Piazza L. da Vinci 32, 20133 Milano, Italy carlo.cercignani@polimi.it

## 1.1 Introduction

The well-posedness of the initial value problem for the Boltzmann equation means that we prove that there is a unique nonnegative solution preserving the energy and satisfying the entropy inequality, from a positive initial datum with finite energy and entropy. However, for general initial data, it is difficult, and until now not known, whether such a well-behaved solution can be constructed globally in time. The difficulty in doing this is obviously related to the nonlinearity of the collision operator and the apparent lack of conservation laws or a priori estimates preventing the solution from becoming singular in finite time.

A complete validity discussion for the Boltzmann equation will automatically contain existence and uniqueness results. Consequently, by Lanford's validity theorem, [18], we already have some existence and uniqueness theorems. Unfortunately, a validity proof involves several hard additional steps beyond existence and uniqueness. Therefore, the Boltzmann equation has been validated rigorously only in a few simple situations (locally in time [18] and globally for a rare gas cloud in all space [16, 17]).

Existence (and in some situations uniqueness) of solutions to the initial-value problem is known for a much larger variety of cases, and it is our purpose in the next section to survey these results.

## 1.2 A survey of the existence theory

The Boltzmann equation reads as follows:

$$\frac{\partial f}{\partial t} + \mathbf{v} \cdot \frac{\partial f}{\partial \mathbf{x}} = Q(f, f) \tag{1.2.1}$$

$$Q(f, f)(\mathbf{x}, \mathbf{v}, t) = \int \int B(\mathbf{n} \cdot (\mathbf{v} - \mathbf{v}_*), |\mathbf{v} - \mathbf{v}_*|) \, (f' f'_* - f f_*) \, \sin \theta \, d\theta d\phi \, d\mathbf{v}_*,$$

$$(1.2.2)$$

where $f = f(\mathbf{x}, \mathbf{v}, t)$ is the probability density of finding a gas molecule at $\mathbf{x}$ with velocity $\mathbf{v}$ at time $t$. We denote by $f_*$ $f(\mathbf{x}, \mathbf{v}_*, t)$, where $\mathbf{v}_*$ is the velocity of a partner in a collision, and $f' = f(\mathbf{x}, \mathbf{v}', t)$, $f'_* = f(\mathbf{x}, \mathbf{v}'_*, t)$, where

$$\mathbf{v}' = \mathbf{v} - \mathbf{n}[\mathbf{n} \cdot (\mathbf{v} - \mathbf{v}_*)]$$
$$\mathbf{v}'_* = \mathbf{v}_* + \mathbf{n}[\mathbf{n} \cdot (\mathbf{v} - \mathbf{v}_*)]. \qquad (1.2.3)$$

Here $\mathbf{n}$ is the unit vector associated with the angles $\theta$ and $\phi$. The quadratic operator $Q(., .)$ is called the collision operator.

The kernel $B$ depends on the molecular model. If the molecules are hard spheres of diameter $\sigma$, then

$$B(\mathbf{n} \cdot (\mathbf{v} - \mathbf{v}_*), |\mathbf{v} - \mathbf{v}_*|) = N\sigma^2 |\mathbf{v} - \mathbf{v}_*| \sin \theta \cos \theta. \qquad (1.2.4)$$

If the molecules are point masses interacting with a force varying as the $n$th inverse power of the distance, then

$$B(\mathbf{n} \cdot (\mathbf{v} - \mathbf{v}_*), |\mathbf{v} - \mathbf{v}_*|) = B(\theta)|\mathbf{V}|^{\frac{n-5}{n-1}}, \qquad (1.2.5)$$

where $B(\theta)$ is a nonelementary function of $\theta$ which for $\theta$ close to $\pi/2$ behaves as the power $-(n+1)/(n-1)$ of $|\pi/2 - \theta|$. In particular, for $n = 5$ one has the Maxwell molecules, for which the dependence on $V = |\mathbf{v} - \mathbf{v}_*|$ disappears. For a detailed explanation of the structure of the collision term, see Ref. 4, 8, or 12.

Sometimes the artifice of cutting the grazing collisions corresponding to small values of $|\theta - \frac{\pi}{2}|$ is used (angle cutoff). In this case one has both the advantage of being able to split the collision term and of preserving a relation of the form (1.2.5) for power-law potentials. This artifice is common in most existence theorems; when it is not stated otherwise, one considers either hard spheres or models with angular cutoff.

When the distribution function of a gas is not dependent on the space variables, the equation is considerably simplified. The collision operator is basically Lipschitz continuous in $L^1_+$ and the equation becomes globally solvable in time. Moreover, uniqueness, asymptotic behavior and a theory of classical solutions have been established. The theory for the spatially homogeneous Boltzmann equation originated in the early 1930s and can be considered rather complete; unfortunately the homogeneous case is hardly of interest for applications. The reader interested in this theory should consult Ref. 8.

The Boltzmann equation has a well-known family of equilibrium solutions, the Maxwellian solutions

$$f = A \exp(-\beta |\mathbf{v} - \mathbf{u}|^2) \quad (A, \beta, \mathbf{u} \quad \text{constants}). \qquad (1.2.6)$$

If the solution is initially sufficiently close to a Maxwellian one, it is possible to prove that a solution can be constructed globally in time, and we have uniqueness and asymptotic behavior. The approach is based on the analysis of the linearized Boltzmann operator, which leads us to a differential inequality of the type

$$\frac{d}{dt}y \leq -ky + y^2,$$

where $y = y(t)$ is some norm of the deviation of the solution from the Maxwellian and $k$ is a positive number. Therefore, if $y(0)$ is sufficiently small, we can control the solution for all times. As we said, the basic ingredient is good control of the linearized Boltzmann operator. This theory is discussed in Refs. 8 and 10.

The case of perturbation of a vacuum is a consequence of the validity result, as mentioned in the previous section, together with the local existence theory. If the initial value is close to a homogeneous distribution, a solution starting from it can be constructed globally in time. Uniqueness and asymptotic behavior can also be proved. The main idea is explained in Ref. 8.

Except for the first one, all these results have a perturbation character. The knowledge of particular solutions helps to construct other solutions which are close to the original ones. The general initial value problem is poorly understood, although a significant and somewhat unexpected step was performed in the late 1980s. Consider an equation similar to the Boltzmann equation for which we have conservation of mass and energy and the $H$-theorem ($\int f \log f d\mathbf{x} d\mathbf{v}$ cannot grow in time). Denote by $f^\epsilon(t)$ the solutions. Here, $\epsilon$ is a regularization parameter such that the solutions formally converge to a solution of the Boltzmann equation in the limit $\epsilon \to 0$. The conservation laws yield the existence of a weak limit denoted by $f(t)$. However, since the collision operator is quadratic in $f$, it cannot be weakly continuous. Thus it does not follow by general arguments that $f(t)$ solves the Boltzmann equation. Nevertheless, some smoothness gained by the streaming operator gives enough compactness to prove that $f(t)$ actually solves the Boltzmann equation in the mild sense.

The method gives neither uniqueness nor energy conservation, but the entropy is seen to decrease along the solution trajectories. The trend to equilibrium in a periodic or specularly reflecting box can be proved, but the asymptotic Maxwellian is not uniquely determined.

Let us begin with some notation and a rather standard definition. Let $\Lambda f$ denote the left-hand side of the Boltzmann equation and

$$f^\#(\mathbf{x}, \mathbf{v}, t) = f(\mathbf{x} + \mathbf{v}t, \mathbf{v}, t).$$

**Definition 2.1** A measurable function $f = f(\mathbf{x}, \mathbf{v}, t)$ on $[0, \infty) \times \Re^3 \times \Re^3$ is a mild solution of the Boltzmann equation for the (measurable) initial value $f_0(\mathbf{x}, \mathbf{v})$ if for almost all $(\mathbf{x}, \mathbf{v})$ $Q_\pm(f, f)^\#(\mathbf{x}, \mathbf{v}, \cdot)$ are in $L^1_{\text{loc}}[0, \infty)$, and if for each $t \geq 0$

$$f^\#(\mathbf{x}, \mathbf{v}, t) = f_0(\mathbf{x}, \mathbf{v}) + \int_0^t Q(f, f)^\#(\mathbf{x}, \mathbf{v}, s)\, ds. \qquad (1.2.7)$$

One of the key ideas used by DiPerna and Lions to prove a general existence theorem was to relax the solution concept even further, such that the bounds provided by the energy conservation and the $H$ theorem could be put to the best use, and then to regain mild solutions via a limit procedure. They called the relaxed solution concept a "renormalized solution" and defined it in the following way.

**Definition 2.2** A function $f = f(\mathbf{x}, \mathbf{v}, t) \in L^1_+(\Re^+_{\mathrm{loc}} \times \Re^3 \times \Re^3)$ is called a renormalized solution of the Boltzmann equation if

$$\frac{Q_\pm(f, f)}{1 + f} \in L^1_{\mathrm{loc}}(\Re_+ \times \Re^3 \times \Re^3) \qquad (1.2.8)$$

and if for every Lipschitz continuous function $\beta : \Re_+ \rightarrow \Re$ which satisfies $|\beta'(t)| \le C/(1+t)$ for all $t \ge 0$ one has

$$\varLambda\beta(f) = \beta'(f)Q(f, f) \qquad (1.2.9)$$

in the sense of distributions.

DiPerna and Lions [12] noticed that renormalization would actually give mild solutions, according to the following lemma.

**Lemma 2.1** *Let* $f \in (L^1_{\mathrm{loc}} \times \Re^3 \times \Re^3)$. *If* $f$ *satisfies* (1.2.8) *and* (1.2.9) *with* $\beta(t) = \ln(1+t)$, *then* $f$ *is a mild solution of the Boltzmann equation. If* $f$ *is a mild solution of the Boltzmann equation and if* $Q_\pm(f, f)/(1 + f) \in L^1_{\mathrm{loc}}(\Re_+ \times \Re^3 \times \Re^3)$, *then* $f$ *is a renormalized solution.*

Their main result is the following.

**Theorem 2.2 (DiPerna and Lions [12])** *Suppose that* $f_0 \in L^1_+(\Re^3 \times \Re^3)$ *is such that*

$$\int_{\Re^3} \int_{\Re^3} f_0(1 + |\mathbf{x}|^2 + |\mathbf{v}|^2)\, d\mathbf{x}\, d\mathbf{v} < \infty$$

*and*

$$\int_{\Re^3} \int_{\Re^3} f_0 |\ln f_0|\, d\mathbf{x}\, d\mathbf{v} < \infty.$$

*Then there is a renormalized solution of the Boltzmann equation such that* $f \in C(\Re_+, L^1(\Re^3, \Re^3))$, $f\big|_{t=0} = f^0$.

The existence theorem of DiPerna and Lions is rightly considered a basic result of the mathematical theory of the Boltzmann equation. Unfortunately, it is far from providing a complete theory, since there is no proof of uniqueness;

in addition, there is no proof that energy is conserved and conservation of momentum can be proved only globally and not locally. More complete results concerning conservation equations can be obtained in the case of solutions depending on just one space coordinate [2, 5] and will be discussed later in this chapter.

## 1.3 The weak form of the collision operator and a useful identity

We introduce the weak form of the collision term, $Q(f, f)$. We shall henceforth use the latter notation for the operator defined by

$$
\int_{[0,T]\times[0,1]\times\mathbb{R}^3} Q(f,f)(x,\mathbf{v},t)\varphi(x,\mathbf{v},t)d\mathbf{v}dxdt
$$

$$
= \frac{1}{2}\int_{[0,T]\times[0,1]\times\mathbb{R}^3\times\mathbb{R}^3\times S^2} B(\mathbf{n}\cdot(\mathbf{v}-\mathbf{v}_*),|\mathbf{v}-\mathbf{v}_*|)
$$

$$
\times (\varphi' + \varphi'_* - \varphi - \varphi_*)ff_*d\mu dt \qquad (1.3.1)
$$

for any test function $\varphi(x,\mathbf{v},t)$ which is twice differentiable as a function of $\mathbf{v}$ with second derivatives uniformly bounded with respect to $x$ and $t$. In Eq. (1.3.1) we have used the notation

$$
d\mu = \sin\theta\, d\theta d\phi\, d\mathbf{v}_* d\mathbf{v} dx. \qquad (1.3.2)
$$

We remark that for classical solutions the above definition is known to be equivalent to that in (1.2.2). The main reason for introducing it is that it may produce weak solutions (as opposed to renormalized solutions in the sense of DiPerna and Lions [12]) even if the collision term is not necessarily in $L^1$. It is surprising that the weak form has not been used before the recent work of the author [2, 5].

For a function $f$ to be a weak solution of the Boltzmann equation, it must satisfy Eq. (1.2.1), where the derivatives in the left-hand side are distributional derivatives and the right-hand side has been defined above.

Now we want to prove that the definition of the weak form of the collision term makes sense for inverse power potentials without introducing an angular cutoff, as first shown in a paper of the author [5]. To this end we consider the following identity:

$$
\int_0^1 ds \int_0^1 dt \frac{\partial^2}{\partial s\partial t}[\varphi(\mathbf{v} + s(\mathbf{v}' - \mathbf{v}) + t(\mathbf{v}_* - \mathbf{v}'))]
$$

$$
= \int_0^1 ds \left\{ \frac{\partial}{\partial s}[\varphi(\mathbf{v} + s(\mathbf{v}' - \mathbf{v}) + (\mathbf{v}_* - \mathbf{v}'))] - \frac{\partial}{\partial s}[\varphi(\mathbf{v} + s(\mathbf{v}' - \mathbf{v}))] \right\}
$$

$$
= \varphi(\mathbf{v}_*) - \varphi(\mathbf{v}') - \varphi(\mathbf{v}'_*) + \varphi(\mathbf{v}). \qquad (1.3.3)
$$

Hence

$$\varphi(\mathbf{v}) + \varphi(\mathbf{v}_*) - \varphi(\mathbf{v}') - \varphi(\mathbf{v}'_*) = \int_0^1 ds \int_0^1 dt \sum_{i,j=1}^3 \frac{\partial^2 \varphi}{\partial v_i \partial v_j} (v'_i - v_i)(v^*_j - v'_j),$$

(1.3.4)

where the argument of $\varphi$ is $\mathbf{v} + s(\mathbf{v}' - \mathbf{v}) + t(\mathbf{v}_* - \mathbf{v}')$. If $K$ is an upper bound for the second derivatives, we obtain the following estimate:

$$|\varphi(\mathbf{v}) + \varphi(\mathbf{v}_*) - \varphi(\mathbf{v}') - \varphi(\mathbf{v}'_*)| \le 9K|\mathbf{v}' - \mathbf{v}||\mathbf{v}^* - \mathbf{v}'| \le |\mathbf{V}||\mathbf{n} \cdot \mathbf{V}|. \quad (1.3.5)$$

Hence if the kernel $B$ diverges for $\theta = \pi/2$, but $B \cos \theta$ is integrable, then the integral with respect to $\theta$ does not diverge. We recall that, if the intermolecular force varies as the $n$th inverse power of the distance, then $B(\mathbf{n} \cdot (\mathbf{v} - \mathbf{v}_*), |\mathbf{v} - \mathbf{v}_*|)$ varies as shown in (1.2.5).

We conclude that for power-law potentials, $B \cos \theta$ behaves as the power $-2/(n-1)$ of $|\pi/2 - \theta|$ and the definition of a weak solution given above makes sense for $n > 3$.

Henceforth we shall consider just Maxwell molecules, for which we state the main result of this section as follows.

**Lemma 3.1** *The following estimate holds:*

$$\left| \int_{\mathbb{R}^3} Q(f,f)(x,\mathbf{v},t)\varphi(x,\mathbf{v},t)d\mathbf{v} \right| \le \beta_0 K \int_{\mathbb{R}^3 \times \mathbb{R}^3} |\mathbf{V}|^2 f f_* d\mathbf{v} d\mathbf{v}_* , \qquad (1.3.6)$$

*where $K$ is un upper bound for the second derivatives of $\phi$ and $\beta_0$ a constant that only depends on molecular parameters.*

## 1.4 Basic estimates

In this section and the next we shall be concerned with the initial value problem for the nonlinear Boltzmann equation when the solution depends on just one space coordinate which might range from $-\infty$ to $+\infty$ or from 0 to 1 (with periodicity boundary conditions); for definiteness we stick to the latter case. Easy modifications, in the vein of Ref. 7, are necessary to deal with the case of different boundary conditions. The $x$-, $y$- and $z$- components of the velocity $\mathbf{v} \in \mathbb{R}^3$ will be denoted by $\xi, \eta$ and $\zeta$ respectively, and the Boltzmann equation reads

$$\frac{\partial f}{\partial t} + \xi \frac{\partial f}{\partial x} = Q(f,f) \qquad (1.4.1)$$

with

$$Q(f,f)(x,\mathbf{v},t) = \int\int B(\mathbf{n}\cdot(\mathbf{v}-\mathbf{v}_*),|\mathbf{v}-\mathbf{v}_*|)\,(f'f'_* - ff_*)\,\sin\theta\,d\theta d\phi\,d\mathbf{v}_*.$$

$$(1.4.2)$$

We now set out to prove the crucial estimates for the solution of the initial value problem and for the collision term. It is safe to assume that we deal with a sufficiently regular solution of the problem, because this can always be enforced by truncating the collision kernel and modifying the collision terms in the way described in earlier work, in particular in Ref. 12. If we obtain strong enough bounds on the solutions of such truncated problems, we can then extract a subsequence converging to a renormalized solution in the sense of DiPerna and Lions; and the bounds which we do get actually guarantee that this solution is then a solution in the weak sense defined above.

Consider now the functional

$$I[f](t) = \int_{x<y}\int_{\mathbb{R}^3}\int_{\mathbb{R}^3}(\xi-\xi_*)f(x,\mathbf{v},t)f(y,\mathbf{v}_*,t)\,d\mathbf{v}_*d\mathbf{v}dxdy, \qquad (1.4.3)$$

where the integral with respect to $x$ and $y$ is over the triangle $0 \le x < y \le 1$. This functional was in the one-dimensional discrete velocity context first introduced by Bony [2]. The use of this functional is the main reason why we have to restrict our work to one dimension; no functional with similar pleasant properties is known, at this time, in more than one dimension (for a discussion of this point see a recent paper of the author [6]). Notice that if we have bounds for the integral with respect to $x$ of $\rho = \int_{\mathbb{R}}^3 f(x,\mathbf{v},t)\,d\mathbf{v}$ and for

$$E(t) = \int_0^1\int|\mathbf{v}|^2 f\,d\mathbf{v}dx,$$

then we have control over the functional $I[f](t)$.

A short calculation with proper use of the collision invariants of the Boltzmann collision operator shows that

$$\frac{d}{dt}I[f] = -\int_{[0,1]}\int_{\mathbb{R}^3}\int_{\mathbb{R}^3}(\xi-\xi_*)^2 f(x,\mathbf{v}_*,t)f(x,\mathbf{v},t)\,d\mathbf{v}d\mathbf{v}_*dx. \qquad (1.4.4)$$

Notice that the first term on the right, apart from the factor $(\xi-\xi_*)^2$, has structural similarity to the collision term of the Boltzmann equation, and the integrand is nonnegative. This is the reason why the functional $I[f]$ is a powerful tool.

After integration from 0 to $T > 0$ and reorganizing,

$$\int_0^T\int_{[0,1]}\int_{\mathbb{R}^3}\int_{\mathbb{R}^3}(\xi-\xi_*)^2 f(x,\mathbf{v}_*,t)f(x,\mathbf{v},t)\,d\mathbf{v}d\mathbf{v}_*dxdt = I[f](0) - I[f](T).$$

$$(1.4.5)$$

According to a previous remark, the right-hand side of (1.4.5) is bounded. Since the total energy is conserved, we have proved the following lemma.

**Lemma 4.1** *If f is a sufficiently smooth solution of the initial value problem given by* (1.4.1) *and* (1.4.2) *with initial value $f_0$, then the integral in Eq.* (1.4.5) *is bounded.*

We now have the following.

**Lemma 4.2** *Under the above assumptions, we have, for the weak solutions of the Boltzmann equation for noncutoff Maxwell molecules:*

$$\int_{\mathbb{R}^3 \times \mathbb{R}^3 \times \mathbb{S}^2 \times [0,T] \times [0,1]} |\mathbf{v} - \mathbf{v}_*|^2 f(x, \mathbf{v}, t) f(x, \mathbf{v}_*, t) B(\theta) dt d\mu < K_0, \qquad (1.4.6)$$

*where $K_0$ is a constant, which only depends on the initial data (and molecular constants).*

In fact, we can take $\varphi = \xi^2$ as a test function and remark that the contribution of the left-hand side is bounded in terms of the initial data because $\xi^2 \leq |\mathbf{v}|^2$. Hence the right-hand side is also bounded. Then:

$$\varphi(\mathbf{v}) + \varphi(\mathbf{v}_*) - \varphi(\mathbf{v}') - \varphi(\mathbf{v}'_*)$$
$$= 2n_1(\xi - \xi_*)\mathbf{n} \cdot (\mathbf{v} - \mathbf{v}_*) - 2n_1^2|\mathbf{n} \cdot \mathbf{V}|^2. \qquad (1.4.7)$$

When computing this integral, we use as polar angles $\theta$ (the angle between $\mathbf{n}$ and $\mathbf{V}$) and $\phi$ (a suitable angle in the plane orthogonal to $\mathbf{V}$) so that the components $n_i$ $(i = 1, 2, 3)$ of $\mathbf{n}$ are given by

$$n_1 = \frac{V_1}{V} \cos \theta - \frac{V_0}{V} \sin \theta \cos \phi$$

$$n_2 = \frac{V_2}{V} \cos \theta + \frac{V_1 V_2}{V V_0} \sin \theta \cos \phi - \frac{V_3}{V} \sin \theta \sin \phi$$

$$n_2 = \frac{V_3}{V} \cos \theta + \frac{V_1 V_3}{V V_0} \sin \theta \cos \phi + \frac{V_2}{V} \sin \theta \sin \phi,$$

where $V_i$ $(i = 1, 2, 3)$ are the components of $\mathbf{V}$ and $V_0 = \sqrt{V_2^2 + V_3^2}$. Then $(V_1 = \xi - \xi_*)$

$$n_1(\xi - \xi_*)\mathbf{n} \cdot \mathbf{V} = \cos \theta(V_1 \cos \theta - V_0 \sin \theta \cos \phi).$$

Then when we integrate with respect to $\phi$, the contribution from the last term disappears and we are left with

$$\int_{[0,T] \times [0,1] \times \mathbb{R}^3} Q(f, f)(x, \mathbf{v}, t)\xi^2 d\mathbf{v} dx dt$$

$$= -\int_{[0,T] \times [0,1] \times \mathbb{R}^3 \times \mathbb{R}^3 \times S^2} B(\theta)\{[n_1(\xi - \xi_*)]^2 - n_1^2|\mathbf{n} \cdot \mathbf{V}|^2\}f f_* d\mu dt.$$
$$(1.4.8)$$

We can separate the contributions from the two terms, since they separately converge and obtain

$$\int_{[0,T]\times[0,1]\times\mathbb{R}^3} Q(f,f)(x,\mathbf{v},t)\xi^2 d\mathbf{v}dxdt$$

$$= -3B_0 \int_{[0,T]\times[0,1]\times\mathbb{R}^3\times\mathbb{R}^3} (\xi-\xi_*)^2 ff_* d\mathbf{v}d\mathbf{v}_* dxdt$$

$$+ B_0 \int_{[0,T]\times[0,1]\times\mathbb{R}^3\times\mathbb{R}^3} |\mathbf{V}|^2 ff_* d\mathbf{v}d\mathbf{v}_* dxdt, \qquad (1.4.9)$$

where if the force between two molecules at distance $r$ is $\kappa r^{-5}$, then

$$B_0 = a\sqrt{\frac{\kappa}{2m^3}} \quad (a = 1.3703\ldots). \qquad (1.4.10)$$

The constant $a$ was first computed by Maxwell [19]; the value given here was computed by Ikenberry and Truesdell [15]. Since we know that the left-hand side of Eq. (1.4.9) is bounded and the first term on the right-hand side is bounded, it follows that the last term is also bounded by a constant depending on initial data (and molecular constants, such as $m$ and $\kappa$).

## 1.5 Existence of weak solutions for noncutoff potentials

In order to prove the existence of a weak solution, we shall assume that this has been proved for Maxwell molecules with an angular cutoff [2]; actually to make the chapter self-contained and the proof more explicit, we shall assume that the proof is available when a cutoff for small relative speed is introduced. In this case, in fact the proof immediately follows from the DiPerna–Lions existence theorem with the estimate of Lemma 4.2; it is enough to remark that a solution exists when we renormalize by division by $1 + \epsilon f$ ($f$ independent of $\epsilon > 0$ and we pass to the limit $\epsilon \to 0$ due to (1.4.6)).

In the noncutoff case we approximate the solution by cutting off the angles close to $\pi/2$ and the small relative speeds. In this way we can obtain a sequence $f_n$ formally approximating the solution $f$ whose existence we want to prove.

**Lemma 5.1** *Let $\{f_n\}$ be a sequence of solutions to an approximating problem. There is a subsequence such that for each $T > 0$*

(i) $\int f_n \, d\mathbf{v} \to \int f \, d\mathbf{v}$ *a.e. and in* $L^1((0,T) \times \mathbb{R}^3)$,
(ii)

$$\int_{\mathbb{R}^3} |\mathbf{V}|^2 f_{n*} d\mathbf{v}_* \to \int_{\mathbb{R}^3} |\mathbf{V}|^2 f_* d\mathbf{v}_*$$

*in* $L^1((0,T) \times \mathbb{R}^3 \times B_R)$ *for all* $R > 0$, *and a.e.*,

(iii)

$$g_n(x,t) = \frac{\int_{\mathbb{R}^3 \times \mathbb{R}^3} |\mathbf{V}|^2 f_n f_{n*} d\mathbf{v} d\mathbf{v}_*}{1 + \int f_n \, d\mathbf{v}} \rightarrow \frac{\int_{\mathbb{R}^3 \times \mathbb{R}^3} |\mathbf{V}|^2 f f_* d\mathbf{v} d\mathbf{v}_*}{1 + \int f \, d\mathbf{v}} = g(x,t)$$

(1.5.1)

*weakly in* $L^1((0,T) \times (0,1))$.

*Proof.* (i) is immediate. (ii) uses an argument well known in the DiPerna–Lions proof with the estimate $\sup_n \int f_n(1 + |\mathbf{v}|^2) \, d\mathbf{v} < \infty$ to reduce the problem to bounded domains with respect to $\mathbf{v}_*$.

For (iii) we use (i) and the fact that $f_n$ converges weakly, but the factor multiplying it in the integral converges a.e. because of (ii).  □

Now we remark that $g_n(x,t)$ converges weakly to $g(x,t)$ and $\rho_n(x,t)$ converges a.e. to $\rho(x,t)$ and the integral $\int \rho_n g_n dx dt$ is uniformly bounded to conclude with the following lemma.

**Lemma 5.2** *Let* $\{f_n\}$ *be a sequence of solutions to an approximating problem. There is a subsequence such that for each* $T > 0$

$$\int_{(0,T) \times (0,1) \times \mathbb{R}^3 \times \mathbb{R}^3} |\mathbf{V}|^2 f_n f_{n*} d\mu dt \rightarrow \int_{(0,T) \times (0,1) \times \mathbb{R}^3 \times \mathbb{R}^3} |\mathbf{V}|^2 f f_* d\mu dt. \quad (1.5.2)$$

We can now prove the basic result.

**Lemma 5.3** *Let* $\{f_n\}$ *be a sequence of solutions to an approximating problem, weakly converging to* $f$. *There is a subsequence such that for each* $T > 0$

$$\int_{(0,T) \times (0,1) \times \mathbb{R}^3} \phi Q_n(f_n, f_n) dt dx d\mathbf{v} \rightarrow \int_{(0,T) \times (0,1) \times \mathbb{R}^3} \phi Q(f, f) dt dx d\mathbf{v},$$

(1.5.3)

*where* $Q_n$ *and* $Q$ *are given by the weak form of the collision operator, as defined in Eq.* (1.3.1).

*Proof.* In fact the integrand on the left-hand side of Eq. (1.5.3) is, thanks to (1.3.6), uniformly bounded by the integrand of Eq. (1.5.2) which weakly converges.  □

Thanks to this result, we can now pass to the limit in the approximating problem to obtain the following.

**Theorem 5.4** *Let* $f_0 \in L^1(\mathbb{R} \times \mathbb{R}^3)$ *be such that*

$$\int f_0(\cdot)(1 + |\mathbf{v}|^2) d\mathbf{v} dx < \infty; \qquad \int f_0 |\ln f_0(.)| d\mathbf{v} dx < \infty. \quad (1.5.4)$$

*Then there is a weak solution* $f(x, \mathbf{v}, t)$ *of the initial value problem* (1.2.1), (1.2.4), *such that* $f \in C(\mathbb{R}_+, L^1(\mathbb{R} \times \mathbb{R}^3))$, $f(., 0) = f_0$.

## 1.6 Conservation of momentum and energy

An immediate consequence of Theorem 5.4 is as follows.

**Theorem 6.1** *The solution whose existence has been proved in the previous section conserves energy globally.*

In fact we can take $\varphi = |\mathbf{v}|^2$ as a test function and this immediately yields the result. The lack of this property is one of the drawbacks of the DiPerna–Lions renormalized solutions.

It is, of course, easy to prove that momentum is conserved globally. But we can prove more.

**Theorem 6.2** *The solution whose existence has been proved in the previous section conserves momentum locally.*

In fact we can take $\varphi = \mathbf{v}g(x)$ as a test function, where $g(x)$ is a smooth periodic function of the space coordinate: the result follows.

## 1.7 Trend to equilibrium

The aim of this section is to discuss the trend to equilibrium. We follow the approach of Desvillettes [11] and Cercignani [9], which is based on a remark by DiPerna and Lions [13] and applies to any domain $\Omega$. The main result is the following.

**Theorem 7.1** *Let $f(\mathbf{x}, \mathbf{v}, t)$ be a DiPerna–Lions solution of the Boltzmann equation, with initial data $f_0(\mathbf{x}, \mathbf{v})$ such that*

$$f_0 \geq 0; \qquad \int_\Omega \int_{\Re^3} f_0(\mathbf{x}, \mathbf{v})(1 + |\mathbf{v}|^2 + |\log f_0(\mathbf{x}, \mathbf{v})|)d\mathbf{x}d\mathbf{v} < +\infty. \quad (1.7.1)$$

*Let $f$ also satisfy general boundary conditions on $\partial\Omega$ compatible with a constant and uniform Maxwellian $M_w$. Then, for every sequence $t_n$ going to infinity, there exist a subsequence $t_{n_k}$ and a local Maxwellian $M(\mathbf{x}, \mathbf{v}, t)$ such that $f_{n_k}(\mathbf{x}, \mathbf{v}, t) = f(\mathbf{x}, \mathbf{v}, t_{n_k} + t)$ converges weakly in $L^1(\Omega \times \Re^3 \times [0, T])$ to $M(\mathbf{x}, \mathbf{v}, t)$ for any $T > 0$. Moreover $M$ satisfies the free transport equation*

$$\frac{\partial M}{\partial t} + \mathbf{v} \cdot \frac{\partial M}{\partial \mathbf{x}} = 0 \qquad (1.7.2)$$

*and the periodicity boundary condition.*

*Proof.* According to the existence proof of DiPerna–Lions [12]

$$\int_0^T \int_{\Re^3} \int_\Omega \int_{S^2} \int_{\Re^3} [f(\mathbf{x}, \mathbf{v}', t)f(\mathbf{x}, \mathbf{v}'_*, t) - f(\mathbf{x}, \mathbf{v}, t)f(\mathbf{x}, \mathbf{v}_*, t)]$$

$$\times \{\log[f(\mathbf{x}, \mathbf{v}', t)f(\mathbf{x}, \mathbf{v}'_*, t)] - \log[f(\mathbf{x}, \mathbf{v}, t)f(\mathbf{x}, \mathbf{v}_*, t)]\}$$

$$\times B(\mathbf{V}, \mathbf{V} \cdot \mathbf{n})d\mathbf{v}_* d\mathbf{n} d\mathbf{x} d\mathbf{v} dt$$

$$+ \sup_t \int_\Omega \int_{\Re^3} f(\mathbf{x}, \mathbf{v}, t)(1 + |\mathbf{v}|^2 + |\log f(\mathbf{x}, \mathbf{v}, t)|)d\mathbf{x}d\mathbf{v} < +\infty. \quad (1.7.3)$$

Thus $f_n(\mathbf{x}, \mathbf{v}, t) = f(\mathbf{x}, \mathbf{v}, t + t_n)$ is weakly compact in $L^1(\Omega \times \Re^3 \times [0, T])$ for any sequence $t_n$ of nonnegative numbers and any $T > 0$. If $t_n \to \infty$, then there exist a subsequence $t_{n_k}$ and a function $M(\mathbf{x}, \mathbf{v}, t)$ in $L^1(\Omega \times \Re^3 \times [0, T])$ such that $f_{n_k}$ converges weakly to $M$ in $L^1(\Omega \times \Re^3 \times [0, T])$ for any $T > 0$. In order to prove that $M$ is a Maxwellian, we remark that, since the first integral in Eq. (1.7.3) is finite, then

$$\int_{t_{n_k}}^{T+t_{n_k}} \int_{\Re^3} \int_\Omega \int_{S^2} \int_{\Re^3} B(\mathbf{V}, \mathbf{V} \cdot \mathbf{n})[f(\mathbf{x}, \mathbf{v}', t)f(\mathbf{x}, \mathbf{v}'_*, t) - f(\mathbf{x}, \mathbf{v}, t)f(\mathbf{x}, \mathbf{v}_*, t)]$$

$$\times \{\log[f(\mathbf{x}, \mathbf{v}', t)f(\mathbf{x}, \mathbf{v}'_*, t)] - \log[f(\mathbf{x}, \mathbf{v}, t)f(\mathbf{x}, \mathbf{v}_*, t)]\}$$

$$\times d\mathbf{v}_* d\mathbf{n} d\mathbf{x} d\mathbf{v} dt \to 0$$

when $k \to \infty$, and thus

$$\int_0^T \int_{\Re^3} \int_\Omega \int_{S^2} \int_{\Re^3} [f_{n_k}(\mathbf{x}, \mathbf{v}', t)f_{n_k}(\mathbf{x}, \mathbf{v}'_*, t) - f_{n_k}(\mathbf{x}, \mathbf{v}, t)f_{n_k}(\mathbf{x}, \mathbf{v}_*, t)]$$

$$\times \{\log[f_{n_k}(\mathbf{x}, \mathbf{v}', t)f_{n_k}(\mathbf{x}, \mathbf{v}'*, t)]$$

$$- \log[f_{n_k}(\mathbf{x}, \mathbf{v}, t)f_{n_k}(\mathbf{x}, \mathbf{v}_*, t)]\}$$

$$\times B(\mathbf{V}, \mathbf{V} \cdot \mathbf{n})d\mathbf{v}_* d\mathbf{n} d\mathbf{x} d\mathbf{v} dt \to 0 \quad (k \to \infty). \quad (1.7.4)$$

But for all smooth nonnegative functions $\phi, \psi$ with compact support:

$$\int_{\Re^3} \int_{S^2} \int_{\Re^3} f_{n_k}(\mathbf{x}, \mathbf{v}', t)f_{n_k}(\mathbf{x}, \mathbf{v}'_*, t)\phi(\mathbf{v})\psi(\mathbf{v}_*)B(\mathbf{V}, \mathbf{V} \cdot \mathbf{n})d\mathbf{v}_* d\mathbf{n} d\mathbf{v}$$

$$\to \int_{\Re^3} \int_{S^2} \int_{\Re^3} M(\mathbf{x}, \mathbf{v}', t)M(\mathbf{x}, \mathbf{v}'_*, t)\phi(\mathbf{v})\psi(\mathbf{v}_*)B(\mathbf{V}, \mathbf{V} \cdot \mathbf{n})d\mathbf{v}_* d\mathbf{n} d\mathbf{v}$$

$$\text{(a.e. in } \Omega \times [0, T] \text{ when } k \to \infty) \quad (1.7.5)$$

and

$$\int_{\Re^3} \int_{S^2} \int_{\Re^3} f_{n_k}(\mathbf{x}, \mathbf{v}, t)f_{n_k}(\mathbf{x}, \mathbf{v}_*, t)\phi(\mathbf{v})\psi(\mathbf{v}_*)B(\mathbf{V}, \mathbf{V} \cdot \mathbf{n})d\mathbf{v}_* d\mathbf{n} d\mathbf{v}$$

$$\to \int_{\Re^3} \int_{S^2} \int_{\Re^3} M(\mathbf{x}, \mathbf{v}, t)M(\mathbf{x}, \mathbf{v}_*, t)\phi(\mathbf{v})\psi(\mathbf{v}_*)B(\mathbf{V}, \mathbf{V} \cdot \mathbf{n})d\mathbf{v}_* d\mathbf{n} d\mathbf{v}$$

$$\text{(a.e. in } \Omega \times [0, T] \text{ when } k \to \infty). \quad (1.7.6)$$

This was proved by DiPerna and Lions [13] with the following kind of argument (we just consider (1.7.6) because the proof for (1.7.5) is analogous). First, remark that for any Borel set $A \in \mathcal{B}(\mathbb{B}_R \times \mathbb{B}_R \times S^2)$ with respect to the measure $d\mu = B(\mathbf{V}, \mathbf{V} \cdot \mathbf{n})d\mathbf{v}_* d\mathbf{n}d\mathbf{v}$, where $\mathbb{B}_R$ is the ball of radius R in velocity space,

$$\int_A f_{n_k}(\mathbf{x}, \mathbf{v}, t)f_{n_k}(\mathbf{x}, \mathbf{v}_*, t)B(\mathbf{V}, \mathbf{V} \cdot \mathbf{n})d\mathbf{v}_* d\mathbf{n}d\mathbf{v}$$

$$\leq M^2\mu(A) + \int_{E_R} f_{n_k}(\mathbf{x}, \mathbf{v}, t)f_{n_k}(\mathbf{x}, \mathbf{v}_*, t)$$

$$\times [\chi(f_{n_k}(\mathbf{x}, \mathbf{v}, t) \geq M) + \chi(f_{n_k}(\mathbf{x}, \mathbf{v}_*, t) \geq M)]$$

$$\times B(\mathbf{V}, \mathbf{V} \cdot \mathbf{n})d\mathbf{v}_* d\mathbf{n}d\mathbf{v}$$

$$\leq M^2\mu(A) + [\psi(M)]^{-1}\overline{K}(t, x)\overline{N}(t, x) \qquad (\forall M > 0), \qquad (1.7.7)$$

where $\chi$ denotes the characteristic function of a set and $\psi(t) \in C([0, \infty))$ is an increasing function such that $\psi \to \infty$ as $t \to \infty$, $\psi(t)(\log t)^{-1} \to 0$ as $t \to \infty$, while

$$\overline{K} = \sup_k \left\{ \left[ \int_{E_R} f_{n_k}(\mathbf{x}, \mathbf{v}, t)f_{n_k}(\mathbf{x}, \mathbf{v}_*, t)[\psi(f_{n_k}(\mathbf{x}, \mathbf{v}, t)) + \psi(f_{n_k}(\mathbf{x}, \mathbf{v}_*, t))] \right. \right.$$

$$\left. \left. \times B(\mathbf{V}, \mathbf{V} \cdot \mathbf{n})d\mathbf{v}_* d\mathbf{n}d\mathbf{v} \right] \left[ 1 + \int_{\Re^3} f_{n_k}d\mathbf{v} \right]^{-1} \right\}$$

$$\overline{N} = \sup_k \left( 1 + \int_{\Re^3} f_{n_k}d\mathbf{v} \right). \qquad (1.7.8)$$

Here $\overline{K}$ and $\overline{N}$ are functions independent of $f_{n_k}$. This proves that the product $f_{n_k}(\mathbf{x}, \mathbf{v}, t)f_{n_k}(\mathbf{x}, \mathbf{v}_*, t)$ is weakly compact in $L^1(E_R, d\mu)$ for almost all $(\mathbf{x}, t) \in \Omega \times (0, T)$. We already know that $f_{n_k}(\mathbf{x}, \mathbf{v}, t)f_{n_k}(\mathbf{x}, \mathbf{v}_*, t)[1 + \int_{\Re^3} f_{n_k}d\mathbf{v}]^{-1}$ converges to the corresponding function, $M(\mathbf{x}, \mathbf{v}, t)M(\mathbf{x}, \mathbf{v}_*, t)[1 + \int_{\Re^3} Md\mathbf{v}]^{-1}$, weakly in $L^1$ for a. a. $(\mathbf{x}, t) \in \Omega \times (0, T)$. But the denominator converges a.e. to $1 + \int_{\Re^3} Md\mathbf{v}$; hence $f_{n_k}(\mathbf{x}, \mathbf{v}, t)f_{n_k}(\mathbf{x}, \mathbf{v}_*, t)$, which has been just shown to be weakly compact, converges weakly to $M(\mathbf{x}, \mathbf{v}, t)M(\mathbf{x}, \mathbf{v}_*, t)$ in $L^1(E_R, d\mu)$ for a. a. $(\mathbf{x}, t) \in \Omega \times (0, T)$. It is then possible to extract a subsequence (which we still denote by $f_{n_k}$), such that

$$\int_{\Re^3} \int_{S^2} \int_{\Re^3} B(\mathbf{V}, \mathbf{V} \cdot \mathbf{n})[f_{n_k}(\mathbf{x}, \mathbf{v}', t)f_{n_k}(\mathbf{x}, \mathbf{v}'_*, t) - f_{n_k}(\mathbf{x}, \mathbf{v}, t)f_{n_k}(\mathbf{x}, \mathbf{v}_*, t)]$$

$$\times \{\log[f_{n_k}(\mathbf{x}, \mathbf{v}', t)f_{n_k}(\mathbf{x}, \mathbf{v}'_*, t)] - \log[f_{n_k}(\mathbf{x}, \mathbf{v}, t)f_{n_k}(\mathbf{x}, \mathbf{v}_*, t)]\}$$

$$\times d\mathbf{v}_* d\mathbf{n}d\mathbf{v} \to 0 \qquad \text{(a.e. in } \Omega \times [0, T] \text{ when } k \to \infty) \qquad (1.7.9)$$

for a dense denumerable set in $C(\Re^3)$ of nonnegative smooth functions $\phi$ and $\psi$. But then the convexity of the function $C(f, g) = (f - g)(\log f - \log g)$ $(\Re_+ \times \Re_+ \to \Re_+)$ implies that we can pass to the limit and obtain

$$[M(\mathbf{x}, \mathbf{v}', t)M(\mathbf{x}, \mathbf{v}'_*, t) - M(\mathbf{x}, \mathbf{v}, t)M(\mathbf{x}, \mathbf{v}_*, t)]$$

$$\times \{\log[M(\mathbf{x}, \mathbf{v}', t)M(\mathbf{x}, \mathbf{v}'*, t)] - \log[M(\mathbf{x}, \mathbf{v}, t)M(\mathbf{x}, \mathbf{v}_*, t)]\}$$
$$\times B(\mathbf{V}, \mathbf{V} \cdot \mathbf{n}) = 0 \qquad \text{(a.e. in } \mathbf{v}_*, n, x, \mathbf{v}_*, t). \qquad (1.7.10)$$

Then, since $C(f, g)$ is nonnegative and $B(\mathbf{V}, \mathbf{V} \cdot \mathbf{n})$ strictly positive:

$$M(\mathbf{x}, \mathbf{v}', t)M(\mathbf{x}, \mathbf{v}'_*, t) = M(\mathbf{x}, \mathbf{v}, t)M(\mathbf{x}, \mathbf{v}_*, t) \qquad \text{(a.e. in } \mathbf{v}, n, x, \mathbf{v}_*, t).$$
$$(1.7.11)$$

Then $M$ is a Maxwellian and, moreover, is, thanks to the property of weak stability, a renormalized solution of the Boltzmann equation, satisfying the boundary conditions. Accordingly $Q(M, M) = 0$ and Eq. (1.7.2) holds. □

Theorem 7.1 tells us that the solutions of the Boltzmann equation with the boundary conditions (1.2.1) behave (in the case of a boundary at constant temperature) as Maxwellians satisfying the free transport equation, Eq. (1.7.2). These Maxwellians have been well known since Boltzmann [7]. They have the following form:

$$M = \exp[a_0 + b_0 \cdot \mathbf{v} + c_0|\mathbf{v}|^2 + d_0|\mathbf{x} - \mathbf{v}t|^2 + e_0 \cdot (\mathbf{x} - \mathbf{v}t) + f_0 \cdot (\mathbf{x} \wedge \mathbf{v})],$$
$$(1.7.12)$$

where $a_0, c_0, d_0 \in \Re$ and $b_0, e_0, f_0 \in \Re^3$ are constants. Now if we impose the condition that $M(\mathbf{x}, ., t)$ is an $L^1$ function for any $t \geq 0$, we see that $c_0$ must be negative and $d_0$ nonpositive. If we exclude from our considerations the cases in which the boundary conditions are compatible with several Maxwellians, i.e., the only acceptable Maxwellian has no drift and constant temperature; this immediately implies that $b_0, d_0, e_0, f_0$ are zero. Hence $M$ is a uniform Maxwellian, which coincides with $M_w$.

L. Arkeryd [1] proved that $f$ actually tends to a Maxwellian in a strong sense for a periodic box, but his argument works in other cases as well; his proof uses techniques of nonstandard analysis and, as such, is outside the scope of this book. Subsequently P.-L. Lions [14] obtained the same result without resorting to nonstandard analysis.

If we now try to apply these results to the case of periodicity boundary conditions, we see that the Maxwellian is not uniquely determined by the above argument, if we stick to DiPerna–Lions solutions. But in the case of the weak solutions discussed in the previous section, conservation laws identify uniquely the momentum and energy and hence the asymptotic Maxwellian.

## 1.8 Concluding remarks

We have surveyed the existence theory of the nonlinear Boltzmann equation with particular attention to a recent result of the author concerning Maxwell molecules, without any truncation on the collision kernel, in the one-dimensional case. To the best of our knowledge, this is the first result for the noncutoff Boltzmann equation. The solution conserves energy globally.

## Acknowledgments

The research described in this chapter was supported by MIUR of Italy.

# References

[1]   L. Arkeryd, On the strong $L^1$ trend to equilibrium for the Boltzmann equation, *Studies in Appl. Math.* **87**, 283–288 (1992).

[2]   M. Bony, Existence globale et diffusion en théorie cinétique discrète. In *Advances in Kinetic Theory and Continuum Mechanics,* R. Gatignol and Soubbarameyer, Eds., 81–90, Springer-Verlag, Berlin (1991).

[3]   C. Cercignani, Global weak solutions of the Boltzmann equation, *Jour. Stat. Phys.* **118**, 333–342 (2005).

[4]   C. Cercignani, *The Boltzmann Equation and Its Applications*, Springer-Verlag, New York (1988).

[5]   C. Cercignani, Weak solutions of the Boltzmann equation without angle cutoff, submitted to *Jour. Stat. Phys.* 2005.

[6]   C. Cercignani, Estimating the solutions of the Boltzmann equation, submitted to *Jour. Stat. Phys.* 2005.

[7]   C. Cercignani and R. Illner, Global weak solutions of the Boltzmann equation in a slab with diffusive boundary conditions, *Arch. Rational Mech. Anal.* **134**, 1–16 (1996).

[8]   C. Cercignani, R. Illner, and M. Pulvirenti, *The Mathematical Theory of Dilute Gases*, Springer-Verlag, New York (1994).

[9]   C. Cercignani, Equilibrium states and trend to equilibrium in a gas according to the Boltzmann equation, *Rend. Mat. Appl.* **10**, 77–95 (1990).

[10]  C. Cercignani, *Slow Rarefied Flow Theory and Application to MEMS*, Birkhäuser, Basel (2006).

[11]  L. Desvillettes, Convergence to equilibrium in large time for Boltzmann and BGK equations, *Arch. Rational Mech. Analysis* **110**, 73–91 (1990).

[12]  R. DiPerna and P. L. Lions, On the Cauchy problem for Boltzmann equations: Global existence and weak stability, *Ann. of Math.* **130**, 321–366 (1989).

[13]  R. DiPerna and P. L. Lions, Global solutions of Boltzmann's equation and the entropy inequality, *Arch. Rational Mech. Anal.* **114**, 47–55 (1991).

[14]  P. L. Lions, Compactness in Boltzmann's equation via Fourier integral operators and applications. I, *Cahiers de Mathématiques de la décision* **9301**, CEREMADE (1993).

[15]  E. Ikenberry and C. Truesdell, On the pressures and the flux of energy in a gas according to Maxwell's kinetic theory. I, *Jour. Rat. Mech. Anal.* **5**, 1–54 (1956).

[16]  R. Illner and M. Pulvirenti, Global validity of the Boltzmann equation for a two- dimensional rare gas in vacuum, *Commun. Math. Phys.* **105**, 189–203 (1986).

[17]  R. Illner and M. Pulvirenti, Global validity of the Boltzmann equation for two- and three-dimensional rare gases in vacuum: Erratum and improved result, *Commun. Math. Phys.* **121**, 143–146 (1989).

[18]  O. Lanford III, in *Time evolution of large classical systems.* Moser, E. J. (ed.). Lecture Notes in Physics **38**, 1–111. Springer-Verlag (1975).

[19]  J. C. Maxwell, On the dynamical theory of gases, *Phil. Trans. Roy Soc. (London)* **157**, 49–88 (1866).

# Results on optimal rate of convergence to equilibrium for spatially homogeneous Maxwellian gases

Ester Gabetta

Dipartimento di Matematica, Università degli Studi di Pavia, Via Ferrata, 1-27100 PAVIA (Italia) `ester.gabetta@unipv.it`

## 2.1 Introduction

This chapter aims to provide a unified presentation of some recent studies (see [CGT99], [CCG00], [BGR05], [CCG05] and [GR05]) on the convergence to equilibrium of the solution of Boltzmann's equation for Maxwellian pseudomolecules and, in particular, of the solution of Kac's analog of Boltzmann's equation. The main feature of these researches, with respect to the other ones, is the pursuit of the optimal rate of exponential convergence both in a weighted $\chi$-metric (see for example [GTW95]) and in the total variation metric for probability measures. For the definition of these metrics see the next section. Now, recall the following basic facts about the aforesaid equations.

The dynamics of spatially homogeneous rarefied gases, in the absence of external force fields, are usually described by the Boltzmann equation:

$$\frac{\partial}{\partial t} f(\boldsymbol{v}, t) = Q(f, f)(\boldsymbol{v}, t). \tag{2.1.1}$$

Here $f(\boldsymbol{v}, t)$ is the probability density for the velocity space distribution of the molecules at time $t$, and $Q$, which represents the effects of binary collisions, has the form

$$Q(f, g)(\boldsymbol{v}) = \int_{\mathbb{R}^3 \times \mathrm{S}^2} B(q, \boldsymbol{q} \cdot \boldsymbol{n}/q) \big[ f(\boldsymbol{v}_1) g(\boldsymbol{w}_1) - f(\boldsymbol{v}) g(\boldsymbol{w}) \big] \, \mathrm{d}\boldsymbol{w} \, \mathrm{d}\boldsymbol{n}. \tag{2.1.2}$$

In expression (2.1.2), $\boldsymbol{n}$ is a unit vector, and $\mathrm{d}\boldsymbol{n}$ denotes *normalized* surface measure on the unit sphere $\mathrm{S}^2$. Moreover $\boldsymbol{q} = \boldsymbol{v} - \boldsymbol{w}$ is the relative velocity, and in $\boldsymbol{q} \cdot \boldsymbol{n}$ the dot denotes the usual inner product in $\mathbb{R}^3$. $Q$ describes the effect of a collision between two identical molecules with *pre-collisional velocities* $\boldsymbol{v}$ and $\boldsymbol{w}$, and *post-collisional velocities* $\boldsymbol{v}_1$ and $\boldsymbol{w}_1$. The vector $\boldsymbol{n}$ parametrizes the set of all kinematically possible post-collisional velocities by

$$v_1 = \frac{1}{2}(v + w + qn),$$

$$w_1 = \frac{1}{2}(v + w - qn).$$

(2.1.3)

The post-collisional velocities are those conserving both momentum and kinetic energy

$$v_1 + w_1 = v + w, \qquad |v_1|^2 + |w_1|^2 = |v|^2 + |w|^2. \qquad (2.1.4)$$

The function $B$ in (2.1.2) is called the *collision kernel* and specifies the rate at which collisions with parameter $n$ occur when the incoming velocities are $v$ and $w$. It is a nonnegative Borel function of its arguments, namely the magnitude of $v - w$ and the angle between $v - w$ and $n$. Notice that $B$ takes into account the nature of the interaction between the molecules. Maxwell found that when this interaction turns out to be inversely proportional to a fifth power of a central force, $B$ depends only on the scattering angle $\vartheta$ in $\cos \vartheta = q \cdot n/q$, and not on $q$ itself. Throughout the present chapter, this condition is assumed to be in force, i.e.,

$$B(q, q \cdot n/q) = B(\cos \vartheta), \qquad (2.1.5)$$

and, in this case, the Boltzmann equation is called the *equation for Maxwellian molecules*. In addition to (2.1.5), the *angular cutoff condition* is postulated here; so without further loss of generality, assume that

$$\int_O^\pi B(\cos \vartheta) \sin \vartheta \; d\vartheta = 1 < +\infty, \qquad (2.1.6)$$

and, then, the molecules are said to be *pseudo-Maxwellian*.

Under the angular cutoff condition, one can split $Q$ into its *gain* and *loss* terms

$$Q(f, g) = Q^+(f, g) - Q^-(f, g), \qquad (2.1.7)$$

where

$$Q^+(f, g)(v) = \int_{\mathbb{R}^3} \int_{S^2} B(\cos \vartheta) f(v_1) g(w_1) \; dn \; dw \qquad (2.1.8)$$

$$Q^-(f, g)(v) = f(v) \int_{\mathbb{R}^3} \left[ \int_{S^2} B(\cos \vartheta) \; d\vartheta \right] g(w) \; dw. \qquad (2.1.9)$$

Consequently (2.1.1) becomes

$$\frac{\partial}{\partial t} f(v, t) + f(v, t) = Q^+(f, f)(v, t). \qquad (2.1.10)$$

Apropos of the adoption of hypotheses (2.1.5) and (2.1.6), the following comments from [CC03] provide a clear illustration of the point:

"It is easier to see things through to the end with Maxwellian molecules, and it is still useful, even today, to look first at the Maxwellian case when investigating any new problem ... With the cutoff condition, it is possible to produce strong solutions that can be studied in greater detail."

In point of fact, with respect to the study of the speed of convergence to the equilibrium, deeper results can be achieved in the further simplified model due to Kac. Recall that Kac ([Ka56], [Ka59]) considered an approach to Boltzmann's theory, for a monatomic dilute gas without mass motion, based on a master equation, by simplifying the original Boltzmann's derivation to an $n$-particle system in one dimension. He obtained the analog of Boltzmann's equation. In such a model the collisions conserve energy but not momentum unless it happens to be zero for the initial data. Consequently all of the kinematically possible collisions $(v, w) \to (v^*, w^*)$ are given by

$$v^* = v \cos \vartheta + w \sin \vartheta \quad \text{and} \quad w^* = -v \sin \vartheta + w \cos \vartheta$$

for $0 \leqslant \vartheta \leqslant 2\pi$.

Regarding the analogy of the Maxwellian equation with cutoff, the expressions (2.1.8) and (2.1.9) of the gain and loss terms in the collision kernel— under the further condition that the collision kernel $B$ is constant and equal to 1—become

$$Q^+(f) = \frac{1}{2\pi} \int_0^{2\pi} \int_{\mathbb{R}} f(v^*) f(w^*) \, dw \, d\vartheta \qquad (2.1.11)$$

and

$$Q^-(f) = \frac{1}{2\pi} \int_0^{2\pi} \int_{\mathbb{R}} f(v) f(w) \, dw \, d\vartheta = f(v), \qquad (2.1.12)$$

respectively. Then the evolution of the probability density $f(v, t)$ for the velocities turns out to be described by the equation

$$\frac{\partial f}{\partial t}(v, t) + f(v, t) = Q^+(f). \qquad (2.1.13)$$

With an abuse of notation, the loss and gain terms, in the Kac setting, are denoted by the same symbols used for the analogous terms in the Maxwellian equation.

It is well known that, with initial data $f_0$, both problem (2.1.1), under (2.1.5)–(2.1.6), and problem (2.1.13) have a unique solution; see, for example, Morgenstern ([Mo54]) and McKean ([MK66]).

As far as the presentation of the subject of this chapter is concerned, it follows the chronological order of publication of the three references mentioned at the beginning of the introduction. So, Section 2.3 contains a concise description of the arguments developed in [CGT99]. Sharp bounds on the rate of exponential convergence to equilibrium are proved there—both in a weighted $\chi$-metric and in the variational metric—under an additional hypothesis about the existence of a finite *Sobolev norm* for the initial data. The remaining assumptions pertain to the existence of finite moments, up to some specific order, and of the Fisher (Linnik) information. Analogous conditions are used in [CCG00] to obtain variational bounds on the error made when the Wild summation for solutions of (2.1.1) is truncated at a certain stage. The relevance of this argument, explained in Section 2.4, is twofold: first, Wild's summation is the only constructive method known for solving Boltzmann's equation; second, the evaluation of the aforesaid error can be used to get a simple proof of the exponentially fast rate of relaxation to equilibrium for Maxwellian molecules. It must be noticed that the analysis exploits the McKean representation of the Wild sum; see McKean ([MK66], [MK67]). Following the viewpoint (see, for example, [IT56]) that, apart from entropy and information measures, it is only the moments of the initial probability distribution that have physical significance, in [CCG05], with reference to the Kac model, it is proved that the existence of finite moments of any order and of the Fisher information suffices to achieve the best rate of convergence. As recalled in Section 5 of [CCG05] a key role in the analysis is played by a decomposition which depends on some basic McKean ideas of a probabilistic nature. When qualitative regularity, of the same type as in [CGT99], of the initial data is not assumed, the above-mentioned decomposition makes it possible to overcome the fact that the evolution does not improve the regularity of the initial data. Necessary preliminaries, for a better understanding of the main results to be reviewed, are collected in Section 2.2. Finally, Section 2.5 indicates some directions for new researches.

## 2.2 Preliminaries

For the sake of clarity and completeness, here one recalls some of the basic concepts and notation that will be used to formulate the results examined in the next sections. These concepts can be split into two different classes: the former pertains to mathematical analysis or mathematical physics, the latter to probability.

### 2.2.1 Analytical and physical preliminaries

In [CGT99] a key role is played by an inequality which yields bounds for Sobolev's space norms of solutions of (2.1.1). For a fixed positive integer $k$, let $H^k$ be the space of all functions $f$ in $L^2(\mathbb{R}^n)$ whose distribution derivatives $\partial^\alpha f$ are $L^2$ functions for $|\alpha| \leqslant k$. Here from now on, for any fixed $p > 0$, $L^p(\mathbb{R}^n)$ will denote as usual the set of all measurable functions $f$ on $\mathbb{R}^n$ for which

$$\|f\|_{L^p(\mathbb{R}^n)} := \left( \int_{\mathbb{R}^n} |f(x)|^p \, \mathrm{d}x \right)^{1/p} < +\infty.$$

Now, for any $f$ in $L^1(\mathbb{R}^n)$, let $\hat{f}$ be its Fourier transform, i.e.,

$$\hat{f}(\xi) := \int_{\mathbb{R}^n} e^{-i\xi x} f(x) \, dx \qquad (\xi \in \mathbb{R}).$$

Then, if $f$ belongs to $H^k$, the quantity

$$\|f\|_{H^k} := \left( \int_{\mathbb{R}^n} |\hat{f}(\xi)|^2 \cdot |\xi|^{2k} \, d\xi \right)^{1/2}$$

is what here is called the *Sobolev space norm* of $f$. It is well known that this norm is equivalent to the one defined by $(\sum_{\alpha \leqslant k} \|\partial^\alpha f\|_2^2)^{1/2}$. For a survey on Sobolev spaces see, for example, Folland ([Fo99]).

One of the main goals of the chapter to be examined in the subsequent sections is that of proving the natural conjecture, due to McKean ([MK66]), that the rate of approach to equilibrium is driven by the spectral properties of a linearized form of the collision operator that we are going to define. Such a definition involves the fundamental concept of *Maxwellian density* on $\mathbb{R}^n$, that is, the function

$$\boldsymbol{v} \mapsto (6\pi T)^{-n/2} \exp\left\{ \frac{-|\boldsymbol{v} - \boldsymbol{u}|^2}{2nT} \right\} =: M(\boldsymbol{v}) \qquad (\boldsymbol{v} \in \mathbb{R}^n),$$

where

$$\boldsymbol{u} = \int_{\mathbb{R}^n} \boldsymbol{v} M(\boldsymbol{v}) \, dv, \quad nT = \int_{\mathbb{R}^n} |\boldsymbol{v} - \boldsymbol{u}|^2 f(\boldsymbol{v}) \, dv,$$

$\boldsymbol{u}$ being bulk velocity and $T$ the temperature. In the present framework, where $n = 1, 3$, one assumes that

$$\int_{\mathbb{R}^n} \boldsymbol{v} f_0(\boldsymbol{v}) = 0, \quad n = \int_{\mathbb{R}^n} |\boldsymbol{v}|^2 f_0(\boldsymbol{v}) \qquad (n = 1, 3). \qquad (2.2.1)$$

Now, suppose that the solution $f$ of (2.1.1), with initial datum $f_0$ satisfying (2.2.1), has the form

$$f(\boldsymbol{v}, t) = M(\boldsymbol{v})\big(1 + \varepsilon h(\boldsymbol{v}, t)\big), \qquad (2.2.2)$$

$M$ being the Maxwellian density with $u = 0$ and $T = 1$, some function for which

$$\int_{\mathbb{R}^n} |h(\boldsymbol{v}, 0)|^2 M(\boldsymbol{v}) \, dv = 1 \qquad (n = 1, 3)$$

and $\varepsilon$ some small number. Correspondingly to (2.2.2), (2.1.1) becomes

$$\frac{\partial}{\partial t} h(\boldsymbol{v}, t) = \mathcal{L}h(\boldsymbol{v}, t) + \varepsilon \frac{1}{M(v)} Q(Mh, Mh)(\boldsymbol{v}) \qquad (2.2.3)$$

with

$$\mathcal{L}h(\boldsymbol{v}) = \frac{1}{M(\boldsymbol{v})}\left(Q^+(M, Mh)(\boldsymbol{v}) + Q^+(Mh, M)(\boldsymbol{v})\right) - \int_{\mathbb{R}^3} Mh(\boldsymbol{v})\,\mathrm{d}v - h(\boldsymbol{v})$$

and $\mathcal{L}$ is just the above-mentioned *linearized collision operator*. As to the spectrum of $\mathcal{L}$, it can be recalled that it has been computed in Wang Chang and Uhlenbeck [WU70]. Notice that $\mathcal{L}$ is self-adjoint and negative semi-definite on $\mathcal{H}$ (the Hilbert space with norm $\|h\|_{\mathcal{H}}^2 = \int_{\mathbb{R}^3} |h(\boldsymbol{v})|^2 M(\boldsymbol{v})\,\mathrm{d}v$), with a five-dimensional null space. The remaining eigenvalues are strictly negative; here, with $\Lambda$ one denotes the absolute value of the first of these eigenvalues when they are arranged in order of increasing magnitude. In other words, $\Lambda$ is the *spectral gap* of $\mathcal{L}$. An explicit evaluation of $\Lambda$ is also given in Section 8 of [CGT99].

Passing to the Kac equation (2.1.13), as explained in Section 2 of [CGT99] starting from a remark of McKean ([MK66]), it can be recalled that $\Lambda$ takes the value $1/4$.

In the introductory section a hint to the entropy and other allied functionals has been given. As a matter of fact, the existence of these functionals forms a constituent part of many statements in the kinetic theory of gases and, in particular, in almost all the results we intend to consider in the following sections. The *entropy* $H(f)$ of a probability density function $f$ on $\mathbb{R}^3$, such that $\int_{\mathbb{R}^3} |v|^2 f(v)\mathrm{d}v < +\infty$, is defined by

$$H(f) = -\int_{\mathbb{R}^3} f(v)\log f(v)\,\mathrm{d}v. \qquad (2.2.4)$$

It was discovered by Boltzmann that $H\big(f(\cdot, t)\big)$ is nondecreasing in time along solutions $f(\cdot, t)$ of (2.1.1). Moreover, it can be constant only if $f(\cdot, t)$ is a Maxwellian distribution. See, for example, [CIP94] and [Vi03].

A different entropy functional is frequently considered in the kinetic literature. It is inspired by the Fisher amount of (statistical) information. Let $f$ be a probability function on $\mathbb{R}^n$ which, besides satisfying the above moment condition, is continuously differentiable; then, this new measure of information, denoted by $L_n$, is defined as

$$L_n(f) = \int_{\mathbb{R}^n} \big|\nabla_v \log f(v)\big|^2 f(v)\,\mathrm{d}v, \qquad (2.2.5)$$

where, as usual, $\nabla$ denotes the gradient vector. It should be recalled that, on the difference of $L_n$, the *Fisher information* aims at quantifying the content of information of a statistical sample drawn to estimate an unknown parameter $\vartheta$ in $\mathbb{R}^d$. More precisely, let $f(\cdot, \vartheta)$ be the probability density function (on $\mathbb{R}^n$) of a sample of size $n$, depending on the parameter $\vartheta$. In this framework, Fisher provided deep motivations for the adoption of the $(d \times d)$ matrix

$$\mathcal{F}(f;\vartheta) = \int_{\mathbb{R}^n} \left(\nabla_\vartheta \log f(x;\vartheta) \cdot (\nabla_\vartheta \log f(x;\vartheta))\right)^T f(x;\vartheta)\ \mathrm{d}x$$

as a measure of statistical information; see Fisher ([Fi22], [Fi25]).

To point out an actual relationship between the measures $L_n$ and $\mathcal{F}$, notice that if $\vartheta$ is a location parameter—i.e., $d = n$ and $f(x;\vartheta) = f(x - \vartheta)$—then $L_n(f)$ turns out to be the same as the trace of the Fisher information matrix $\mathcal{F}(f;\vartheta)$.

In particular in the one-dimensional case (see equation (2.1.13)), $L_n(f)$ becomes the well-known *Linnik's information* (see [Li59]):

$$L_1(f) =: L(f) = \int_{\mathbb{R}} \frac{1}{f(v)} \left(f'(v)\right)^2\ \mathrm{d}v.$$

As to the connection between entropy and Linnik information, remember that

$$\lim_{\delta \to 0^+} \frac{2}{\delta}\left\{H(f_\delta) - H(f)\right\} = L(f),$$

where $f_\delta$ is the density one obtains by convolving $f$ with the one-dimensional Maxwell density with variance $\delta$.

### 2.2.2 Distances between probability measures and McKean trees: an overview

As recalled repeatedly in foregoing paragraphs, the ultimate goal of the papers to be reviewed is the study of convergence to equilibrium in the $L^1$-norm for probability density functions. Such an object is achieved through the study of the convergence of probability measures in a weaker norm. Moreover, the proofs are based, in part, on some suitable discrete probabilistic structures. Then, it seems right to provide a concise description both of the distances to be used and of the McKean binary tree graphs, before tackling the main subject.

Given two probability measures $P$ and $Q$ on the same measurable space $(\Omega, \mathcal{F})$, the *variational distance* between $P$ and $Q$ is defined as

$$d_1(P,Q) = \sup_{A \in \mathcal{F}} |P(A) - Q(A)|.$$

In particular, if $\Omega = \mathbb{R}^n$, if $\mathcal{F}$ is the Borel class on $\mathbb{R}^n$ and if $P$ and $Q$ are absolutely continuous with respect to the Lebesgue measure on $\mathbb{R}^n$ with density functions $p$ and $q$, respectively, then $d_1(P,Q)$ reduces to

$$d_1(P,Q) = \frac{1}{2}\int_{\mathbb{R}^n} |p(x) - q(x)|\ \mathrm{d}x$$

$$= \frac{1}{2}\|p - q\|_{L^1(\mathbb{R}^n)}.$$

Now denote the Fourier transforms of $p$ and $q$ by $\hat{p}$ and $\hat{q}$ respectively, and assume that conditions

$$\int_{\mathbb{R}^n} |x|^4 \left( p(x) + q(x) \right) \, \mathrm{d}x < +\infty, \quad \int_{\mathbb{R}^n} \pi_i(x)^k \left( p(x) - q(x) \right) \, \mathrm{d}x = 0 \quad (2.2.6)$$

hold true for $\pi_i(x) = x_i$ if $x = (x_1, \dots, x_n) \in \mathbb{R}^n$, $i = 1, \dots, n$, $k = 1, 2, 3$. Then from elementary properties of the characteristic functions the quantity

$$\|\hat{p} - \hat{q}\| := \sup_{\xi \in \mathbb{R}^n} \frac{|\hat{p}(\xi) - \hat{q}(\xi)|}{|\xi|^4} \qquad (2.2.7)$$

turns out to be well defined. In Section 14.2 of [Ra91], it is designated by the term *weighted $\chi$-metric* and it has been diffusely utilized in kinetic theory starting from the publication of [GTW95]. This metric—when it is well defined—is topologically weaker than $d_1$.

After saying how one intends to study the relaxation to the equilibrium of the solution of (2.1.1) or of (2.1.13), it is worth recalling the sole representation at our disposal of that solution, i.e., the Wild series. See Wild ([Wi51]). More precisely, what is needed here is the McKean version ([MK66], [MK67]) of this series. To make these concepts precise start by defining integral (2.1.8) to be the *Wild convolution* of $f$ and $g$, $f \circ g$ in symbols. Accordingly, operation (2.1.11) will be indicated by the same symbol. Then, recall that Wild proved that the solution of the equation at issue, when $f_0$ is the initial density, is given by

$$f(v, t) = \sum_{n \geqslant 1} \mathrm{e}^{-t} \left( 1 - \mathrm{e}^{-t} \right)^{n-1} Q_n^+(v; f_0)(v),$$

where, following McKean, $Q_n^+$ can be written as

$$Q_n^+(v; f_0) = \sum_{\gamma \in G(n)} p_n(\gamma) C_\gamma(v; f_0) \quad (n = 1, 2, \dots), \qquad (2.2.8)$$

where $C_\gamma$ denotes the $n$-fold Wild convolution performed according to a procedure which can be described in a complete way by means of the "structure" of a tree $\gamma$ belonging to the set $G(n)$ of all McKean's trees with $n$ leaves.

These trees are characterized by the fact that each node has either zero or two "children": a "left child" and a "right child." To illustrate this circumstance, two elements of $G(8)$ are visualized in Figure 2.1. In each tree $\gamma$ in $G(n)$, leaves—i.e., shaded circles—can be labelled from left to right by the integers from 1 to $n$, according to a natural left-to-right order. Imagine now that $f_0$ is set in each of the leaves of $\gamma$ and assume that the deepest level in $\gamma$ is $i$. At this stage, to go from $\gamma$ to the corresponding convolution $C_\gamma$, find the leftmost pair of leaves at level $i$, erase this pair of leaves which makes the former "parent" node a leaf, and write down $Q_1 := Q(f_0, f_0)$ in the new leaf. After erasing all the leaves at level $i$, in this way, proceed to erase pairs of leaves at level $(i - 1)$ and, for any pair, write down

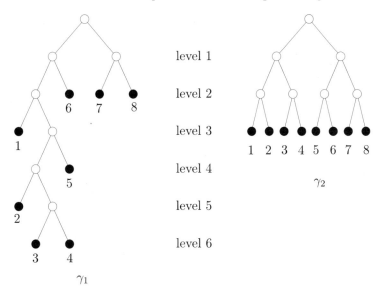

level 1

level 2

level 3

level 4

level 5

level 6

**Fig. 2.1.** Shaded (unshaded) circles stand for leaves (nodes).

$$Q_2 = Q\big(Q_l(f_0, f_0), Q_r(f_0, f_0)\big) \tag{2.2.9}$$

in the leaf which replaces the corresponding "parent" node, where $l$ ($r$, respectively) can be 0 or 1 according to whether the left (right, respectively) leaf of the pair was a leaf of $\gamma$ or the "parent" node of a pair erased in the previous step, with the proviso that $Q_0 := f_0$. Then, after erasing all the leaves at level $(i-1)$, proceed to erase pairs of leaves at level $(i-2)$ in the same way and, for any pair, write down (2.2.9) in the leaf which replaces the "parent" node of the pair at issue, keeping in mind that both $l$ and $r$ can belong to $\{0, 1, 2\}$. Once this has been done until only the "root" of $\gamma$ is left, one has written $C_\gamma$ in the "root." This process can be described by a proper use of brackets. As an example, the $C_\gamma$ associated with graph $\gamma_1$ in Figure 2.1 can be represented as

$$\left(\Big(f_0 \circ \big((f_0 \circ (f_0 \circ f_0)) \circ f_0\big)\Big) \circ f_0\right) \circ (f_0 \circ f_0)$$

and, analogously,

$$\big((f_0 \circ f_0) \circ (f_0 \circ f_0)\big) \circ \big((f_0 \circ f_0) \circ (f_0 \circ f_0)\big)$$

gives the convolution corresponding to tree $\gamma_2$ in the same figure. From a well-known result from enumerative combinatorics it turns out that the number of elements of $G(n)$ is the Catalan number $\binom{2n-2}{n-1}/n$; see, for example, [Co70].

It follows from the previous description that the level reached by leaves of a tree plays an important role in determining convolutions. To quantify this fact, the concepts of depth of a leaf and depth of a tree are very useful. By *depth of a leaf* of $\gamma$ in $G(n)$ here one means the number of generations which separate the leaf from the root of $\gamma$. For example, the depth of leaf 2 in $\gamma_1$ of Figure 2.1 is 5. From now on the depth of leaf $j$ in $\gamma$ will be denoted by $\delta_j(\gamma)$. The quantities

$$\delta_{(1)}(\gamma) := \min\{\delta_1(\gamma), \dots, \delta_n(\gamma)\}$$

and

$$\delta_{(n)}(\gamma) := \max\{\delta_1(\gamma), \dots, \delta_n(\gamma)\}$$

are said to be the *depth* and the *height*, respectively, *of tree* $\gamma$. Tree $\gamma_1$ has depth 2 and height 6; in graph $\gamma_2$, depth and height are the same ($= 3$) and the graph is said to be balanced.

As to the coefficients $p_n$ in (2.2.8), it suffices that $p_n(\gamma)$ is the (conditional) probability that a distinguished random walk on $G$ passes through $G(n)$ at the $n$th step. So, given $n$, $\delta_j$ and $\delta_{(1)}$, $\delta_{(n)}$ can be viewed as random variables with respect to such a random walk. Clearly, $p_n(\cdot)$ induces a probability distribution for depths of leaves and trees. In studying the probability distribution of a leaf, through probability generating functions as in [GR05], one encounters the expectation of the random variable

$$\sum_{j=1}^{n} \xi^j x^{\delta_j},$$

where $x$ and $\xi$ are arbitrary numbers. This expectation, denoted by $V_n(x, \xi)$, has an explicit form given by

$$V_n(x, \xi) = \sum_{j=1}^{n} \xi^j \frac{1}{(n-j)!(j-1)!} \sum_{d \geqslant 0} \sum_{k} |s(n-j, k)| \cdot |s(j-1, d-k)|,$$

where $s(n, k)$ stands for a Stirling number of the first kind; see [GR05]. From a different viewpoint, $V_n$ was introduced, with $\xi = 1$ and $x = c/2$ with $c$ in $(0, 1)$, in [CCG00] to get a concise measure of the location of the joint distribution of the depths of the leaves of trees in $G(n)$. In fact, $V_n(c/2, 1)$ is the same as the quantity $T(n)$ as defined in [CCG00]. It is clear that small values of $T(n)$ indicate that the distribution of $(\delta_1, \dots, \delta_n)$ is located around great values for the depths of leaves and this, in turn, entails analytical regularity for the Wild convolutions. This point will be discussed in Section 2.4.

## 2.3 Optimal exponential convergence results under Sobolev regularity and moments

An early result about the explicit determination of the rate of trend, in total variation, toward equilibrium of the solution of a kinetic equation was obtained by McKean in 1966 (see [MK66]) who dealt with the Kac equation.

He stated, under the hypotheses of *boundness* both of the *third moment* and the *Linnik information* of the initial datum,

$$\|f - M\|_{L^1} \leqslant ct^{3/2}e^{-\lambda t} \qquad \left(\lambda := \frac{2}{9}\left(-\frac{8}{3\pi} + 1\right) \approx 0,04\right),$$

i.e., the geometrically fast approach to the Maxwellian at the rate $\lambda$. Moreover he suggested that this bound could be improved, as recalled in Subsection 2.2.1.

Previously Ikenberry and Truesdell ([IT56]) had proven that all moments of the solution of this spatially homogeneous Maxwell equation, that exist initially, converge exponentially to the corresponding ones of the equilibrium distribution. More recent works on the convergence to equilibrium, in the case of intermolecular forces harder than Maxwellian ones and in the case of pseudo-Maxwellian molecules, are due to Arkeryd ([Ar88]) and Wennberg ([We93]) respectively. These authors based their arguments on the spectral theory of the linearized collision operator $\mathcal{L}$ (see Subsection 2.2.1). On the one hand, they obtained exponential convergence to equilibrium under the additional hypothesis that the initial distribution belongs to a suitable neighborhood of the equilibrium itself. On the other hand, they did not explicitly compute the spectrum of $\mathcal{L}$ in the specific space where they studied the subject. Notice that their space is different from the one evocated in Subsection 2.2.1.

In view of these remarks, Carlen, Gabetta and Toscani ([CGT99]) proposed to provide a proof for McKean's conjecture both for the Kac and the Maxwell models. The main problem was the one of showing that $\mathcal{L}$ governs the rate of approach to equilibrium in the spatially homogeneous case for initial data far from equilibrium. This fact has been ascertained by Cercignani, Lampis and Sgarra ([CLS88]) when the initial distribution belongs to a sufficiently small neighborhood of the limiting distribution, by determining an upper bound for the nonlinear term in (2.1.10). Direct extensions of the CLS argument to the case in which initial data are far from equilibrium are made difficult by the determination of uniform bound (in $t$) for some specific neighborhood of the Maxwellian distribution. Therefore [CGT99] tried to bound the $H^m$ norm (see Subsection 2.2.1) of solutions of (2.1.1) uniformly in time. This has been done on the basis of the fact that bounds for the total variation distance can be expressed as geometric means of $H^m$ and $\|\cdot\|$ bounds (see Subsection 2.2.2). So, to prove that the total variation distance decays exponentially to zero it would be enough to show that the $\|\cdot\|$ term converges to zero exponentially as $t \to \infty$ and the $H^m$ norm is uniformly bounded with respect to time. Moreover, in view of (2.1.10) the latter circumstance would follow, under the hypothesis that $\|f_0\|_{H^m}$ is finite, from the combination of the following occurrences: (a) $\|Q^+(f)\|_{H^m}^2$ is bounded when $\|f\|_{H^m}^2$ is finite, $f$ being any probability density function with finite second moments; (b) if $\|f_0\|_{H^m}$ is finite, then the solution of (2.1.10) can be bounded uniformly with respect to time and, therefore, from (a) it follows that $\|Q^+(f(\cdot,t))\|_{H^m}$ turns out to be uniformly bounded.

For the sake of clarity, recall that Section 4 of [CGT99] establishes the aforesaid bound for the total variation distance, in terms of geometric means, as

$$\|f - M_f - S\|_{L^1} \leqslant C_\varepsilon \|f - M_f - S\|^{1-\varepsilon} \|f - M_f - S\|_{H_m}^\varepsilon, \qquad (2.3.1)$$

where: $M_f$ denotes the Maxwellian density with the same mean and variance as $f$; $S$ is a subtraction term that is inserted in such a way that $(f - M_f - S)$ satisfies conditions (2.2.6) in order to get a definition of $\|f - M_f - S\|$ consistent with the framework described in Section 2.2.2.

As to the $\| \cdot \|$ term in (2.3.1), Theorem 1.3 in [CGT99] establishes exponential decay when $f(\cdot, t)$ is the solution of (2.1.10), the pseudo-Maxwellian molecules equation, in the following terms.

**Theorem 2.3.1** *Let $f_0$ be any probability density with $\int_{\mathbb{R}^3} |\boldsymbol{v}|^4 f_0(\boldsymbol{v}) \, \mathrm{d}^3 v < +\infty$ and let $\varepsilon > 0$ be given. Then there are constants $B$ and $C$ and a function $S(\cdot, t)$ such that*

$$\|f(\cdot, t) - M_f(\cdot) - S(\cdot, t)\| \leqslant Bte^{-t(1-\varepsilon)\Lambda} \|f(\cdot, 0) - M_f(\cdot) - S(\cdot, 0)\|$$

*for all $t \geqslant 0$, and with $\|f(\cdot, 0) - M_f(\cdot) - S(\cdot, 0)\| < \infty$, such that for all $m$*

$$e^{t\Lambda}(\|S(\cdot, t)\|_{L^1} + \|S(\cdot, t)\|_{H^m}) \leqslant C$$

*for all $t \geqslant 0$.*

The meaning of $\Lambda$ has been explained in Subsection 2.2.1.

In the ambit of the Kac equation (2.1.13), the previous result can be improved according to Theorem 3.1 in [CGT99].

**Theorem 2.3.2** *Let $f_0$ be a probability density with $\int_{\mathbb{R}} |\boldsymbol{v}|^4 f_0(\boldsymbol{v}) \, \mathrm{d}v < +\infty$. Then there are constants $B$ and $C$ and a function $S(\cdot, t)$ such that*

$$\|f(\cdot, t) - M_f(\cdot) - S(\cdot, t)\| \leqslant Bte^{-t\Lambda} \|f(\cdot, 0) - M_f(\cdot) - S(\cdot, 0)\|$$

*for all $t \geqslant 0$ and with $\|f(\cdot, 0) - M_f(\cdot) - S(\cdot, 0)\| < \infty$ such that for all $m$*

$$e^{t\Lambda}(\|S(\cdot, t)\|_{L^1} + \|S(\cdot, t)\|_{H^m}) \leqslant C$$

*for all $t \geqslant 0$. Here, $\Lambda = 1/4$ is the spectral gap in the linearized collision operator for the Kac model.*

Now, as for the $\| \cdot \|_{H^m}$ term in (2.3.1), by an easy argument it can be seen that the circumstance mentioned in point (b) comes true; see p. 525 in [CGT99]. Harder is the problem of proving (a) and, then, of bounding the $\| \cdot \|_{H^m}$ term in (2.3.1) by

$$\|f - M_f - S\|_{H^m} \leqslant 2\|Q^+(f, f)\|_{H^m} + \|f\|_{H^m} + \|M + S\|_{H^m}.$$

The crucial point is solved by Theorem 7.1 in [CGT99] for equation (2.1.10), in the following terms.

**Theorem 2.3.3** *Let $f$ be any probability density with finite second moments on $\mathbb{R}^3$ such that $\|f\|_{H^m}$ is finite. Then there is a constant $C(m, T)$ so that*

$$\|Q^+(f)\|_{H^m}^2 \leqslant (1/2)\|f\|_{H^m}^2 + C(m, T)$$

*whenever*

$$\|f - M_f\|_{L^1}^2 \leqslant 2^{-(m+5)}.$$

There are several useful variants of this result. For example, one can mention a version which can be called *entropic* since it involves the entropy defined by (2.2.8).

**Theorem 2.3.4** *Let $f$ be any probability density on $\mathbb{R}^3$ with finite second moments such that $\|f\|_{H^m}$ is finite. Then there is a constant $C(m, T)$ so that*

$$\|Q^+(f)\|_{H^m}^2 \leqslant (1/2)\|f\|_{H^m}^2 + C(m, T)$$

*whenever*

$$\big(H(M_f) - H(f)\big) \leqslant 2^{-(m+4)}.$$

Another variant is based on the measure of information $L_3(f)$ defined by (2.2.5).

**Theorem 2.3.5** *Let $f$ be any probability density on $\mathbb{R}^3$ with finite second moments and such that $\|f\|_{H^m}$ is finite. Then there is a constant $C(m, T, L_3(f))$ so that*

$$\|Q^+\|_{H^m}^2 \leqslant (1/2)\|f\|_{H^m}^2 + C(m, T, L_3(f)).$$

Analogous results also hold true, with slight improvements, for the one-dimensional case (Kac equation); see Section 2 of [CGT99]. Note that in the last theorem, $L_3(f)$ is replaced by the Linnik functional according to the remarks made in Subsection 2.2.1.

Now, the previous statements can be combined to get the main results. In particular, for the Kac equation we have the following.

**Theorem 2.3.6** *Let $f_0$ be any probability density on the real line with unit variance, finite fourth moment $m_4$ and finite Linnik functional $L(f_0)$. Then, for any $\varepsilon > 0$ there is a fixed constant $m$, depending only on $\varepsilon$, so that if*

$$\int_{\mathbb{R}} f(v)|v|^{2m} \, dv + \|f\|_{H^m} \leqslant K,$$

*there is a universal, computable constant $C$ depending only on $\varepsilon$, $m_4$, $L(f_0)$ and $K$ so that the solution of the Kac equation $f(\cdot, t)$ with initial data $\mathring{f}_0(\cdot)$ satisfies*

$$\|f(\cdot, t) - M_f(\cdot)\|_{L^1} \leqslant C e^{-(1-\varepsilon)\Lambda t},$$

*where $\Lambda$ is the spectral gap in the linearized collision operator $\mathcal{L}$ for the Kac equation: i.e., $\Lambda = 1/4$. Moreover, increasing $m$ we obtain the same result if the $L_1$ norm is replaced by any $H^k$ norm.*

Its extension to Maxwellian collisions proceeds in the same form.

**Theorem 2.3.7** *Let $f_0(\boldsymbol{v})$ be the initial data for (2.1.10). Suppose that the bulk velocity $\boldsymbol{u} = 0$, and the temperature $T = 1$. Let $\varepsilon > 0$ be given. Then there is a number $n$ depending only on $\varepsilon$ so that whenever*

$$\int_{\mathbb{R}^3} |\boldsymbol{v}|^{2n} f_0(\boldsymbol{v}) \, \mathrm{d}^3 v + \|f\|_{H^m} < \infty$$

*then it holds that*

$$\|f(\cdot, t) - M\|_{L^1} \leqslant C_\varepsilon e^{-(1-\varepsilon)\Lambda t}.$$

*Here $\Lambda$ is the spectral gap of the linearized collision operator and $C_\varepsilon$ is a computable constant.*

## 2.4 Sharp exponential convergence exclusively based on moments

It is time to analyse convergence to equilibrium without assuming smoothness hypotheses—such as the Sobolev regularity in Section 2.3—on the initial data, besides existence of moments. The essence of the problem is that, if $f_0$ does not possess a certain degree of smoothness, then, in general, neither will $Q_n^+(f_0, f_0)$, no matter how large $n$ is. So, it would be strategic to be able to decompose the probability density $Q_n^+$ as a convex combination of two densities and to show that the weight of the one which attracts the initial irregularity becomes negligible as $n$ goes to infinity. A decomposition of this kind has been advanced in [CCG05] on the basis of the correspondence between Wild convolutions and McKean graphs overviewed in Subsection 2.2.2. To start, for every positive integer $n$, consider the class $\mathcal{B}_{n,k}$ of all trees with $n$ leaves (see notation in the aforesaid subsection) and with depth not smaller than $k$, i.e.,

$$\mathcal{B}_{n,k} := \{\gamma \in G(n) \; : \; \delta_{(1)}(\gamma) \geqslant k\}.$$

Through Lemma 1.4 in [CCG00], the same authors show in [CCG05] that the probability of $\mathcal{B}_{n,k}$—induced by the coefficients $p_n(\gamma)$ in (2.2.8)—goes to one as $n$ diverges. More precisely, the statement at issue can be formulated according to the final part of [GR05] as follows:

*If $p_{n,k}$ stands for the probability of the complement of $\mathcal{B}_{n,k}$ $(\mathcal{U}_{n,k})$, then*

$$0 \leqslant p_{n,k} \leqslant C \frac{(\log n)^k}{n}$$

*holds for a suitable constant $C$, for every $n \geqslant 2$. Compare also [BGR05].*

Now consider the decomposition of $Q_n^+(f_0, f_0)$ given by

$$Q_n^+(f_0, f_0) = (1 - p_{n,k})B_{n,k}(f_0) + p_{n,k}U_{n,k}(f_0) \qquad (2.4.1)$$

with

$$B_{n,k}(f_0) = \frac{1}{1 - p_{n,k}} \sum_{\gamma \in \mathcal{B}_{n,k}} p_n(\gamma)C_\gamma(\cdot; f_0). \qquad (2.4.2)$$

As stated by Theorem 3.1 in [CCG05], the distribution $B_{n,k}(f_0)$ incorporates a regularity of the same type as the one assumed, by hypothesis, in [CGT99], i.e.,

*If $f_0$ has finite Linnik information $L(f_0)$, then there is a constant $C$ depending only on $k$ and $L(f_0)$ for which*

$$\|B_{n,k}(f_0)\|_{H^{k/2}(\mathbb{R})} \leqslant C.$$

In view of this fact, in [CCG05] the set $\mathcal{B}_{n,k}$ ($\mathcal{U}_{n,k}$, respectively) is called the set of the "beautiful" ("ugly" respectively) trees.

Next, since $p_{n,k}$ becomes negligible as $n$ diverges, in order that the new approach may be successful it would suffice to show that the "discrepancy" between $U_{n,k}$ and the Maxwell distribution is bounded in a sense to be specified. In the chapter we are overviewing here, discrepancy is measured through the weighted $\chi$-metric defined by (2.2.7), while the upper bound is fixed in terms of a specific functional $\Phi$ which plays an important role throughout the chapter. Such a functional is defined for probability densities $f$ with finite fourth moment; let $m_k(f)$ be the $k$th moment of $f$, $k = 1, 2, 3, 4$, with $m_2(f) = 1$. Set $\mu(f) := (m_1{}^2(f) + m_3{}^2(f))^{1/2}$ and, for any $\xi$ in $\mathbb{R}$, $\hat{p}_f(\xi) := (im_1(f)\xi - im_3(f)\xi^3/6)\psi(|\xi|)$, where $\psi$ is a $C^\infty$ monotone decreasing function on $[0, +\infty)$ such that for $r \leqslant L_1$, $\psi(r) = 1$, while $\psi(r) = 0$ for $r \geqslant L_2$, and where $0 < L_1 < L_2$ are specific constants. Indicate by $P_f$ the inverse image of $\hat{p}_f$ and define $\Phi(f)$ by

$$\Phi(f) = \|(f - P_f) - M\| + k\mu(f),$$

where $M$ denotes, here and through the rest of the chapter, the univariate Maxwell density with $T = 1$ and $k$ is a specific constant.

Going back to the problem of bounding the ugly piece of (2.4.1), we are now in a position to report the remaining part of Theorem 3.1 [CCG05]:

*If $f_0$ has finite Linnik information $L(f_0)$ and finite fourth moment, with $m_2(f_0) = 1$, then*

$$\|U_{n,k}(f_0) - M\| \leqslant \Phi(f_0).$$

Some interpolation inequalities, proved in Section 4 of [CCG05], show that the $\|\cdot\|_{L^1(\mathbb{R})}$ norm (used to define the variational distance between probabilities) can be majorized in the form of a product of the $\|\cdot\|$ norm and terms which, in view of the properties, above described, of the pieces of decomposition (2.4.1)–(2.4.2), turn out to be bounded. So, to conclude, it remains to prove that the $\|\cdot\|$ norm behaves well. In fact, by the definition of $\Phi$, one gets

$$\|Q_n^+(f_0, f_0) - M\| = \Phi\big(Q_n^+(f_0, f_0)\big). \tag{2.4.3}$$

Moreover, $\Phi$ is a "strictly convolution-convex" functional, which entails

$$\Phi\big(Q_n^+(f_0)\big) = \Phi\left(\sum_{\gamma \in G(n)} p_n(\gamma) C_\gamma(f_0)\right)$$
$$\leqslant \sum_{\gamma \in G(n)} p_n(\gamma) \Phi\big(C_\gamma(f_0)\big)$$
$$\leqslant \sum_{\gamma \in G(n)} p_n(\gamma) \Phi(f_0) \sum_{j=1}^n \left(\frac{1-\Lambda}{2}\right)^{\delta_j(\gamma)}, \tag{2.4.4}$$

where $\Lambda$ is the spectral gap as defined in Subsection 2.2.1; see Section 2 in [CCG05] and Lemma 1.5 in [CCG00]. Now, it should be observed that

$$T(n) := \sum_{\gamma \in G(n)} p_n(\gamma) \sum_{j=1}^n \left(\frac{\Lambda}{2}\right)^{\delta_j(\gamma)}$$

is the expectation of the random variable $W(\cdot) := \sum_{j=1}^n \left(\frac{\Lambda}{2}\right)^{\delta_j(\cdot)}$, with respect to the probability law induced by the coefficients $p_n(\cdot)$. According to Lemma 1.4 in [CCG00], $T(n)$ goes to zero as $n$ diverges. Notice that, as mentioned in Subsection 2.2.2, $T(n)$ is the same as $V_n(c/2, 1)$. So, combining of (2.4.3) and (2.4.4) with Proposition 8 in [GR05] yields

*For every initial data $f_0$ with finite fourth moment, one has*

$$\|Q_n^+(f_0, f_0) - M\| \leqslant An^{-\Lambda}.$$

This, in turn, due to the above-mentioned interpolation inequalities, gives the main Theorem in [CCG05].

**Theorem 2.4.1** *Let $f_0$ be a probability density, with finite Linnik information, possessing moments of every order. Then, for any $\varepsilon > 0$, there is a finite constant $C$ depending only on the moments of $f_0$ and on $\varepsilon$, so that*

$$\|Q_n^+(f_0, f_0) - M\|_{L^1(\mathbb{R})} \leqslant Cn^{\varepsilon - \Lambda}.$$

With this result in hand, one can write

$$\|f(t, \cdot) - M\|_{L^1} = \left\|\sum_{n=1}^\infty e^{-t}(1 - e^{-t})^{n-1}\{Q_n^+(f_0, f_0) - M\}\right\|_{L^1}$$

$$\leqslant \sum_{n \geqslant 1} e^{-t}(1 - e^{-t})^{n-1} \|Q_n^+(f_0, f_0) - M\|_{L^1}$$

$$\leqslant c \sum_{n \geqslant 1} e^{-t}(1 - e^{-t})^{n-1} n^{-\Lambda + \varepsilon}$$

$$= c e^{-t} \sum_{n \geqslant 1} \frac{(1 - e^{-t})^n}{(1 + n)^{\Lambda - \varepsilon}} = c e^{-t} L\big(1 - e^{-t}, \Lambda - \varepsilon, 1\big),$$

where L is Lerch's transcendent. Finally, from the well-known limiting properties of $L(z, 1, \alpha)$ as $z \to 1$ (see, for example, [MOS66], p. 33), one gets

$$c e^{-t} L\big(1 - e^{-t}, \Lambda - \varepsilon, 1\big) \sim c \Gamma(1 - \Lambda + \varepsilon) e^{-t(\Lambda - \varepsilon)} \qquad t \to +\infty,$$

$\Gamma$ standing, as usual, for the Euler gamma function. This can be formalized in the following.

**Theorem 2.4.2** *Let $f_0$ be a probability density, with finite Linnik information such that $\int_{\mathbb{R}} v^2 f_0(v) \, dv = 1$, and possessing moments of every order. Let $\varepsilon > 0$, and $C$ be the constant in Theorem 2.4.1 and let $f(\cdot, t)$ be the solution of (2.1.13) with initial data $f_0$; then*

$$\|f(\cdot, t) - M\|_{L^1(\mathbb{R})} \leqslant C \, \Gamma(1 + \Lambda + \varepsilon) e^{-t(\Lambda - \varepsilon)}.$$

## 2.5 Further research

There are at least two lines of research that could be viewed as the natural developments of the subject analysed in the previous sections. The former could point to extending the concluding result in the last section to pseudo-Maxwellian molecules. In other words, one could try to deduce the thesis of the main result reviewed in Section 2.4 for equation (2.1.10), without assuming Sobolev regularity for the initial data.

The latter line of research could be directed towards the analysis of how far from necessity the hypothesis of existence of moments of every order and of Linnik information is. A study of this kind could start by determining necessary and sufficient conditions for convergence to equilibrium in the topology of weak convergence of probability measures, and continue with the assessment of the rate of convergence, established as a function of the maximum moment order admitted, with respect to the various measures of discrepancy such as: uniform metric for distribution functions (Kolmogorov); Wasserstein (Kentorovich) metrics or, more properly, dissimilarity metrics (Gini); variational distance. This way of proceeding could indicate the minimum moment order which guarantees the optimal rate of convergence. At the state of things, this part of the program would profit by an interpretation of the solution of the equations of interest as the law of a sum of a (random) number of random numbers, which follows from the McKean representation of the Wild sum.

With these elements in hand one could rely on some adaptation (necessary because of the lack, in the present case, of independence of the summands) of the classical central limit theorem in order to "initialize" the program sketched above.

## Acknowledgments

The author would like to thank Eugenio Regazzini for stimulating discussions and helpful suggestions.

This work was performed under the auspices of the *National Group for Mathematical Physics* of the *Istituto Nazionale di Alta Matematica* and was partially supported by *Italian Ministery of University* (MIUR National Project "Mathematical Problems of Kinetic Theories," Cofin2004) and the IMATI (CNR-Pavia, Italy).

# References

[Ar88]  L. Arkeryd: *Stability in $L^*$ for the spatially homogeneous Boltzmann equation*. Arch. Rat. Mech. Anal. 103, 151–167 (1988).

[BGR05]  F. Bassetti, E. Gabetta, E. Regazzini: *On the depth of the trees in the McKean representation of Wild's sums*. IMATI Publication N.27-PV (2005); to appear in Transport Theory and Stat. Phys.

[CC03]  E. A. Carlen, M. C. Carvalho: *Probabilistic methods in kinetic theory*. Riv. Mat. Univ. Parma (7) $2^*$, 101–149 (2003).

[CCG00]  E. A. Carlen, M. C. Carvalho, E. Gabetta: *Central limit theorem for Maxwellian molecules and truncation of the Wild expansion*. Comm. Pure Appl. Math., 53, 370–397 (2000).

[CCG05]  E. A. Carlen, M. C. Carvalho, E. Gabetta: *On the relation between rates of relaxation and convergence of Wild sums for solutions of the Kac equation*. J. Func. Anal. 220, 362–387 (2005).

[CGT99]  E. A. Carlen, E. Gabetta, G. Toscani: *Propagation of smoothness and the rate of exponential convergence to equilibrium for a spatially homogeneous Maxwellian gas*. Commun. Math. Phys. 305, 521–546 (1999).

[CIP94]  C. Cercignani, R. Illner, M. Pulvirenti: *The Mathematical Theory of Dilute Gases*. Springer, New York (1994).

[CLS88]  C. Cercignani, M. Lampis, C. Sgarra: *$L^2$-stability near equilibrium of the solution of the homogeneous Boltzmann equation in the case of Maxwellian molecules*. Meccanica 23, 15–18 (1988).

[Co70]  L. Comtet: *Analyse Combinatoire*, Presses Universitaire de France, Paris (1970).

[Fi22]  R. A. Fisher: *On the mathematical foundations of theoretical statistics*. Philos. Trans. R. Soc. London. A222, 309–368. CP18 (1922).

[Fi25]  R. A. Fisher: *Theory of statistical estimation*. Proc. Cambridge Philos. Soc. 22, 700–725, CP42 (1925).

[Fo99]  G. B. Folland: *Real Analysis: Modern Techniques and Their Applications (2nd ed.)*, Wiley, New York (1999).

[GR05]     E. Gabetta, E. Regazzini: *Some new results for McKean's graphs with applications to Kac's equation.* IMATI Publications N.28-PV (2005); to appear in Jour. Stat. Phys.

[GTW95]   E. Gabetta, G. Toscani, B. Wennberg: *Metrics for probability distributions and the trend to equilibrium for solutions of the Boltzmann equation.* J. Stat. Physics, 81, nos. 5/6, 901–934 (1995).

[IT56]     E. Ikenberry, C. Truesdell: *On the pressure and the flux of energy according to Maxwell's kinetic energy.* I. J. Rat. Mech. Anal. 5, 1–54 (1956).

[Ka56]     M. Kac: *Foundations of kinetic theory.* Proceedings of the third Berkeley Symposium on Mathematical and Statistical Problems. J. Neyman, ed. University of California, vol. 3, 171–197 (1956).

[Ka59]     M. Kac: *Probability and Related Topics in Physical Sciences.* (Lectures in Applied Mathematics) Interscience Publishers, Ltd, London (1959).

[Li59]     Yu. V. Linnik: *An information-theoretic proof of the central limit theorem with the Lindeberg condition.* Theory of Prob. and its Applications, vol. IV, no. 3, 288–299 (1959).

[MK66]     H. P. McKean Jr.: *Speed of approach to equilibrium for Kac's caricature of a Maxwellian gas.* Arch. Rat. Mech. Anal. 21, 343–367 (1966).

[MK67]     H. P. McKean Jr.: *An exponential formula for solving Boltzmann's equation for a Maxwellian gas.* J. Combin. Theory, 2, 358–382 (1967).

[Mo54]     D. Morgenstern: *General existence and uniqueness proof for spatially homogeneous solutions of the Maxwell-Boltzmann equation in the case of Maxwellian molecules.* Proc. Nat. Acad. Sci. 40, 719–721 (1954).

[MOS66]   W. Magnus, F. Oberhettinger, R. P. Soni: *Formulas and Theorems for the Special Functions of Mathematical Physics,* Springer, Berlin (1966).

[Ra91]     S. T. Rachev: *Probability Metrics and the Stability of Stochastic Modules,* Wiley, Chichester (1991).

[Vi03]     C. Villani: *Topics on Optimal Transportation.* Am. Math. Soc., Providence, RI (2003).

[We93]     B. Wennberg: *Stability and exponential convergence in $L^p$ for the spatially homogeneous Boltzmann equation.* Nonlinear Anal. 20, no. 8, 935–964 (1993).

[Wi51]     I. Wild: *On Boltzmann's equation in the kinetic theory of gases.* Proc. Cambridge Philos. Soc. 47, 602–609 (1951).

[WU70]     C. S. Wang Chang, G. E. Uhlenbeck: *The Kinetic Theory of Gases.* In: *Studies in Statistical Mechanics V,* eds. J. de Boer and G. E. Uhlenbeck, Amsterdam: North-Holland Publishing Co., 1970.

# Nonresonant velocity averaging and the Vlasov–Maxwell system

François Golse

Université Paris 7 & Laboratoire Jacques-Louis Lions Boîte courrier 187, F75252 Paris cedex 05 golse@math.jussieu.fr

## 3.1 Mean-field equations of Vlasov type

The Vlasov equation governs the number density in single-particle phase space of a large particle system (typically a rarefied ionized gas or plasma), subject to some external force field (for instance the Lorentz force acting on charged particles). Most importantly, collisions between particles are neglected in the Vlasov equation, unlike the case of the Boltzmann equation. Hence the only possible source of nonlinearity in the Vlasov equation for charged particles is the self-consistent electromagnetic field created by charges in motion: each particle is subject to the electromagnetic field created by all the particles other than itself.

The Vlasov equation reads

$$\partial_t \mathbf{f} + \mathrm{div}_x(\mathbf{v}(\xi)\mathbf{f}) + \mathrm{div}_\xi(\mathbf{F}(t,x)\mathbf{f}) = 0, \tag{3.1.1}$$

where $\mathbf{f} \equiv \mathbf{f}(t,x,\xi) \in M_N(\mathbb{R})$ is the diagonal matrix of number densities for the system of particles considered. Specifically, there are $N$ different species of particles in the system, and

$$\mathbf{f}(t,x,\xi) = \begin{pmatrix} f_1(t,x,\xi) & 0 & 0 \\ 0 & \ddots & 0 \\ 0 & 0 & f_N(t,x,\xi) \end{pmatrix},$$

where $f_j(t,x,\xi)$ is the density of particles of the $j$th species located at the position $x \in \mathbb{R}^3$ with momentum $\xi \in \mathbb{R}^3$ at time $t$. Likewise

$$\mathbf{v}(\xi) = \begin{pmatrix} v_1(\xi) & 0 & 0 \\ 0 & \ddots & 0 \\ 0 & 0 & v_N(\xi) \end{pmatrix},$$

where $v_j(\xi)$ is the velocity of particles of the $j$th species with momentum $\xi$, while

$$\mathbf{F}(t, x) = \begin{pmatrix} F_1(t, x) & 0 & 0 \\ 0 & \ddots & 0 \\ 0 & 0 & F_N(t, x) \end{pmatrix},$$

where $F_j(t, x)$ is the force field at time $t$ and position $x$ acting on particles of the $j$th species. The divergence operators act entrywise on their arguments, meaning that

$$\text{div}_x(\mathbf{v}(\xi)\mathbf{f}(t, x, \xi))$$

$$= \begin{pmatrix} \text{div}_x(v_1(\xi)f_1(t, x, \xi)) & 0 & 0 \\ 0 & \ddots & 0 \\ 0 & 0 & \text{div}_x(v_N(\xi)f_N(t, x, \xi)) \end{pmatrix},$$

while

$$\text{div}_\xi(\mathbf{F}(t, x)\mathbf{f}(t, x, \xi))$$

$$= \begin{pmatrix} \text{div}_\xi(F_1(t, x)f_1(t, x, \xi)) & 0 & 0 \\ 0 & \ddots & 0 \\ 0 & 0 & \text{div}_\xi(F_N(t, x)f_N(t, x, \xi)) \end{pmatrix}.$$

Henceforth

$$\mathbf{m} = \begin{pmatrix} m_1 & 0 & 0 \\ 0 & \ddots & 0 \\ 0 & 0 & m_N \end{pmatrix}, \quad \mathbf{q} = \begin{pmatrix} q_1 & 0 & 0 \\ 0 & \ddots & 0 \\ 0 & 0 & q_N \end{pmatrix},$$

where $m_j$ and $q_j$ are respectively the mass and the charge of particles of the $j$th species.

## The relativistic Vlasov–Maxwell model

This is the fundamental model for relativistic particles with strong electromagnetic coupling. Hence, denoting by $c$ the speed of light in a vacuum,

$$\mathbf{v}(\xi) = \nabla_\xi \mathbf{e}(\xi), \quad \mathbf{F}(t, x) = E(t, x)\mathbf{q} - \tfrac{1}{c}B(t, x) \times \nabla_\xi \mathbf{e}(\xi)\mathbf{q}, \tag{3.1.2}$$

where

$$\mathbf{e}(\xi) = (\mathbf{m}^2 c^4 + c^2|\xi|^2\mathbf{I})^{1/2}, \tag{3.1.3}$$

while $E \equiv E(t, x) \in \mathbb{R}^3$ and $B \equiv B(t, x) \in \mathbb{R}^3$ are respectively the electric and the magnetic field at time $t$ and position $x$. They are governed by the system of Maxwell's equations

$$\begin{aligned} \text{div}_x B = 0, \quad & \partial_t B + c\,\text{curl}_x E = 0, \\ \text{div}_x E = \rho, \quad & \partial_t E - c\,\text{curl}_x B = -j. \end{aligned} \tag{3.1.4}$$

The charge density $\rho$ is defined as

$$\rho(t, x) = \int_{\mathbb{R}^3} \text{trace}(\mathbf{q}\mathbf{f}(t, x, \xi))d\xi, \tag{3.1.5}$$

while the current density $j \equiv j(t, x) \in \mathbb{R}^3$ is given by

$$j(t, x) = \int_{\mathbb{R}^3} \text{trace}(\mathbf{q}\mathbf{v}(\xi)\mathbf{f}(t, x, \xi))d\xi. \tag{3.1.6}$$

The main mathematical problem concerning the Vlasov–Maxwell system is the question of global existence and uniqueness of smooth solutions of the Cauchy problem, which remains open at the time of this writing.

Since the time-dependent vector field

$$\mathbb{R}^3 \times \mathbb{R}^3 \ni (x, \xi) \mapsto (v_j(\xi), F_j(t, x)) \in \mathbb{R}^3 \times \mathbb{R}^3$$

is divergence free for each $j = 1, \ldots, N$, the quantity

$$\|f_j(t, \cdot, \cdot)\|_{L^p(\mathbb{R}^3 \times \mathbb{R}^3)} = \text{Const.}$$

is an invariant of the motion for each $j = 1, \ldots, N$ and each $1 \le p \le \infty$.

The total energy of the particle system is also an invariant of the motion. It reads

$$\frac{1}{2} \iint_{\mathbb{R}^3 \times \mathbb{R}^3} \text{trace}(\mathbf{e}(\xi)\mathbf{f}(t, x, \xi))dxd\xi + \frac{1}{2} \int_{\mathbb{R}^3} (|E|^2 + |B|^2)(t, x)dx = \text{Const.}$$

Since all the matrices $\mathbf{m}$, $\mathbf{e}(\xi)$ and $\mathbf{f}(t, x, \xi)$ have nonnegative entries, there are no cancellations in the expression above, so that the energy conservation implies a priori estimates of the form

$$\iint_{\mathbb{R}^3 \times \mathbb{R}^3} \sqrt{m_j^2 c^4 + c^2 |\xi|^2} f_j(t, x, \xi)dxd\xi \le \text{Const.}$$

and

$$\int_{\mathbb{R}^3} (|E|^2 + |B|^2)(t, x)dx \le \text{Const.}$$

for the Vlasov–Maxwell system.

## 3.2 A kinetic formulation of the Maxwell equations

Henceforth, we consider a classical solution $(\mathbf{f}, E, B)$ of the relativistic Vlasov–Maxwell system (3.1.1), (3.1.2), (3.1.4), (3.1.5), (3.1.6), with initial data

$$\mathbf{f}\big|_{t=0} = \mathbf{f}^{in}, \quad E\big|_{t=0} = E^{in}, \quad B\big|_{t=0} = B^{in}, \tag{3.2.1}$$

where $\mathbf{f}^{in}$, $E^{in}$ and $B^{in}$ are at least of class $C^\infty$ in all their variables, and satisfy the compatibility conditions

$$\operatorname{div}_x E^{in} = \int_{\mathbb{R}^3} \operatorname{trace}(\mathbf{q}\mathbf{f}^{in})d\xi\,, \quad \operatorname{div}_x B^{in} = 0\,. \tag{3.2.2}$$

For simplicity, we shall assume moreover that $f^{in}$, $E^{in}$ and $B^{in}$ are compactly supported.

It will be especially convenient to represent the electromagnetic field in terms of the distribution of Liénard–Wiechert potentials created by each one of the moving charged particles in the system considered. For a classical presentation of Liénard–Wiechert potentials, see for instance §14.1 in [14], or §63 in [16]. Here, we propose a slightly different (and yet equivalent) formulation of that notion.

Define $\mathbf{u} \equiv \mathbf{u}(t, x, \xi) \in M_N(\mathbb{R})$ to be the solution of

$$\begin{aligned}\Box_{t,x}\mathbf{u} = \mathbf{f}\,, \quad t > 0\,, \ x, \xi \in \mathbb{R}^3\,,\\ \mathbf{u}\big|_{t=0} = \partial_t\mathbf{u}\big|_{t=0} = 0\,,\end{aligned} \tag{3.2.3}$$

where $\Box_{t,x} = \partial_t^2 - c^2\Delta_x$ is the d'Alembert operator. The $j$th diagonal entry of $\mathbf{u}(t, x, \xi)$ is exactly the distribution of Liénard–Wiechert potentials created at time $t$ by particles of the $j$th species distributed under $f_j^{in} \equiv f_j^{in}(x, \xi)$ initially.

Define then the electromagnetic potential

$$\begin{aligned}\mathbf{\Phi}(t, x) = \int_{\mathbb{R}^3} \mathbf{q}\mathbf{u}(t, x, \xi)d\xi\,,\\ \mathbf{A}(t, x) = \int_{\mathbb{R}^3} \mathbf{q}\mathbf{v}(\xi)\mathbf{u}(t, x, \xi)d\xi\,.\end{aligned} \tag{3.2.4}$$

We also define (a diagonal matrix of) vector potentials $\mathbf{A}^0 \equiv \mathbf{A}^0(t, x)$ so that

$$\Box_{t,x}\mathbf{A}^0 = 0 \tag{3.2.5}$$

with the following compatibility conditions:

$$\operatorname{div}_x \mathbf{A}^0\big|_{t=0} = 0\,, \quad \operatorname{div}_x \partial_t\mathbf{A}^0\big|_{t=0} = -\int_{\mathbb{R}^3} \mathbf{q}\mathbf{f}^{in}(x, \xi)d\xi\,, \tag{3.2.6}$$

as well as

$$\operatorname{curl}_x \operatorname{trace}(\mathbf{A}^0)\big|_{t=0} = B^{in}\,, \quad \operatorname{trace}(\partial_t A^0)\big|_{t=0} = -E^{in}\,. \tag{3.2.7}$$

One easily checks that $(\mathbf{\Phi}, \mathbf{A}^0 + \mathbf{A})$ is the electromagnetic potential leading to the electromagnetic field $(E, B)$ by the formulas

$$E = -\partial_t \operatorname{trace}(\mathbf{A}^0 + \mathbf{A}) - \nabla_x\mathbf{\Phi}\,, \quad B = \operatorname{curl}_x \operatorname{trace}(\mathbf{A}^0 + \mathbf{A})$$

and satisfying the Lorentz gauge entrywise:

$$\partial_t \boldsymbol{\Phi} + \mathrm{div}_x(\mathbf{A}^0 + \mathbf{A}) = 0 .$$

Hence one can replace the Vlasov–Maxwell system (3.1.1), (3.1.2), (3.1.4), (3.1.5), (3.1.6) with the equivalent system (3.1.1), (3.2.3) with the following formula for the Lorentz force:

$$\mathbf{F}(t,x) = -\tfrac{1}{c}\,\mathrm{curl}_x\,\mathrm{trace}\left(\mathbf{A}^0 + \int_{\mathbb{R}^3} \mathbf{q}\mathbf{v}(\xi)\mathbf{u}(t,x,\xi)d\xi\right) \times \nabla_\xi \mathbf{e}(\xi)\mathbf{q}$$
$$- \mathrm{trace}\left(\partial_t \mathbf{A}^0 + \partial_t \int_{\mathbb{R}^3} \mathbf{q}\mathbf{v}(\xi)\mathbf{u}(t,x,\xi)d\xi + \nabla_x \int_{\mathbb{R}^3} \mathbf{q}\mathbf{u}(t,x,\xi)d\xi\right)\mathbf{q} . \tag{3.2.8}$$

Observe that the (diagonal matrix of) vector potentials $\mathbf{A}^0$ can be chosen as smooth as the initial data $(f^{in}, E^{in}, B^{in})$, i.e., of class $C^\infty$ in all its variables, since the wave equation (3.2.5) propagates the regularity of the initial data (3.2.6), (3.2.7) of $\mathbf{A}^0$. Hence the only possibility for a finite time blow-up of classical solutions of the relativistic Vlasov–Maxwell model would therefore come from the **f-u** coupling in the system (3.1.1), (3.2.3). In the next section, we shall analyze carefully some conditional smoothing mechanisms for such systems.

## 3.3 Nonresonant velocity averaging for transport+wave systems

Throughout this section, we set $c = 1$. We are concerned with coupled transport+wave systems of the form

$$\Box_{t,x}u(t,x,\xi) = f(t,x,\xi)$$
$$(\partial_t + v(\xi)\cdot\nabla_x)f(t,x,\xi) = P(D_\xi)g(t,x,\xi), \tag{3.3.1}$$

where $P(D_\xi)$ is a differential operator of order $m \geq 0$. Specifically, we are interested in the local regularity in $(t,x)$ of averages of $u$ of the form

$$\bar{U}(t,x) = \int_{\mathbb{R}^D} u(t,x,\xi)\phi(\xi)d\xi .$$

Observe that the expression of the Lorentz force in (3.2.8) involves precisely averages of **u** of this type, instead of **u** itself.

For simplicity, we consider first regularity estimates in $L^2$-based Sobolev spaces. Assume that $f, u, g \in L^2_{loc}(\mathbb{R} \times \mathbb{R}^D \times \mathbb{R}^D)$ while $v \in C^\infty(\mathbb{R}^D; \mathbb{R}^D)$ and $\phi \in C^\infty_c(\mathbb{R}^D)$.

Under the full-rank condition

$$\mathrm{rank}\,\nabla_\xi v(\xi) = D \quad \text{for each } \xi \in \mathrm{supp}\,\phi, \tag{3.3.2}$$

the classical velocity averaging lemma implies that

$$\int_{\mathbb{R}^D} f(t,x,\xi)\phi(\xi)d\xi \in H_{loc}^{1/2(m+1)}(\mathbb{R} \times \mathbb{R}^D).$$

On the other hand

$$\Box_{t,x}U = \int_{\mathbb{R}^D} f\phi d\xi$$

and the usual energy estimate for the wave equation on $u$, obtained by multiplying each side of that equation by $\partial_t U$, leads to

$$\partial_t \frac{1}{2}\left(|\partial_t U|^2 + |\nabla_x U|^2\right) - \mathrm{div}_x(\partial_t U \nabla_x U) = \partial_t U \int_{\mathbb{R}^D} f d\xi.$$

After localizing in $(t,x)$ and integrating in $x$, this clearly shows that $U$ gains one derivative in $L^2$ in each variable $t,x$ over the average of $f$:

$$\int_{\mathbb{R}^D} f(t,x,\xi)\phi(\xi)d\xi.$$

Hence

$$U \in H_{loc}^{1+\frac{1}{2(m+1)}}(\mathbb{R} \times \mathbb{R}^D).$$

Going back to our formulation (3.1.1), (3.2.3), we see that, in order for the vector field

$$\mathbb{R}^3 \times \mathbb{R}^3 \ni (v_j(\xi), F_j(t,x)) \in \mathbb{R}^3 \times \mathbb{R}^3$$

to generate a unique characteristic flow, $F_j$ should be locally Lipschitz continuous in $x$ uniformly in $t$, for each $j = 1, \dots, N$. In terms of the distribution of Liénard–Wiechert potentials $\mathbf{u}$, this amounts precisely to controlling second-order derivatives of averages of $\mathbf{u}$, as can be seen from (3.2.8). Unfortunately, the strategy based on the classical velocity averaging lemma as explained above (in the most favorable $L^2$ setting) fails to gain that much regularity—in fact, $m = 1$ in the Vlasov equation, so that the best one can hope for with this method is a gain of $1 + \frac{1}{4}$ derivatives in $(t,x)$, which is not enough to allow us to define characteristics for the Vlasov–Maxwell system.

However, this approach to the regularity question leaves aside an important feature of the Vlasov–Maxwell system. The electromagnetic field consists of waves that propagate at the speed of light, whereas the charged particles, all of which have positive mass, move at a lesser speed. Indeed, the speed of a particle of mass $m$ and momentum $\xi$ is

$$\frac{c^2|\xi|}{\sqrt{m^2c^4 + c^2|\xi|^2}} < c.$$

The fact that the speed of propagation in the wave equation (3.2.3) is larger than the speed of particles $v(\xi)$ in the Vlasov equation (3.1.1) leads to a new regularizing mechanism, which we now explain.

Consider the system (3.3.1). We shall call this system *nonresonant* if

$$|v(\xi)| < 1, \quad \text{for each } \xi \in \mathbb{R}^D. \tag{3.3.3}$$

**Theorem 3.3.1 (Bouchut-Golse-Pallard [2])** *Let $u, f, g$ in $L^2_{loc}(\mathbb{R}_t \times \mathbb{R}^D_x \times \mathbb{R}^D_\xi)$ satisfy (3.3.1) with $v \in C^\infty(\mathbb{R}^D; \mathbb{R}^D)$. Assume that this system is nonresonant. Then, for each $\phi \in C^\infty_c(\mathbb{R}^D_\xi)$, one has*

$$U(t, x) = \int_{\mathbb{R}^D} u(t, x, \xi)\phi(\xi)d\xi \in H^2_{loc}(\mathbb{R} \times \mathbb{R}^D).$$

This result generalizes the fact that the operator $\Box_{t,x}$ is microlocally elliptic on the null space of the transport operator whenever $|v(\xi)| < 1$.

There is also an interesting difference with the strategy based on the usual velocity averaging lemma described above. Indeed, this new method leads to a gain of 2 derivatives on momentum averages of $u$ in the nonresonant case—without gaining more than $1 + \frac{1}{2(m+1)}$ derivatives in $(t, x)$ on momentum averages of $f$ itself.

To see the importance of the nonresonance condition (3.3.3), we briefly sketch the proof of Theorem 3.3.1.

*Proof (Sketch of the proof).* Set $T^\pm_\xi = \partial_t \pm v(\xi) \cdot \nabla_x$ and consider the second-order differential operator

$$Q_\xi = \Box_{t,x} - \lambda T^-_\xi T^+_\xi.$$

First, one checks that

$$\begin{aligned}
Q_\xi u = f - \lambda T^-_\xi \Box^{-1}_{t,x} D^m_\xi g &= f - \lambda D^m_\xi \Box^{-1}_{t,x} T^-_\xi g - \lambda \Box^{-1}_{t,x}[T^-_\xi, D^m_\xi]g \\
&= f - \lambda D^m_\xi \Box^{-1}_{t,x} T^-_\xi g - \lambda \Box^{-1}_{t,x} D^m_\xi v(\xi) \cdot \nabla_x g \\
&= a + d^m_\xi b \in L^2_{loc}(dtdxd\xi) + D^m_\xi L^2_{loc}(dtdxd\xi).
\end{aligned}$$

Here, we have denoted by $\Box^{-1}_{t,x}$ the operator defined by $\Box^{-1}_{t,x}\psi = \Psi$ where $\Psi$ is the solution of the Cauchy problem

$$\Box_{t,x}\Psi = 0, \quad x \in \mathbb{R}^D, \; t > 0,$$
$$\Psi\big|_{t=0} = 0,$$
$$\partial_t\Psi\big|_{t=0} = \psi.$$

Next, we observe that, for $\xi \in \operatorname{supp}\phi$ and $\lambda$ such that

$$\sup_{\xi \in \operatorname{supp}\phi} |v(\xi)| < \lambda < 1,$$

the operator $Q_\xi$ is elliptic for each $\xi \in \operatorname{supp}\phi$.

More precisely, denoting by $q_\xi(\omega, k)$ the symbol of $Q_\xi$, one has

$$\sup_{\xi \in \operatorname{supp}\phi} \left| D^m_\xi\left(\frac{1}{q_\xi(\omega, k)}\right) \right| \le \frac{C_m}{\omega^2 + |k|^2},$$

where $C_m$ may depends on $m$ but is *uniform* in $\xi$. Then

$$\int_{\mathbb{R}^D} \hat{u}\phi(\xi)d\xi = \int \frac{\hat{a}}{q_\xi(\omega,k)}\phi(\xi)d\xi + (-1)^m \int_{\mathbb{R}^D} D_\xi^m \left(\frac{\phi(\xi)}{q_\xi(\omega,k)}\right) \hat{b}d\xi$$

with $\hat{a}$ and $\hat{b} \in L^2_{\omega,k,\xi}$ have $H^2$-decay in $\omega, k$.

*Remarks*

(a) First, one easily checks that none of the assumptions in Theorem 3.3.1 can be dispensed with.

(b) That one gains 2 derivatives is special to the $L^2$-case, since $\Box^{-1}_{t,x}$ gains 1 derivative in $(t,x)$ by the energy estimate for the wave equation.

In $L^p$ with $1 < p < \infty$, $\Box^{-1}_{t,x}$ gains $1-(D-1)|\frac{1}{2}-\frac{1}{p}|$ derivatives in $(t,x)$—see for instance [18], [19])—whenever $|\frac{1}{2} - \frac{1}{p}| \le \frac{1}{D-1}$.

Using this result and the Mihlin-Hörmander theorem on $L^p$ multipliers—see for instance Theorem 3 on p. 96 of [20]—the same method as above shows that

$$\int_{\mathbb{R}^D} u(t,x,\xi)\phi(\xi)d\xi \in W^{1+\gamma,p}_{loc}(\mathbb{R} \times \mathbb{R}^D) \text{ with } \gamma = 1 - (D-1)\left|\frac{1}{2} - \frac{1}{p}\right|.$$

This result, due to C. Pallard [17] suggests a gain of 1 derivative in $L^1$ or $L^\infty$ in space dimension 3. The regularity statement above is still true in these limiting cases of nonresonant velocity averaging; however, the proof rests on the explicit formula for $\Box^{-1}_{t,x}$—the forward fundamental solution of the d'Alembert operator—in physical (instead of Fourier) space. In the case of 3 space dimensions, this fundamental solutions turns out to be a measure, and thus behaves nicely with $L^\infty$ data. The proof also uses a "division lemma" discussed in the next section. See [17] for a complete proof of the $L^p$ variant of nonresonant velocity averaging, including the aforementioned limiting cases.

## 3.4 Applications to the Vlasov–Maxwell system

### 3.4.1 A conditional regularity result

R. DiPerna and P.-L. Lions [4] have proved that the Cauchy problem for the Vlasov–Maxwell system has globally defined renormalized solutions for initial data with finite energy. However, their method does not allow defining characteristic curves for the Vlasov equation; i.e., trajectories for the charged particles governed by the Vlasov–Maxwell system. Their analysis, written in the case of the classical Vlasov–Maxwell system, i.e., for

$$\mathbf{v}(\xi) = \mathbf{m}^{-1}\xi,$$

obviously applies to the relativistic Vlasov–Maxwell system considered here, which is somewhat more consistent on physical grounds, since the Maxwell equations are themselves a relativistic model.

By using both the standard velocity averaging argument for large momenta and the nonresonant velocity averaging method for momenta below some threshold $R$, one arrives at the following conditional result, upon optimizing in $R$.

**Theorem 3.4.1 (Bouchut-Golse-Pallard [2])** *Consider the relativistic Vlasov–Maxwell system* (3.1.1), (3.1.2), (3.1.4), (3.1.5), (3.1.6) *with initial condition* (3.2.1). *Assume that the initial data satisfy*

$$0 \le \mathbf{f}^{in} \in L^\infty(\mathbb{R}^3 \times \mathbb{R}^3) \,, \quad E^{in} \text{ and } B^{in} \in H^1_{loc}(\mathbb{R}^3)$$

*with the compatibility condition*

$$\operatorname{div} B^{in} = 0 \,, \quad \operatorname{div} E^{in} = \int_{\mathbb{R}^3} \operatorname{trace}(\mathbf{q}\mathbf{f}(t,x,\xi))d\xi$$

*and the finite energy condition*

$$\iint_{\mathbb{R}^3 \times \mathbb{R}^3} \operatorname{trace}(\mathbf{e}(\xi)\mathbf{f}^{in}(x,\xi))d\xi dx + \frac{1}{2}\int_{\mathbb{R}^3}(|E^{in}|^2 + |B^{in}|^2)(x)dx < \infty \,.$$

*Let* $(f, E, B)$ *be a renormalized solution of the Vlasov–Maxwell system with those initial data. If the macroscopic energy density satisfies*

$$\int_{\mathbb{R}}^3 \operatorname{trace}(\mathbf{e}(\xi)\mathbf{f}^{in}(t,x,\xi))d\xi \in L^p_{loc}(\mathbb{R}_+ \times \mathbb{R}^3) \text{ with } \frac{3}{2} < p \le 2,$$

*then the electromagnetic field has Sobolev regularity*

$$(E, B) \in H^s_{loc}(\mathbb{R}^*_+ \times \mathbb{R}^3) \text{ with } s < \frac{4p-6}{4p+3} \,.$$

See [2] for a proof of this result. The theorem above falls short of providing the amount of regularity on the electromagnetic field that one would need in order to define a characteristic flow, even in a generalized sense—see [1], [5]. Perhaps, its main interest is to indicate the relevance of the idea of nonresonant velocity averaging in the context of the Vlasov–Maxwell system. Most likely, further ideas are needed in order to apply the method of nonresonant velocity averaging to the Vlasov–Maxwell system with more convincing output.

### 3.4.2 A new proof of the Glassey–Strauss theorem

Consider the Cauchy problem for the Vlasov–Maxwell system (3.1.1), (3.1.2), (3.1.4), (3.1.5), (3.1.6) with initial condition (3.2.1). As mentioned above, R. DiPerna and P.-L. Lions [4] have proved that this Cauchy problem has globally defined renormalized solutions for initial data with finite energy.

However, such solutions are not known to be uniquely defined by their initial data. Besides, one would expect that the regularity of initial data propagates, so that it seems reasonable to seek classical solutions, with $(f, E, B)$ at least of class $C^1$. The benefit of dealing with classical solutions is twofold: first, such solutions are uniquely defined by their initial data. Furthermore, one can define characteristic curves of the Vlasov equation (3.1.1) for $C^1$ solutions by a simple application of the Cauchy–Lipschitz theorem.

Unfortunately, global existence of classical solutions of the Vlasov–Maxwell system for any $C^1$ initial data of arbitrary size with good enough decay property at infinity remains an open problem. The best result in that direction is the following theorem.

**Theorem 3.4.2 (R. Glassey-W. Strauss [13])** *Let* $\mathbf{f} \in C^1([0, T) \times \mathbb{R}^3 \times \mathbb{R}^3)$ *and* $E, B \in C^1([0, T) \times \mathbb{R}^3)$ *be a solution of the Vlasov–Maxwell system (3.1.1), (3.1.2), (3.1.4), (3.1.5), (3.1.6) with initial condition (3.2.1). Assume that* $\mathbf{f}^{in} \in C_c^1(\mathbb{R}^3 \times \mathbb{R}^3)$ *and that* $E^{in}, B^{in} \in C_c^2(\mathbb{R}^3)$ *with*

$$\mathrm{div}_x\, E^{in} = \int_{\mathbb{R}^3} \mathrm{trace}(\mathbf{q}\mathbf{f}^{in})d\xi\,, \quad \mathrm{div}_x\, B^{in} = 0.$$

*If*

$$\overline{\lim}_{\,t \to T^-} \|\mathbf{f}(t)\|_{Lip_{x,\xi}} + \|(E, B)(t)\|_{Lip_x} = +\infty,$$

*then*

$$\overline{\lim}_{\,t \to T^-} R_{\mathbf{f}}(t) = +\infty,$$

*where*

$$R_{\mathbf{f}}(t) = \inf\{r > 0 \,|\, \mathbf{f}(t, x, \xi) = 0 \text{ for each } x \in \mathbb{R}^3 \text{ and } |\xi| > r\}\,.$$

The original proof of this result is fairly hard to read in detail, although the general strategy is very clearly explained in [13]. For this reason, together with the considerable interest in the result itself, there have been some attempts at finding alternatives to the original proof. One is due to S. Klainerman and G. Staffilani [15]; although perhaps not very much simpler than the original proof, it is based on a completely new and different idea which may be of great interest in further understanding the Vlasov–Maxwell system.

The proof of the Glassey–Strauss theorem sketched below originates from [3]. The underlying strategy is essentially the same as in the original proof; however, it is much simpler in two very different respects. A first, considerable simplification over [13] comes from the representation of the electromagnetic field $(E, B)$ in terms of the number densities $\mathbf{f}$: whereas the original argument led the reader through complicated manipulations of integrals involving vector analysis, using the kinetic formulation of the Maxwell equations in terms of the distribution of Liénard–Wiechert potentials as in Section 3.2 reduces that burden to performing similar manipulations on *scalar* solutions of the wave equation.

But the most important part of the Glassey–Strauss analysis was a subtle decomposition of the vector fields corresponding to space- and time-derivatives into their projection on the wave cone and the free streaming operator. This decomposition was then used in the representation of the electromagnetic field with several integrations by parts to smooth out the singularities of the integral kernels involved.

In our analysis, this last step is replaced by a "division lemma" bearing on the fundamental solution of the d'Alembert operator, which is vaguely reminiscent of the classical "Preparation Theorem." The main advantage of this argument is that it does not depend at all on the explicit form of the fundamental solution, and remains the same for other space dimensions—whereas the 2-dimensional analogue of the Glassey–Strauss analysis required a different argument, since the 2-dimensional fundamental solution of the d'Alembert operator is not concentrated on the wave cone in even space dimensions.

**The division lemma**

At variance with the Glassey–Strauss analysis, our argument uses only the following symmetries of the d'Alembert operator. Denote the Lorentz boosts on $\mathbb{R}_t \times \mathbb{R}_x^D$ by

$$L_j = x_j \partial_t + t \partial_{x_j} , \qquad j = 1, \ldots, D .$$

We recall that these Lorentz boosts commute with the d'Alembert operator $\Box_{t,x}$ on $\mathbb{R}_t \times \mathbb{R}_x^D$:

$$[\Box_{t,x}, L_j] = 0 , \quad j = 1, \ldots, D .$$

Let $Y$ be the forward fundamental solution of $\Box_{t,x}$, i.e.,

$$\Box_{t,x} Y = \delta_{(0,0)} , \quad \operatorname{supp} Y \subset \mathbb{R}_+ \times \mathbb{R}^D$$

—for instance, in space dimension $D = 3$, one has

$$Y(t, x) = \mathbf{1}_{t \geq 0} \frac{\delta(t - |x|)}{4\pi t} .$$

Then, since $L_j$ commutes with $\Box_{t,x}$, one finds that

$$\Box_{t,x} L_j Y = L_j \delta_{(0,0)} = 0 , \quad \operatorname{supp} L_j Y \subset \mathbb{R}_+ \times \mathbb{R}^D ,$$

whence, by the uniqueness of the solution to the Cauchy problem for the wave equation,

$$L_j Y = 0 , \quad j = 1, \ldots, D .$$

**Lemma 3.4.3** *Let $D \geq 2$. For each $\xi \in \mathbb{R}^D$, there exists $b_{ij}^k \equiv b_{ij}^k(t, x, \xi)$ in $C^\infty$ on $\mathbb{R}^{D+1} \setminus 0$ and homogeneous of degree $-k$ in $(t, x)$ such that*

*(i) the homogeneous distribution $b_{ij}^2 Y$ of degree $-D - 1$ on $\mathbb{R}^{D+1} \setminus 0$ has null residue at the origin, and*

(ii) *there exists an extension of* $b_{ij}^2 Y$ *as a homogeneous distribution of degree* $-D-1$ *on* $\mathbb{R}^{D+1} \setminus 0$, *still denoted* $b_{ij}^2 Y$, *that satisfies*

$$\partial_{ij} Y = T^2(b_{ij}^0 Y) + T(b_{ij}^1 Y) + b_{ij}^2 Y , \quad i, j = 0, \dots, D.$$

Here $T$ denotes the advection operator $T = \partial_t + v(\xi) \cdot \nabla_x$.

*Remark.* The null residue condition reads

$$\int_{\mathbb{S}^2} b_{ij}^2(1, y) d\sigma(y) = 0 \quad \text{if } D = 3,$$

$$\int_{|y| \le 1} b_{ij}^2(1, y) \frac{dy}{\sqrt{1-|y|^2}} = 0 \quad \text{if } D = 2.$$

In the first formula, $d\sigma$ designates the surface element on the unit sphere.

*Proof (Sketch of the proof).* Observe that

$$\sum_{j=1}^{D} v_j(\xi) L_j = v(\xi) \cdot x \partial_t + t v(\xi) \cdot \nabla_x = (v(\xi) \cdot x - t)\partial_t + tT.$$

Since $L_j Y = 0$ for $j = 1, \dots, D$, one has

$$(t - v(\xi) \cdot x)\partial_t Y = tTY.$$

Furthermore, since supp $\partial_t Y \cap \{t - v(\xi) \cdot x = 0\} = \{(0,0)\}$,

$$\partial_t Y - a_0^0 TY = 0.$$

Indeed, $Y$ is a homogeneous distribution of degree $1 - D$ on $\mathbb{R}^{D+1}$, so that $\partial_t Y - a_0^0 TY$ is a homogeneous distribution of degree $-D$ on $\mathbb{R}^{D+1} \setminus 0$. It has therefore a unique extension to $\mathbb{R}^{D+1}$ as a distribution of degree $-D$; since this distribution is supported at the origin, it is a linear combination of $\delta_{(0,0)}$ and its derivatives. Because $\delta_{(0,0)}$ is homogeneous of degree $-D-1$ on $\mathbb{R}^{D+1}$, this linear combination must be 0. Hence

$$\partial_t Y = T(a_0^0 Y) - (Ta_0^0)Y.$$

One finds analogous formulas for $\partial_{x_j} Y$ with $j = 1, \dots, D$ by combining the formula above with the fact that $L_j Y = 0$ for $j = 1, \dots, D$.

Statement (ii) is obtained by iterating the argument above once in each variable.

As for statement (i), observe that $b_{ij}^k Y$ is a homogeneous distribution of degree $1 - k - D$ on $\mathbb{R}^{D+1} \setminus 0$. Hence, whenever $k = 0, 1$, $b_{ij}^k Y$ has a unique extension as a homogeneous distribution of degree $1 - k - D$ on $\mathbb{R}^{D+1}$. Since

$$\beta_{ij}^2 Y = \partial_{ij} Y - T^2(b_{ij}^0 Y) - T(b_{ij}^1 Y)$$

and the right-hand side is a homogeneous dsitribution on $\mathbb{R}^{D+1}$, the left-hand side is a homogeneous distribution of degree $-1 - D$ on $\mathbb{R}^{D+1} \setminus 0$ that has a homogeneous extension to $\mathbb{R}^{D+1}$. Hence, it has null residue at $(0,0)$: see for instance §3 in chapter 3 of [6].

**Application to the Glassey–Strauss theorem**

We use the division lemma above to estimate the first-order derivatives of the electromagnetic field. This amounts to estimating the second-order derivatives of the momenum averages of the distribution of Liénard–Wiechert potentials:

$$\partial_{ij} \int m(\xi)\mathbf{u}(t, x, \xi)d\xi = \sum_{k=0}^{2} \int m(\xi) \left( b_{ij}^{k-l} Y \star T^l(\mathbf{1}_{t\geq 0}\mathbf{f})(t, x, \xi) \right) d\xi.$$

Here, $m$ denotes any $C^\infty$ function with compact support that coincides with either 1 or each component of $v(\xi)$ on the $\xi$-support of $f$.

The idea is to use the Vlasov equation to compute $T^l(\mathbf{1}_{t\geq 0}\mathbf{f})$ and integrate by parts to bring the $\xi$-derivatives to bear on $b_{ij}^{k-l}$ and $m$.

In fact, the worst term is for $l = 0$:

$$\int m(\xi) \left( b_{ij}^2 Y \star (\mathbf{1}_{t\geq 0}\mathbf{f})(t, x, \xi) \right) d\xi \,.$$

By using the null residue condition, we write this term in the form

$$\int m(\xi) \int_\epsilon^t \int_{\mathbb{S}^2} b_{ij}^2(1, \omega, \xi)\mathbf{f}(t-s, x-s\omega, \xi)\frac{d\sigma(\omega)}{4\pi s}dsd\xi$$
$$+ \int m(\xi) \int_0^\epsilon \int_{\mathbb{S}^2} b_{ij}^2(1, \omega, \xi)\frac{\mathbf{f}(t-s, x-s\omega, \xi)-\mathbf{f}(t, x, \omega)}{4\pi s}d\sigma(\omega)dsd\xi.$$

If the size $R_\mathbf{f}(t)$ of the $\xi$-support of $\mathbf{f}$ is bounded on $[0, T)$, i.e., if

$$\overline{\lim}_{t\to T^-} R_\mathbf{f}(t) < +\infty,$$

this term is bounded by

$$C(1 + \ln_+(t\|\nabla_x\mathbf{f}\|_{L^\infty})).$$

Hence, the Lipschitz semi-norm $N(t) = \|\nabla_{x,\xi}\mathbf{f}(t, \cdot, \cdot)\|_{L^\infty}$ satisfies a logarithmic Gronwall inequality of the form

$$N(t) \leq N(0) + \int_0^t (1 + \ln_+ N(s))N(s)ds\,, \quad t \in [0, T)\,.$$

Therefore, $N$ is uniformly bounded on $[0, T]$, which implies in turn that the fields $(E, B) \in L^\infty([0, T]; W^{1,\infty}(\mathbb{R}^3))$.

# References

[1]    Ambrosio, L., Transport equation and Cauchy problem for *BV* vector fields. *Invent. Math.* **158** (2004), 227–260.

[2]   Bouchut, F., Golse, F., Pallard, C., Nonresonant smoothing for coupled wave+transport equations and the Vlasov–Maxwell system, *Revistà Mat. Iberoam.* **20** (2004), 865–892.

[3]   Bouchut, F., Golse, F., Pallard, C., Classical solutions and the Glassey–Strauss theorem for the 3D Vlasov–Maxwell system, *Arch. for Rational Mech. and Anal.* **170** (2003), 1–15.

[4]   DiPerna, R.J., Lions, P.-L., Global weak solutions of the Vlasov–Maxwell system, *Comm. on Pure and Appl. Math.* **42** (1989), 729–757.

[5]   DiPerna, R.J., Lions, P.-L., Ordinary differential equations, transport theory and Sobolev spaces. *Invent. Math.* **98** (1989), 511–547.

[6]   Gel'fand, I.M., Shilov, G.E., *Generalized functions. Vol. 1. Properties and operations.* Academic Press, New York-London, 1964.

[7]   Glassey, R., Strauss, W., High velocity particles in a collisionless plasma. *Math. Methods Appl. Sci.* **9** (1987), 46–52.

[8]   Glassey, R., Strauss, W., Absence of shocks in an initially dilute collisionless plasma. *Comm. Math. Phys.* **113** (1987), 191–208.

[9]   Glassey, R., Strauss, W., Large velocities in the relativistic Vlasov–Maxwell equations. *J. Fac. Sci. Univ. Tokyo Sect. IA Math.* **36** (1989), 615–627.

[10]  Glassey, R.T., Schaeffer, J.W., Global existence for the relativistic Vlasov–Maxwell system with nearly neutral initial data. *Comm. Math. Phys.* **119** (1988), 353–384.

[11]  Glassey, R.T., Schaeffer, J.W., The "two and one-half-dimensional" relativistic Vlasov Maxwell system. *Comm. Math. Phys.* **185** (1997), 257–284.

[12]  Glassey, R.T., Schaeffer, J.W., The relativistic Vlasov–Maxwell system in two space dimensions. I, II. *Arch. for Rational Mech. and Anal.* **141** (1998), 331–354 & 355–374.

[13]  Glassey, R.T., Strauss, W.A., Singularity formation in a collisionless plasma could occur only at high velocities. *Arch. for Rational Mech. and Anal.* **92** (1986), 59–90.

[14]  Jackson, J.D., *Classical Electrodynamics* Wiley, New York, 1975.

[15]  Klainerman, S., Staffilani, G., A new approach to study the Vlasov–Maxwell system, *Comm. on Pure and Appl. Anal.* **1** (2002), 103–125.

[16]  Landau, L.D., Lifshitz, E.M., *Cours de Physique Théorique. Vol. 2: Théorie des champs*, Editions Mir, Moscow, 1970.

[17]  Pallard, C., Nonresonant smoothing in Sobolev spaces for coupled wave+transport equations, *Bull. Sci. Math.* **127** (2003), 705–718.

[18]  Peral, J., $L^p$ estimates for the wave equation, *J. Funct. Anal.* **36** (1980), 114–145.

[19]  Seeger, A.; Sogge, C.D.; Stein, E.M. Regularity properties of Fourier integral operators. *Ann. of Math.* (2) **134** (1991), 231–251.

[20]  Stein, E.M., *Singular Integrals and Differentiability Properties of Functions*, Princeton Univ. Press, Princeton, NJ, 1970.

# Part II

Modeling Applications, Inverse and
Computational Problems in Quantum Kinetic
Theory

# 4

# Multiband quantum transport models for semiconductor devices

Luigi Barletti,[1] Lucio Demeio[2] and Giovanni Frosali[3]

[1] Dipartimento di Matematica "U. Dini" Università degli Studi di Firenze, Viale Morgagni 67/A, I-50134 Firenze, Italy `barletti@math.unifi.it`
[2] Dipartimento di Scienze Matematiche Università Politecnica delle Marche, Via Brecce Bianche 1, I-60131 Ancona, Italy `demeio@dipmat.unian.it`
[3] Dipartimento di Matematica "G.Sansone" Università di Firenze, Via S. Marta 3, I-50139 Firenze, Italy `giovanni.frosali@unifi.it`

## 4.1 Introduction

The modeling of semiconductor devices, which is a very active and intense field of research, has to keep up with the speed at which the fabrication technology proceeds; the devices of the last generations have become increasingly smaller, reaching a size so small that quantum effects dominate their behaviour. Quantum effects such as resonant tunneling and other size-quantized effects cannot be described by classical or semiclassical theories; they need a full quantum description [Fre90, JAC92, KKFR89, MRS90, RBJ91, RBJ92]. A very important feature, which has appeared in the devices of the last generation and which requires a full quantum treatment, is the presence of the interband current: a contribution to the total current which arises from transitions between the conduction and the valence band states. Resonant interband tunneling diodes (RITDs) are examples of semiconductor devices which exploit this phenomenon; they are of big importance in nanotechnology for their applications to high-speed and miniaturized systems [YSDX91, SX89]. In the band diagram structure of these diodes there is a small region where the valence band edge lies above the conduction band edge (valence quantum well), making interband resonance possible.

So far, most of the existing literature has been devoted to quantum transport models where only conduction band electrons contribute to the current flow and under the parabolic band approximation, with only a small region of the Brillouin zone near the minimum of the band being populated. In bipolar models, the contribution of the valence band (the current due to the holes) is also included at the macroscopic level. Quantum models which include the interband resonance process are called "multiband models" and have largely been formulated and analyzed only in the last five to ten years. Like other models for semiconductor devices, they can essentially be divided into two classes: Schrödinger-based models and Wigner function-based (or density matrix-based) models. The former aim at the calculation of the wave function for the system or device under study, and contain no statistics. The latter involve electron statistics or transport theory concepts.

Hydrodynamic models have also been formulated and discussed [Gar94]; again, for the most part only single-band hydrodynamics has received attention. Only recently, multiband hydrodynamic models, based on the multiband kinetic models mentioned above, have appeared.

In this review chapter, we describe the multiband models that have recently been formulated in both classes. Attention is given to the definitions of the relevant quantities which characterize each model and to the advantages and disadvantages of each model compared to others. The technical details of the derivations of the various models, as well as the rigorous proofs of consistency and existence of the solutions, are diverted directly to the papers where the models have been described.

This chapter is organized as follows: in Section 4.2 we briefly recall the Bloch theory of electrons moving in a periodic potential; Section 4.3 is devoted to the envelope function theory; in Section 4.4 we deal with the multiband models based on the Schrödinger equation; Section 4.5 contains the statistical kinetic models based on the Wigner function approach and in Section 4.6 we give an outline of the hydrodynamic models.

## 4.2 The Schrödinger equation and the wave function in a periodic potential

The starting point of any theoretical description of a quantum system is the Schrödinger equation, which we now discuss for a periodic Hamiltonian [RS72, MRS90].

We consider an ensemble of electrons moving in a semiconductor crystal. The electrostatic potential generated by the crystal ions is represented by a periodic potential $V_{\mathcal{L}}(x)$, the periodicity being described as follows:

$$V_{\mathcal{L}}(x + a) = V_{\mathcal{L}}(x), \qquad \forall a \in \mathcal{L},$$

where $\mathcal{L}$ is the periodic lattice of the crystal. The quantum dynamics of a single electron is, therefore, generated by the Hamiltonian

$$H = H_0 + V(x), \tag{4.2.1}$$

where $H_0 = p^2/2m + V_{\mathcal{L}}(x)$ is the periodic part of the Hamiltonian, which contains the kinetic energy and the periodic potential. Also, $p = -i\hbar\nabla$ is the momentum operator, $m$ is the electron mass, $\hbar$ is Planck's constant over $2\pi$ and $V(x)$ is the potential due to external fields, such as barriers or bias. The periodic Hamiltonian $H_0$ has a complete system of generalized eigenfunctions $b_n(x, k)$, called *Bloch waves*, where the "pseudomomentum" or "crystal momentum" variable $k$ runs over the *Brillouin zone*. This is defined as the centered fundamental domain of the reciprocal lattice $\mathcal{L}^*$, i.e.,

$$B = \left\{ k \in \mathbb{R}^3 \,\middle|\, k \text{ is closer to } 0 \text{ than to any other point of } \mathcal{L}^* \right\}.$$

The Bloch waves satisfy the generalized eigenvalue equation

$$H_0 b_n(x, k) = \epsilon_n(k) b_n(x, k), \tag{4.2.2}$$

(or $H_0 \mid nk\rangle = \epsilon_n(k) \mid nk\rangle$ in Dirac's notation), where the generalized eigenfunctions $\epsilon_n(k)$ are the *energy bands* of the crystal. Accordingly, the integer $n$ is called the "band-index."

Using Dirac's notation, we choose the following normalization of the Bloch functions:

$$\langle nk \mid n'k'\rangle = |B| \, \delta_{nn'} \, \delta(k - k'), \tag{4.2.3}$$

so that any wave function $\Psi$ can be decomposed as

$$\Psi(x) = \sum_n \int_B \sigma_n(k) \, b_n(x, k), \tag{4.2.4}$$

where

$$\sigma_n(k) = \int_{\mathbb{R}^3} dx \, \bar{b}_n(x, k) \, \Psi(x). \tag{4.2.5}$$

It is well known that the Bloch waves can be written in the form

$$b_n(x, k) = e^{ik \cdot x} u_n(x, k), \tag{4.2.6}$$

where $u_n(x, k)$, called *Bloch functions*, are $\mathcal{L}$-periodic in $x$ and have the property that $\{u_n(\cdot, k) \mid n \in \mathbb{N}\}$ is an orthonormal basis of $L^2(C)$ for any fixed $k \in B$, where $C$ denotes the fundamental cell of the direct lattice $\mathcal{L}$. In particular,

$$\int_C \bar{u}_n(x, k) \, u_{n'}(x, k) \, dx = \delta_{nn'}, \qquad k \in B. \tag{4.2.7}$$

The electron population of the semiconductor material is partitioned into the energy bands of the Hamiltonian. The highest occupied energy band usually contains only a small electron population and therefore it has many unoccupied states; this is the conduction band. The states of all other (lower energy) bands are instead fully occupied and form the valence bands. In the older devices, based on resonant tunneling, only the electrons of the conduction band contribute to the flow of the current across the device. In some of the devices of the last generation, instead, the resonant tunneling occurs between states belonging to different bands, so that the carrier population of the valence band also contributes to the flow of current. For the description of these devices, multiband models must be used.

## 4.3 Envelope function theory

The wave function of an electron moving under the action of a periodic potential, which we have described in the previous section, is a fast-oscillating object (both in time and space) and is therefore not well suited for numerical computations. A widely used methodology is that of smoothing out these fast oscillations, thus leading to the "envelope function" approach. Envelope functions can be introduced in basically two different ways, one due to Wannier [Wan62] (called the Wannier–Slater envelope functions) and one due to Luttinger and Kohn [LK55] (called the Luttinger–Kohn envelope functions). The Luttinger–Kohn envelope functions are the building blocks of the Kane model, which will be described in the next section. Here, we introduce the definitions and outline the most important properties of both kinds of envelope functions.

### 4.3.1 Wannier–Slater envelope functions

The Wannier–Slater (W-S) envelope functions [Wan62] are defined as follows:

$$f_n(x) = \frac{1}{(2\pi)^{3/2}} \int_B \sigma_n(k)\, e^{ix \cdot k} dk, \qquad (4.3.1)$$

where $\sigma_n(k)$ is given by (4.2.5). Note that the W-S envelope functions are inverse Fourier transforms to which fast oscillations due to the periodic potential have been removed. In other words, each envelope function $f_n$ has the property that its Fourier transform is supported in the Brillouin zone $B$. The W-S envelope functions are easily expressed in terms of the wave function by introducing "continuous-index Wannier functions"

$$a_n(x, x') = \frac{1}{(2\pi)^{3/2}} \int_B b_n(x, k) e^{-ix' \cdot k} dk. \qquad (4.3.2)$$

Using (4.2.5), (4.3.1) and (4.3.2) we get

$$f_n(x) = \int_{\mathbb{R}^3} \bar{a}_n(x', x)\Psi(x')\, dx', \qquad (4.3.3)$$

and, conversely,

$$\Psi(x) = \sum_n \frac{1}{|B|} \int_{\mathbb{R}^3} a_n(x, x') f_n(x')\, dx'. \qquad (4.3.4)$$

To better understand the meaning of the W-S envelope functions, consider the (discrete-index) Wannier functions [Wan62], which are the Fourier components of the Bloch waves with respect to $k$:

$$a_n(x - \lambda) = \int_B b_n(x, k) e^{-ik \cdot \lambda} = \int_B u_n(x, k) e^{ik \cdot (x - \lambda)}, \qquad (4.3.5)$$

where $\lambda$ is a point of the periodic lattice $\mathcal{L}$. The most important property of the Wannier functions is that they are localized at the sites of the lattice, with an exponential decay away from those sites. The Bloch waves, on the contrary, are delocalized and maintain their highly oscillatory behaviour throughout $\mathbb{R}$. The Wannier functions, like the Bloch waves, form a complete, generalized, orthonormal basis and any wave function can be expanded as

$$\Psi(x) = \sum_n \sum_{\lambda \in \mathcal{L}} f_n(\lambda) a_n(x - \lambda),$$

where

$$f_n(\lambda) = \int_B \frac{dk}{|B|} \sigma_n(k) e^{ik \cdot \lambda}.$$

If the length scale of the crystal lattice is small with respect to the macroscopic scale described by the variable $x$, the Brillouin zone becomes very large and the Fourier coefficients $f_n(\lambda)$ can be replaced by the continuous Fourier transform, yielding definition (4.3.1).

The dynamics of the W-S envelope functions can be deduced from (4.3.3) and (4.3.4), and from the Schrödinger equation

$$i\hbar \frac{\partial}{\partial t} \Psi(x, t) = H\Psi(x, t),$$

where $H$ is the Hamiltonian operator (4.2.1). This yields (see [Bar03b] for the details of the derivation)

$$i\hbar \frac{\partial}{\partial t} f_n(x, t) = \widetilde{\epsilon}_n(-i\nabla) f_n(x, t) + \sum_{n'} \int_{\mathbb{R}^3} V_{nn'}^{\mathrm{WS}}(x, x') f_{n'}(x', t)\, dx'. \quad (4.3.6)$$

Here,

$$V_{nn'}^{\mathrm{WS}}(x, x') = \frac{1}{|B|} \int_{\mathbb{R}^3} \bar{a}_n(y, x) V(y) a_n(y, x')\, dy \quad (4.3.7)$$

are matrix elements of the external potential with respect to the continuous-index Wannier functions and $\widetilde{\epsilon}_n(-i\nabla)$ are pseudo-differential operators associated to the energy bands with a cut-off outside the Brillouin zone, namely

$$\widetilde{\epsilon}_n(-i\nabla) f_n(x, t) = \frac{1}{(2\pi)^3} \int_{\mathbb{R}^6} \mathbb{1}_B(k) \, \epsilon_n(k) \, f_n(x') \, e^{ik \cdot (x-x')} dx' dk.$$

where $\mathbb{1}_B$ is the characteristic function of the Brillouin zone $B$.

### 4.3.2 Luttinger–Kohn envelope functions

A general definition of envelope functions in the sense of Luttinger and Kohn [Bur92, LK55] may be given as follows. Let $\{v_n(x) \mid n \in \mathbb{N}\}$ be $\mathcal{L}$-periodic functions that form an orthonormal basis of $L^2(C)$. Then, the Luttinger–Kohn (L-K) envelope functions of a wave function $\Psi$, with respect to the basis $v_n(x)$, are functions $F_n(x)$ such that

(i) $\Psi(x) = \sum_n F_n(x) v_n(x)$;

(ii) the $F_n$ are slowly varying with respect to the lattice periodicity, namely

$$\mathrm{supp}(\hat{F}_n) \subset B, \qquad n \in \mathbb{N}, \tag{4.3.8}$$

where $\hat{F}_n$ denotes the Fourier transform of $F_n$.

Usually the basis functions $v_n$ are chosen to be the Bloch functions $u_n(x, k)$ evaluated at $k = 0$, so that

$$\Psi(x) = \sum_n F_n(x) u_n(x, 0),$$

but, of course, other choices are possible. It can be proved that the L-K envelope functions are uniquely determined by the two conditions (i) and (ii) and that the Parseval-like equality

$$\|\Psi\|^2 = \frac{1}{|C|} \sum_n \|F_n\|^2 \tag{4.3.9}$$

holds. It is not difficult to see that the L-K envelope functions are easily expressed in terms of the wave function as follows:

$$f_n(x) = \int_B \frac{dk}{|B|^{1/2}} \int_{\mathbb{R}^3} dy \, \overline{\mathcal{X}}_n(y, k) \, e^{ik \cdot x} \, \Psi(y), \tag{4.3.10}$$

where

$$\mathcal{X}_n(y, k) = \frac{1}{|B|^{1/2}} v_n(y) e^{ik \cdot y}, \quad y \in \mathbb{R}^3, \; k \in B, \; n \in \mathbb{N} \tag{4.3.11}$$

is a (generalized) Luttinger–Kohn basis [LK55]. By using the above relations it is possible to deduce the dynamics of L-K envelope functions. In the case $v_n(x) = u_n(x, 0)$ we have [Wen99]

$$i\hbar \frac{\partial}{\partial t} F_n(x, t) = \epsilon_n(0) F_n(x, t) - \frac{\hbar^2}{2m} \Delta F_n(x, t) - \frac{\hbar^2}{m} \sum_{n'} K_{nn'} \cdot \nabla F_{n'}(x, t)$$

$$+ \sum_{n'} \int_{\mathbb{R}^3} V_{mn'}^{\mathrm{LK}}(x, x') F_{n'}(x', t) \, dx'. \tag{4.3.12}$$

Here, $\epsilon_n(0)$ is the $m$th energy band evaluated at $k = 0$ and

$$K_{nn'} = \int_C u_n(x, 0) \nabla u_{n'}(x, 0) \, dx = -K_{n'n} \tag{4.3.13}$$

are the matrix elements of the gradient operator between Bloch functions (which, we recall, are real valued). The matrix elements of the external potential are given by

$$V_{nn'}^{LK}(x, x') = \frac{1}{(2\pi)^3} \int_B dk \int_{\mathbb{R}^3} dy \int_B dk'$$
$$\times \left\{ e^{ik \cdot x} \overline{\mathcal{X}}_n(y, k) \, V(y) \, \mathcal{X}_{n'}(y, k') \, e^{-ik' \cdot x'} \right\}, \quad (4.3.14)$$

where, of course, $u_n$ has to be used in definition (4.3.11) in place of $v_n$.

## 4.4 Pure-state multiband models

The equations of envelope function dynamics, equations (4.3.6) and (4.3.12), are still too complicated for modeling purposes and, therefore, they should be considered as starting points for building simpler models rather then models *per se*.

First of all we note that, if the external potential is slowly varying with respect to the lattice period, then the $\mathcal{L}$-periodic function $\overline{u_n}(y, 0) \, u_{n'}(y, 0)$

in (4.3.14) (see definition (4.3.11)) can be substituted by its average on a periodic cell. Hence, we can write

$$V_{nn'}^{LK}(x, x') \approx \frac{1}{|B| \, |C| \, (2\pi)^3} \int_C \overline{u}_n(z) \, u_{n'}(z) \, dz$$
$$\times \int_B dk \int_{\mathbb{R}^3} dy \int_B dk' \left\{ e^{ik' \cdot x} e^{-iy \cdot (k-k')} e^{-ik' \cdot x'} V(y) \right\}$$

and so, using (4.2.7), $|C| \, |B| = (2\pi)^3$, and $B \approx \mathbb{R}^3$,

$$V_{nn'}^{LK}(x, x') \approx \delta_{nn'} \delta(x - x') V(x - x'). \quad (4.4.1)$$

In other words, if the potential $V$ is smooth enough, the complicated potential term in eq. (4.3.12) can be approximated by the simple multiplication by $V(x)$ of each $F_n$. The same property holds for $V_{nn'}^{WS}(x, x')$ (see definition (4.3.7)) and the proof is similar.

Another typical approximation is the effective mass dynamics. This can be easily deduced from the Wannier–Slater equations (4.3.6) by simply substituting the energy-band function $\epsilon_n(k)$ with its parabolic approximation near a stationary point (that we assume to be always $k = 0$ for the sake of simplicity). This, together with the approximation (4.4.1) yields a completely decoupled dynamics of the form

$$i\hbar \frac{\partial}{\partial t} f_n(x, t) = -\frac{\hbar^2}{2} \nabla \cdot \mathbb{M}_n^{-1} \nabla f_n(x, t) + V(x) \, f_n(x, t),$$

where $\mathbb{M}$ is the effective mass tensor:

$$\mathbb{M}_n^{-1} = \nabla \otimes \nabla \, \epsilon_n(k) \, |_{k=0}.$$

The effective mass model is widely used in semiconductor modeling and it has been rigorously studied, as an asymptotic dynamics, in Refs. [AP05], [BLP78] and [PR96]. However, if *interband effects* have to be included, then we have to go beyond the effective mass approximation and include at least two coupled bands.

### 4.4.1 The two-band Kane model

A simple multiband model was introduced by Kane [Kan56] in the early 1950s in order to describe the electron transport with two allowed energy bands separated by a forbidden region. The Kane model is a simple two-band model capable of including one conduction band and one valence band and it is formulated as two coupled Schrödinger-like equations for the conduction-band and valence-band envelope functions [BFZ03]. The coupling term is treated by the $k \cdot P$ perturbation method [Wen99], which gives the solutions of the single electron Schrödinger equation in the neighborhood of the bottom of the conduction band and the top of the valence bands, where most of the electrons and holes, respectively, are concentrated. The Kane model is very important for modeling of RITD devices, and is widely discussed in the the literature [SX89, YSDX91].

From our point of view, the Kane model can be viewed as an approximate evolution equation for L-K envelope functions arising from equation (4.3.12) when using the following approximations:

1. the external potential kernel (4.3.14) is substituted by the local and diagonal approximation (4.4.1);
2. only two bands (conduction and valence) are included;
3. the bottom of the conduction band $E_c = \epsilon_c(0)$ and top of the valence band $E_v = \epsilon_v(0)$ are viewed as functions of the position $x$ (this allows one to model band heterostructures).

Thus, using the indices $c$ for conduction and $v$ for valence, we have a two-term L-K envelope function expansion

$$\Psi(x) = \Psi_c(x)u_c(x) + \Psi_v(x)u_v(x)$$

of the wave function $\Psi$ and the following evolution equations for $\Psi_c$ and $\Psi_v$:

$$i\hbar\frac{\partial}{\partial t}\Psi_c(x,t) = (E_c + V)(x)\Psi_c(x,t) - \frac{\hbar^2}{2m}\Delta\Psi_c(x,t) - \frac{\hbar^2}{m}K \cdot \nabla\Psi_v(x,t),$$

$$i\hbar\frac{\partial}{\partial t}\Psi_v(x,t) = (E_v + V)(x)\Psi_v(x,t) - \frac{\hbar^2}{2m}\Delta\Psi_v(x,t) + \frac{\hbar^2}{m}K \cdot \nabla\Psi_c(x,t),$$

$$(4.4.2)$$

which is the two-band Kane model. Note that the quantity $K$, called the Kane momentum, is given by

$$K = K_{cv} = -K_{vc} = \int_C u_c(x)\,\nabla u_v(x)\,dx$$

(see (4.3.13) and recall that the Bloch functions $u_c$ and $u_v$ are real valued). A word of caution on the notation: $\Psi_c$ and $\Psi_v$ are not really band projections (spectral projections) of the wave function, not only because of the envelope function approximation but also because the Hamiltonian operator defined by the right-hand side of equation (4.4.2) is *not* diagonal, even in the absence of external potentials. The identification of $\Psi_c$ and $\Psi_v$ with spectral projections is only approximately true for $k \approx 0$.

The Kane model in the Schrödinger-like form (4.4.2) has been recently studied by J. Kefi, [Kef03], and in the Wigner equation form by Borgioli, Frosali and Zweifel [BFZ03].

### 4.4.2 The Morandi–Modugno multiband model

In this section we briefly introduce the multiband envelope function model, introduced recently by Modugno and Morandi (M-M); for the complete derivation of the model we refer the reader to [MM05].

The starting point is the W-S envelope function dynamics (4.3.6). When the potential $V$ is smooth enough, we can approximate the matrix elements $V_{nn'}^{\mathrm{WS}}$ in the same way as we deduced equation (4.4.1), obtaining

$$V_{nn'}^{\mathrm{WS}}(x,x') \approx \frac{1}{|B|\,(2\pi)^{3/2}} \int_{\mathbb{R}^3} dk \int_{\mathbb{R}^3} dk' \left\{ e^{ik\cdot x} B_{nn'}(k,k')\, \hat{V}(k-k')\, e^{-ik'\cdot x'} \right\},$$

(4.4.3)

where

$$B_{nn'}(k,k') = \frac{1}{|C|} \int_C \bar{u}_n(z,k)\, u_{n'}(z,k')\, dz.$$

(4.4.4)

By using the eigenvalue equation (4.2.2) one obtains

$$B_{nn'}(k,k') = \frac{1}{|C|} \frac{\hbar}{m} (k-k') \frac{P_{nn'}(k,k')}{\Delta E_{nn'}(k,k')}, \qquad \text{for } n \neq n',$$

where

$$P_{nn'}(k,k') = \int_C \bar{u}_n(x,k)(-i\hbar\nabla)u_{n'}(x,k')\, dx$$

(4.4.5)

and

$$\Delta E_{nn'}(k,k') = \epsilon_n(k) - \epsilon_{n'}(k') - \frac{\hbar^2}{2m}(k^2 - k'^2).$$

Moreover, as can be deduced from equation (4.2.3), the diagonal terms are simply given by

$$B_{nn}(k,k') = \frac{|B|}{(2\pi)^3} = \frac{1}{|C|}.$$

(4.4.6)

Using (4.4.3), (4.4.4) and (4.4.6) in equation (4.3.6) (and recalling that $B \approx \mathbb{R}^3$) we get

$$i\hbar \frac{\partial}{\partial t} f_n(x,t) = \epsilon_n(-i\nabla) f_n(x,t) + V(x) f_n(x,t)$$

$$+ \frac{\hbar}{m} \sum_{n'\neq n} \int_{\mathbb{R}^3} dk \int_{\mathbb{R}^3} dk' \frac{e^{ik\cdot x}}{(2\pi)^3} \frac{P_{nn'}(k,k')}{\Delta E_{nn'}(k,k')} \hat{V}(k-k') \hat{f}_{n'}(k',t),$$

where a diagonal part and a nondiagonal part of the dynamics can be clearly distinguished. Assuming, for the sake of simplicity, that the stationary point of each band is $k = 0$ and that the crystal momentum $k$ remains small during the whole evolution, we can expand the term $P_{nn'}/\Delta E_{nn'}$, which characterizes the interband coupling, to first order in $k$ and $k'$. After some manipulations by means of standard perturbation techniques, we get the multiband equation

$$i\hbar \frac{\partial}{\partial t} f_n(x,t) = \epsilon_n(-i\nabla) f_n(x,t) + V(x) f_n(x,t)$$

$$- \frac{i\hbar}{m} \nabla V(x) \cdot \sum_{n'\neq n} \frac{P_{nn'}}{\Delta E_{nn'}} f_{n'}(x,t) - \frac{\hbar}{m} \nabla V(x) \cdot \sum_{n'\neq n} \frac{M^*_{nn'}}{\Delta E_{nn'}} \nabla f_{n'}(x,t)$$

$$- \frac{\hbar}{m} \sum_{n'\neq n} \frac{M_{nn'}}{\Delta E_{nn'}} \left[ \nabla^2 V(x) f_{n'}(x,t) + \nabla V(x) \cdot \nabla f_{n'}(x,t) \right], \quad (4.4.7)$$

where we set $P_{nn'} \equiv P_{nn'}(0,0)$, $\Delta E_{nn'} \equiv \Delta E_{nn'}(0,0)$ and

$$M_{nn'} = \frac{\hbar}{m} \sum_{n''\neq n'} \frac{P_{nn''} P_{n''n'}}{E_n - E_{n''}}, \quad M^*_{n'n} = \frac{\hbar}{m} \sum_{n''\neq n'} \frac{P_{nn''} P_{n''n'}}{E_{n'} - E_{n''}}$$

are effective mass terms. A simple two-band model can be built using the following assumptions:

1. only two bands ($c$ and $v$) are included;
2. the energy band operator $\epsilon_n(-i\nabla)$ is substituted by its parabolic approximation (effective mass energy band);
3. the interband terms of order greater than 2 in $k$ are neglected (this amounts to neglecting
   terms proportional to the matrices $M_{nn'}$);
4. the bottom of the conduction band and the top of the valence band are functions of the position $x$ (as in the two-band Kane model).

This yields

$$i\hbar \frac{\partial}{\partial t} \Phi_c(x,t) = (E_c + V)(x) \Phi_c(x,t) - \frac{\hbar^2}{2} \nabla \cdot \mathrm{M}_c^{-1} \nabla \Phi_c(x,t)$$

$$- \frac{i\hbar}{m E_g(x)} \nabla V(x) \cdot P \Phi_v(x,t),$$

$$\tag{4.4.8}$$

$$i\hbar \frac{\partial}{\partial t} \Phi_v(x,t) = (E_v + V)(x) \Phi_v(x,t) - \frac{\hbar^2}{2} \nabla \cdot \mathrm{M}_v^{-1} \nabla \Phi_v(x,t)$$

$$- \frac{i\hbar}{m E_g(x)} \nabla V(x) \cdot P \Phi_c(x,t),$$

where $E_g(x) = E_c(x) - E_v(x)$ is the band gap. In contrast to the Kane model (4.4.2), in the the M-M model (4.4.8) the envelope functions $\Phi_c$ and $\Phi_v$ are true band functions, to the extent that in the absence of external potentials ($V = 0$) the dynamics is diagonal.

## 4.5 Statistical multiband models: density matrix and Wigner function

We now turn our attention to the multiband models that make use of statistical concepts, mainly of the Wigner function approach [Wig32, MRS90, BJ99, JBBB01, Bar03a]. A multiband model involving the density matrix was already introduced by Krieger and Iafrate [KI87, IK86] by taking matrix elements of the density operator between Bloch states. Subsequently, a number of multiband models based on Wigner function approach were developed. In [Bar03b, Bar04a, BD02] envelope functions were used to construct the multiband Wigner function; in [BFZ03] a Wigner version of the Kane model was introduced; in [DBBBJ02, DBBJ02, DBJ03a, DBJ03b] the multiband Wigner function was obtained by using the Bloch-state representation of the density matrix.

We recall that statistical states in quantum mechanics are described either in terms of the density operator $\rho$ or the Wigner function $f(x, p)$, [Fey72]. The density operator is usually defined by a statistical mixture of states, say $\{\Psi_j \mid j \in \mathbb{N}\}$, where $\Psi_j(x)$ are the wave functions that characterize each state of the mixture. If $\lambda_j \geq 0$ is the probability distribution of the states, then $\sum_j \lambda_j = 1$ and the density operator is given by

$$\rho = \sum_j \lambda_j \mid \Psi_j \rangle\langle \Psi_j \mid \tag{4.5.1}$$

in Dirac's notation, and the density matrix in the space representation is given by

$$\rho(x, x') = \sum_j \lambda_j \Psi_j(x)\overline{\Psi}_j(x') = \sum_j \lambda_j \langle x \mid \Psi_j\rangle\langle\Psi_j \mid x'\rangle. \tag{4.5.2}$$

The Wigner function $f(x, p)$ is defined by the Wigner–Weyl transform of the density operator, that is

$$f(x, p) = \int \frac{d\eta}{(2\pi\hbar)^3} \, \rho\left(x + \frac{\eta}{2}, x - \frac{\eta}{2}\right) e^{-ip\eta/\hbar}. \tag{4.5.3}$$

In the theoretical models based on the solution of the Schrödinger equation (pure states), the calculation of the current across the device, $j(x)$, follows the standard quantum mechanical definition

$$J(x) = -\frac{\hbar}{m} \text{Im} \left( \Psi(x) \nabla \overline{\Psi}(x) \right). \tag{4.5.4}$$

In the statistical models, instead, the current is expressed in terms of the density matrix or in terms of the Wigner function. In the first case the current is

$$J(x) = -\frac{i\hbar}{2m} \left( \nabla_x - \nabla_{x'} \right) \rho(x, x')|_{x=x'}, \tag{4.5.5}$$

and, using the Wigner function, by

$$J(x) = \frac{1}{m} \int pf(x, p) \, dp, \tag{4.5.6}$$

an expression which is, remarkably, identical to the classical expression for the current in statistical systems. It can be easily shown that, in the case of pure states, these two expressions coincide with (4.5.4).

### 4.5.1 Wigner function-based statistical models

A suitable partition of the Wigner function among the energy bands can be obtained by using the completeness of the Bloch states in equation (4.5.3). We adopt hereafter Dirac's notation and consider, for the sake of simplicity, the one-dimensional case only. By defining the coefficients $\Phi_{mn}$ and the integral kernel $W_{mn}$,

$$\Phi_{mn}(k, k', x, p) = \int \int \frac{d\eta}{2\pi\hbar} \left\langle x + \frac{\eta}{2} \middle| nk \right\rangle \left\langle nk' \middle| x - \frac{\eta}{2} \right\rangle e^{-ip\eta/\hbar} \tag{4.5.7}$$

$$W_{mn}(x, p, x', p') = \int_{B^2} dk dk' \Phi_{mn}(k, k', x, p) \Phi_{mn}^*(k, k', x', p'), \tag{4.5.8}$$

the Wigner function can be written as a sum of projections over the Floquet subspaces of the energy bands (see [DBBJ02] for details):

$$f(x, p) = \sum_{mn} f_{mn}(x, p), \tag{4.5.9}$$

where

$$f_{mn}(x, p) = \int_{B^2} dk dk' \rho_{mn}(k, k') \Phi_{mn}(k, k', x, p). \tag{4.5.10}$$

By expressing $\rho$ as a function of $f$, we can write $f_{mn} = \mathcal{P}_{mn} f$, where $\mathcal{P}_{mn}$ is the linear integral operator

$$\left( \mathcal{P}_{mn} f \right)(x, p) \equiv \frac{1}{2\pi\hbar} \int \int dx' dp' W_{mn}(x, p, x', p') f(x', p').$$

Here, $\rho_{mn}(k, k') = \langle mk \mid \rho \mid nk' \rangle$ are the matrix elements of the density operator in the Bloch-state representation, and the linear integral operator $\mathcal{P}_{mn}$ is a projection operator and yields the Wigner projections $f_{mn}$ from the total Wigner function $f$.

The time evolution of the Wigner function is given by the sum of the time evolutions of the band projections,

$$i\hbar \frac{\partial f}{\partial t}(x, p, t) = \sum_{mn} i\hbar \frac{\partial f_{mn}}{\partial t}(x, p, t),$$

given by [DBBJ02]

$$i\hbar \frac{\partial f_{mn}}{\partial t} = \sum_{\mu \in \mathcal{L}} \left[ \widehat{\epsilon}_m(\mu) f_{mn}\left(x + \frac{\mu}{2}, p, t\right) - \widehat{\epsilon}_n(\mu) f_{mn}\left(x - \frac{\mu}{2}, p, t\right) \right] e^{ip\mu/\hbar}$$

$$+ \int \int dx' d\eta \widehat{W}_{mn}(x, p, x', -\eta) \delta V(x', \eta) \widehat{f}(x', \eta, t), \quad (4.5.11)$$

where $\widehat{W}_{mn}$ is the Fourier transform of $W_{mn}$ with respect to the momentum variable:

$$\widehat{W}_{mn}(x, p, x', \eta) = \frac{1}{2\pi\hbar} \int dp' W_{mn}(x, p, x', p') e^{ip'\eta/\hbar}. \quad (4.5.12)$$

Equation (4.5.11) is the equation that governs the time evolution of the Floquet projections $f_{mn}$ of the Wigner function for an ensemble of electrons moving in a semiconductor crystal in the presence of external fields and allowing for energy bands of arbitrary shape. The first term, containing the sum over the lattice vectors, refers to the action of the periodic potential of the crystal lattice, while the last term, written in the form of an integral operator, refers to the action of the external or self-consistent fields acting on the electrons. The first term, as shown in [DBBJ02], reduces to the usual free-streaming operator in the case of a single parabolic band; for this reason we shall refer to this term as the streaming term, while the second term will be called the force term, in analogy with the corresponding force term of the Boltzmann equation. These equations show that, in the absence of external fields, different bands remain dynamically uncoupled and each contribution to the Wigner function evolves independently. In the case $V(x) \equiv 0$, these equations were already written by Markowich, Mauser and Poupaud [MRS90, MMP94] for a single band. It can be shown that, in the case of a single parabolic band, eq. (4.5.11) reduces to the usual Wigner equation in the effective mass approximation

$$\frac{\partial f}{\partial t} + \frac{p}{m^*} \frac{\partial f}{\partial x} + \frac{i}{\hbar} \Theta[\delta V] f = 0,$$

where $m^*$ is the (one-dimensional) electron effective mass in the selected band and

$$(\Theta[\delta V]f)(x,p) = \frac{1}{2\pi\hbar}\int_{\mathbb{R}^2} e^{-i(p-p')\xi/\hbar}\,\delta V(x,\xi)\,f(x,p')\,d\xi\,dp' \qquad (4.5.13)$$

is a pseudo-differential operator with symbol

$$\delta V(x,\xi) = V\left(x+\frac{\xi}{2}\right) - V\left(x-\frac{\xi}{2}\right). \qquad (4.5.14)$$

A multiband model for electron transport in semiconductors, based on the density matrix approach, was introduced by Krieger and Iafrate in [KI87, IK86]. They considered a statistical ensemble of electrons moving under the action of an external time-dependent electric field. Here, we briefly summarize this model in a simplified form. Their model is obtained by expanding the density matrix elements in Bloch functions:

$$\rho(y,z) = \sum_{mn}\int_{B^2} dk\,dk'\,\rho_{mn}(k,k')b_m(k,y)\bar{b}_n(k',z), \qquad (4.5.15)$$

where $\rho_{mn}(k,k') = \langle mk \mid \rho \mid nk'\rangle$ are the already-introduced matrix elements between Bloch functions, whose evolution is given by

$$i\hbar\frac{\partial\rho_{mn}(k,k')}{\partial t} = [\epsilon_m(k)-\epsilon_n(k')]\,\rho_{mn}(k,k')$$

$$+ \sum_l \int_B dk''\,[V_{ml}(k,k'')\rho_{ln}(k'',k') - V_{ln}(k'',k')\rho_{ml}(k,k'')]. \qquad (4.5.16)$$

Here, $V_{mn}(k,k') = \langle mk \mid V \mid nk'\rangle$ are the matrix elements of the external potential in the Bloch representation. The main source of difficulty with this approach lies exactly in these matrix elements, which are ill defined for most potentials of practical interest.

The Wigner function formalism has also been used by Buot and Jensen [Buo74, Buo76, Buo86, BJ90] to formulate multiband models within the framework of the lattice Weyl transform, in which a noncanonical definition of the Wigner function, based on a discrete Fourier transform, was introduced. This definition of the Wigner function makes use of the Wannier functions introduced by (4.3.5). Let $\{\mid m\lambda\rangle, m \in \mathbb{N}, \lambda \in \mathcal{L}\}$ be the states corresponding to the Wannier functions (see equation (4.3.5)); here, $\mathcal{L}$ is the direct lattice and the vectors $\lambda$ are elements of the direct lattice. We can consider matrix elements of the density operator $\rho$ in the Wannier representation,

$$\rho_{mn}(\lambda,\mu) = \langle m\lambda \mid \rho \mid n\mu\rangle$$

with $\lambda,\mu \in \mathcal{L}$. A Wigner function is then introduced by

$$f_{mn}(k,\lambda) = \mathcal{N}\sum_{v\in\mathcal{L}}\rho_{mn}(\lambda+v,\lambda-v)e^{2ikv}, \qquad (4.5.17)$$

where $\mathcal{N}$ is a normalization factor, $\lambda \in \mathcal{L}$ is a lattice vector and $k \in B$. This definition of the Wigner function is sometimes called the discrete Wigner–Weyl transform, and it has a similar structure to that of the definition given in (4.5.3). There are however some important differences: the Wigner function is only defined on the lattice points; it is defined by a Fourier series, rather than the Fourier transform; it is a function of the crystal momentum, which has not been integrated over. According to (4.5.6), the current density is then given by

$$J(\lambda) = \sum_{mn} \int_B \frac{p}{m} f_{mn}(k, \lambda), \qquad (p = \hbar k)$$

and is also defined on the lattice points.

## 4.5.2 Reduced Wigner–Bloch–Floquet models

Equations (4.5.11) are the most general time evolution equations that can be written for the Floquet projections of the multiband Wigner function in the presence of external fields and in the absence of collisions. The action of the periodic potential is described by the first term, which contains the Fourier coefficients of the energy bands, and which reduces to the usual free-streaming operator in the parabolic band approximation. The second term describes the action of the external potential. We note that, while the first term requires only the knowledge of the energy band functions, the second term requires the knowledge of the Bloch eigenfunctions of the material of interest. Therefore, the model equations (4.5.11) are very hard to solve in full generality in practical applications, and the derivation of a set of simplified models is needed. In the following subsections, we outline some of the reduced models which have been derived within the Bloch–Floquet approach.

### Two-band model in the parabolic band approximation without external fields

It is interesting to consider a simple two-band model in the parabolic band approximation and without external fields, in order to study the off-diagonal Floquet projections of the Wigner function, which arise in this case. In a two-band model, the Wigner function and its evolution equation are given by equations (4.5.9) and (4.5.11) without external fields, and with $m = 0, 1$ and $n = 0, 1$. The Wigner function is given by the sum of four contributions, $f_{00}$, $f_{01}$, $f_{10}$ and $f_{11}$. It can be seen easily from equations (4.5.7), (4.5.10) and (4.5.8) that $f_{01} = \overline{f}_{10}$, while $f_{00}$ and $f_{11}$ are real. Each of the four contributions evolves according to equations (4.5.11). In the parabolic band approximation, the differential equations for $f_{00}$ and $f_{11}$ are

$$\frac{\partial f_{00}}{\partial t} + \frac{p - \hbar k_0}{m_0} \frac{\partial f_{00}}{\partial x} = 0, \qquad \frac{\partial f_{11}}{\partial t} + \frac{p - \hbar k_1}{m_1} \frac{\partial f_{11}}{\partial x} = 0, \qquad (4.5.18)$$

where $m_0$ and $m_1$ are the effective masses for band 0 and band 1 respectively and $k_0$ and $k_1$ are the values of the crystal momentum at which band 0 and band 1 attain their minimum. The evolution equations for $f_{01}$ and $f_{10} = \overline{f}_{01}$ have instead a different structure. A simple calculation shows that

$$i\hbar \frac{\partial f_{01}}{\partial t} = \left\{ \left[ \epsilon_0(k_0) + \frac{(p - \hbar k_0)^2}{2m_0} \right] - \left[ \epsilon_1(k_1) + \frac{(p - \hbar k_1)^2}{2m_1} \right] \right\} f_{01}(x, p)$$
$$- \frac{i\hbar}{2} \left( \frac{p - \hbar k_0}{m_0} + \frac{p - \hbar k_1}{m_1} \right) \frac{\partial f_{01}}{\partial x} - \frac{1}{8} \left( \frac{\hbar^2}{m_0} - \frac{\hbar^2}{m_1} \right) \frac{\partial^2 f_{01}}{\partial x^2}, \quad (4.5.19)$$

which follows from equation (4.5.11) after expanding $f_{mn}(x \pm \eta/2, p, t)$ in Taylor series about $\mu = 0$ and using parabolic profiles for the two bands. By introducing the frequencies

$$\omega_{01} = (\epsilon_0(k_0) - \epsilon_1(k_1))/\hbar$$
$$\Omega_{01}(p) = \omega_{01} + (p - \hbar k_0)^2/(2m_0\hbar) - (p - \hbar k_1)^2/(2m_1\hbar)$$

and the new function

$$g_{01}(x, p, t) = f_{01}(x, p, t)e^{i\Omega_{01}(p)t},$$

equation (4.5.19) can be cast in the more elegant form

$$\frac{\partial g_{01}}{\partial t} + \frac{1}{2} \left( \frac{p - \hbar k_0}{m_0} + \frac{p - \hbar k_1}{m_1} \right) \frac{\partial g_{01}}{\partial x} - \frac{i\hbar}{8} \left( \frac{1}{m_0} - \frac{1}{m_1} \right) \frac{\partial^2 g_{01}}{\partial x^2} = 0.$$
$$(4.5.20)$$

Note that in the definition of the Wigner function (4.5.9) $f_{01}$ and $f_{10}$ appear only in the combination $f_{01} + f_{10}$, consistently with the Wigner function being real. Equation (4.5.19) shows that the time evolution of $f_{01}$ is given by three contributions: an oscillatory term, a free-streaming term and a diffusive term with imaginary diffusion coefficient (Schrödinger-like term). The frequency of the oscillatory term, $\Omega_{01}$, is proportional to the difference of the total energies of the particles of the two bands; the velocity of the free streaming term is an average of the relative velocities of the particle with respect to the two minima and the imaginary diffusion coefficient vanishes when the two effective masses are equal.

Equations (4.5.18) and (4.5.20) completely describe the time evolution of all the components of the Wigner function in a two-band model with the parabolic band approximation and in the absence of external fields. Note that these evolution equations are uncoupled.

## Multiband model in the Luttinger–Kohn approximation

As we have already seen in Section 4.3.2, the Luttinger–Kohn model [LK55] considers the carrier populations near minima (or maxima) of the energy bands and it is therefore to be used in conjunction with the parabolic band approximation. For the Bloch states near the minimum (or maximum) of the band, the Bloch functions $u_n(x, k)$ are replaced with the set of functions $u_n(x, k_n)$, i.e., the Bloch functions at the bottom (or top) of the band, here assumed at $k = k_n$. The functions $e^{ikx}u_n(x, k_n)$, after a suitable normalization, also form a complete set (see Ref. [LK55] and see also Section 4.3.2) and any wave function can be expanded in their basis. In this section, we use the Luttinger–Kohn basis for expressing the Floquet projections $f_{mn}$ of the Wigner function and for writing the evolution equations. The action of the free Hamiltonian is treated in the parabolic band approximation.

If the $n$th band has an extremum at $k = k_n$, we can approximate the Bloch waves as

$$\langle x \mid nk \rangle = b_n(x, k) \approx u_n(x, k_n) e^{ikx}. \tag{4.5.21}$$

Since the functions $u_n(x, k_n)$ are periodic functions with period $a$, we can introduce their Fourier expansion,

$$u_n(x, k_n) = \sum_{n'=-\infty}^{\infty} \widehat{U}_{n'}^n e^{iK_{n'}x},$$

where $K_n = 2\pi n/a$ are vectors of the reciprocal lattice with $K_{-n} = -K_n$. After evaluating the coefficients $\Phi_{mn}$ and the integral kernel $W_{mn}$ in this basis, and after carrying out the integration over the momentum variables $k$ and $k'$, one obtains for the Floquet projection $f_{mn}$ of the Wigner function:

$$f_{mn}(x, p) = 4\pi \sum_{m'n'm''n''} \widehat{U}_{m'}^m \widehat{U}_{n'}^{n*} \widehat{U}_{m''}^{m*} \widehat{U}_{n''}^n e^{i(K_{m'} - K_{n'})x}$$

$$\times \mathcal{H}\left( \frac{\pi}{a} - \left| \frac{p}{\hbar} - \frac{K_{m'} + K_{n'}}{2} \right| \right)$$

$$\times \int dx' \frac{\sin 2[\pi/a - |p/\hbar - (K_{m'} + K_{n'})/2|](x - x')}{x - x'}$$

$$\times f\left( x', p - \hbar \frac{K_{m'} + K_{n'} - K_{m''} - K_{n''}}{2} \right) e^{-i(K_{m''} - K_{n''})x'},$$

where the integrals are performed over the whole real line and $\mathcal{H}$ is the Heaviside function. The evolution equations have been formulated for the case of two energy bands in the parabolic band approximation. If $m_0$ and $m_1$ are the effective masses for band 0 and band 1 respectively and $k_0$ and $k_1$ are the values of the crystal momentum at which band 0 and band 1 attain their minimum, we have that

$$\frac{\partial f_{00}}{\partial t} + \frac{p - \hbar k_0}{m_0} \frac{\partial f_{00}}{\partial x} + \frac{i}{\hbar} (\overline{\Theta}_{00} f)(x, p) = 0 \tag{4.5.22}$$

$$\frac{\partial f_{11}}{\partial t} + \frac{p - \hbar k_1}{m_1} \frac{\partial f_{11}}{\partial x} + \frac{i}{\hbar} (\overline{\Theta}_{11} f)(x, p) = 0, \tag{4.5.23}$$

$$i\hbar \frac{\partial f_{01}}{\partial t} = \left\{ \left[ \epsilon_0(k_0) + \frac{(p - \hbar k_0)^2}{2m_0} \right] - \left[ \epsilon_1(k_1) + \frac{(p - \hbar k_1)^2}{2m_1} \right] \right\} f_{01}(x, p)$$

$$- \frac{i\hbar}{2} \left[ \frac{p - \hbar k_0}{m_0} + \frac{p - \hbar k_1}{m_1} \right] \frac{\partial f_{01}}{\partial x} - \frac{1}{8} \left( \frac{\hbar^2}{m_0} - \frac{\hbar^2}{m_1} \right) \frac{\partial^2 f_{01}}{\partial x^2}$$

$$+ (\overline{\Theta}_{01} f)(x, p), \tag{4.5.24}$$

where $\overline{\Theta}_{mn}$ is an operator acting on the whole Wigner function $f$ and, recalling definition (4.5.13), is given by

$$(\overline{\Theta}_{mn} f)(x, p, t)$$

$$= \int \int dx' d\eta \widehat{W}_{mn}(x, p, x', -\eta) \delta V(x', \eta) \widehat{f}(x', \eta, t)$$

$$= 4\pi \sum_{m'n'm''n''} \widehat{U}_{m'}^m \widehat{U}_{n'}^{n*} \widehat{U}_{m''}^{m*} \widehat{U}_{n''}^n e^{i(K_{m'} - K_{n'})x} \mathcal{H} \left( \frac{\pi}{a} - \left| \frac{p}{\hbar} - \frac{K_{m'} + K_{n'}}{2} \right| \right)$$

$$\times \int dx' e^{-i(K_{m''} - K_{n''})x'} \frac{\sin 2[\pi/a - |p/\hbar - (K_{m'} + K_{n'})/2|](x - x')}{x - x'}$$

$$\times \int d\eta \delta V(x', \eta) e^{-i(p - (K_{m'} + K_{n'} - K_{m''} - K_{n''})/2)\eta/\hbar} \widehat{f}(x', \eta, t). \tag{4.5.25}$$

## A two-band model with empty-lattice eigenfunctions

A different simplification of the transport equations can be obtained by using the Bloch functions of the "empty lattice," that is, periodic plane waves. Here, we consider only the two lowest energy bands, given by

$$\epsilon_0(k) = \frac{\hbar^2 k^2}{2m} \tag{4.5.26}$$

$$\epsilon_1(k) = \frac{\hbar^2}{2m} [\mathcal{H}(k)(k - K)^2 + \mathcal{H}(-k)(k + K)^2], \tag{4.5.27}$$

with $K = 2\pi/a$ and $m$ the bare electron mass, and whose eigenfunctions are

$$\Psi_{0k}(x) = \langle x \mid 0k \rangle = \frac{1}{\sqrt{2\pi}} e^{ikx} \tag{4.5.28}$$

$$\Psi_{1k}(x) = \langle x \mid 1k \rangle = \frac{1}{\sqrt{2\pi}} (\mathcal{H}(k) e^{-iKx} + \mathcal{H}(-k) e^{iKx}) e^{ikx}. \tag{4.5.29}$$

By using this basis in the definition (4.5.10) of the multiband Wigner function, one obtains for the band projections $f_{mn}$ (see [DBJ03b] for the details):

$$f_{00}(x, p) = \frac{1}{\pi} \mathcal{H} \left( \frac{K}{2} - \left| \frac{p}{\hbar} \right| \right) \int \frac{\sin 2(K/2 - |p/\hbar|)(x - x')}{x - x'} f(x', p) dx'$$

$$\tag{4.5.30}$$

$$f_{01}(x,p) = \frac{1}{\pi} \int \left[ \widetilde{\mathcal{H}} \left( -\frac{3\hbar K}{4}, p, 0 \right) e^{i(\alpha_1 + \alpha_2 + K)(x-x')} \frac{\sin(\alpha_2 - \alpha_1)(x-x')}{x-x'} \right.$$

$$\left. + \widetilde{\mathcal{H}} \left( 0, p, \frac{3\hbar K}{4} \right) e^{i(\alpha_3 + \alpha_4 - K)(x-x')} \frac{\sin(\alpha_4 - \alpha_3)(x-x')}{x-x'} \right]$$

$$\times f(x',p) dx' \tag{4.5.31}$$

$$f_{11}(x,p) = \frac{1}{\pi} \int \left[ \widetilde{\mathcal{H}} \left( -\hbar K, p, -\frac{\hbar K}{2} \right) \frac{\sin 2(K/4 - |p/\hbar + 3K/4|)(x-x')}{x-x'} \right.$$

$$+ \widetilde{\mathcal{H}} \left( \frac{\hbar K}{2}, p, \hbar K \right) \frac{\sin 2(K/4 - |p/\hbar - 3K/4|)(x-x')}{x-x'}$$

$$\left. + 2\mathcal{H} \left( \frac{K}{4} - \left| \frac{p}{\hbar} \right| \right) \frac{\sin 2(K/4 - |p/\hbar|)(x-x')}{x-x'} \cos \frac{3}{2} K(x-x') \right]$$

$$\times f(x',p) dx', \tag{4.5.32}$$

where the function $\widetilde{\mathcal{H}}(a,x,b) \equiv \mathcal{H}(x-a)\mathcal{H}(b-x)$ has been introduced, and

$$\alpha_1(p) = -\frac{K}{2} + \left| \frac{p}{\hbar} + \frac{K}{2} \right| \qquad \alpha_2(p) = \frac{K}{4} - \left| \frac{p}{\hbar} + \frac{K}{4} \right|$$

$$\alpha_3(p) = -\frac{K}{4} + \left| \frac{p}{\hbar} - \frac{K}{4} \right| \qquad \alpha_4(p) = \frac{K}{2} - \left| \frac{p}{\hbar} - \frac{K}{2} \right|.$$

The time evolution of the Floquet projections of the Wigner function is given by

$$\frac{\partial f_{00}}{\partial t} + \frac{p}{m} \frac{\partial f_{00}}{\partial x} + \frac{i}{\hbar} (\overline{\Theta}_{00} f)(x,p) = 0$$

$$\frac{\partial f_{11}}{\partial t} + \frac{p}{m} \frac{\partial f_{11}}{\partial x} + \frac{i}{\hbar} (\overline{\Theta}_{11} f)(x,p) = 0,$$

$$\frac{\partial f_{01}}{\partial t} + \frac{p}{m} \frac{\partial f_{01}}{\partial x} + \frac{i}{\hbar} (\overline{\Theta}_{01} f)(x,p) = 0,$$

where $\overline{\Theta}$ is an operator acting on the total Wigner function $f$ and is given by

$$(\overline{\Theta}_{00} f)(x,p) = \frac{1}{\pi} \mathcal{H} \left( \frac{\pi}{a} - \left| \frac{p}{\hbar} \right| \right) \int \frac{\sin 2(\pi/a - |p/\hbar|)(x-x')}{x-x'}$$

$$\times \int \delta V(x',\eta) \widehat{f}(x',\eta,t) e^{-ip\eta/\hbar} \, d\eta \, dx'$$

$$(\overline{\Theta}_{01} f)(x,p) = \frac{1}{\pi} \int \left[ \widetilde{\mathcal{H}} \left( -\frac{3\hbar K}{4}, p, 0 \right) e^{i(\alpha_1 + \alpha_2 + K)(x-x')} \frac{\sin(\alpha_2 - \alpha_1)(x-x')}{x-x'} \right.$$

$$\left. + \widetilde{\mathcal{H}} \left( 0, p, \frac{3\hbar K}{4} \right) e^{i(\alpha_3 + \alpha_4 - K)(x-x')} \frac{\sin(\alpha_4 - \alpha_3)(x-x')}{x-x'} \right]$$

$$\times \int \delta V(x',\eta) \widehat{f}(x',\eta,t) e^{-ip\eta/\hbar} \, d\eta \, dx'$$

$$(\overline{\Theta}_{11}f)(x,p) = \frac{1}{\pi}\int\left[\widetilde{\mathcal{H}}\left(-\hbar K, p, -\frac{\hbar K}{2}\right)\frac{\sin 2(K/4 - |p/\hbar + 3K/4|)(x - x')}{x - x'}\right.$$

$$+\widetilde{\mathcal{H}}\left(\frac{\hbar K}{2}, p, \hbar K\right)\frac{\sin 2(K/4 - |p/\hbar - 3K/4|)(x - x')}{x - x'}$$

$$\left.+2\mathcal{H}\left(\frac{K}{4} - \left|\frac{p}{\hbar}\right|\right)\frac{\sin 2(K/4 - |p/\hbar|)(x - x')}{x - x'}\cos\frac{3}{2}K(x - x')\right]$$

$$\times\int\delta V(x', \eta)\widehat{f}(x', \eta, t)e^{-ip\eta/\hbar}\,d\eta\,dx'.$$

Equations (4.5.30)–(4.5.32) show that the Floquet projections of the Wigner function given by this model are functions with compact support and cover different portions of the phase space. The support of the projection $f_{00}$ on the lower band, for example, corresponds to the first Brillouin zone; the supports of the other projections are larger and extend beyond the first Brillouin zone. The equations of this two-band model are very hard to approach numerically, because of the presence of convolution integrals of highly oscillatory functions.

### 4.5.3 Envelope function-based statistical models

An alternative approach to statistical models based on the Wigner picture starts from an envelope function model, such as the Kane model (4.4.2) or the M-M model (4.4.8). Then the Wigner transformation (4.5.3) is applied directly to the envelope functions ($\Psi_c$ and $\Psi_v$ in the former case, $\Phi_c$ and $\Phi_v$ in the latter).

For a two-band model we need a $2 \times 2$ matrix of Wigner functions (Wigner matrix), defined as the component-wise Wigner transform:

$$w_{ij}(x,p) = (\mathcal{W}\rho_{ij})(x,p), \qquad i,j \in \{c, v\},$$

where $\mathcal{W}$ denotes the Wigner transformation (4.5.3) and $\rho_{ij}$ is an envelope function density matrix (i.e., in the pure-state case, it is given by $\rho_{ij}(x,x') = \Psi_i(x)\overline{\Psi}_j(x')$ for the Kane model and by $\rho_{ij}(x,x') = \Phi_i(x)\overline{\Phi}_j(x')$ for the M-M model). The self-adjointness of the density operator implies the Hermiticity of the Wigner matrix for any fixed $(x, p)$:

$$\rho_{ij}(x,x') = \overline{\rho_{ji}(x',x)} \implies w_{ij}(x,p) = \overline{w_{ji}(x,p)}.$$

The evolution equation for the Wigner matrix in the case of the Kane model (4.4.2) is

$$\left(\frac{\partial}{\partial t} + \frac{p}{m}\cdot\nabla_x + \frac{i}{\hbar}\Theta[V_{cc}]\right)w_{cc} = -\frac{i\hbar K}{2m}\cdot\nabla_x(w_{cv} - w_{vc}) - \frac{K\cdot p}{m}(w_{cv} + w_{vc})$$

$$\left(\frac{\partial}{\partial t} + \frac{p}{m}\cdot\nabla_x + \frac{i}{\hbar}\Theta[V_{cv}]\right)w_{cv} = \frac{i\hbar K}{2m}\cdot\nabla_x(w_{cc} + w_{vv}) + \frac{K\cdot p}{m}(w_{cc} - w_{vv})$$

$$\left(\frac{\partial}{\partial t} + \frac{p}{m} \cdot \nabla_x + \frac{i}{\hbar}\Theta[V_{vc}]\right) w_{vc} = -\frac{i\hbar K}{2m} \cdot \nabla_x(w_{cc} + w_{vv}) + \frac{K \cdot p}{m}(w_{cc} - w_{vv})$$

$$\left(\frac{\partial}{\partial t} + \frac{p}{m} \cdot \nabla_x + \frac{i}{\hbar}\Theta[V_{vv}]\right) w_{vv} = -\frac{i\hbar K}{2m} \cdot \nabla_x(w_{cv} - w_{vc}) + \frac{K \cdot p}{m}(w_{cv} + w_{vc}),$$

$$(4.5.33)$$

where we set

$$V_{ij}(x,\xi) = (E_i + V)\left(x + \frac{\xi}{2}\right) - (E_j + V)\left(x - \frac{\xi}{2}\right), \quad i,j \in \{c,v\}, \quad (4.5.34)$$

and the pseudo-differential operator, in the present three-dimensional case, is given by

$$(\Theta[\phi]f)(x,p) = \frac{1}{(2\pi\hbar)^3} \int_{\mathbb{R}^6} e^{-i(p-p')\cdot\xi/\hbar} \phi(x,\xi) f(x,p') \, d\xi \, dp'. \quad (4.5.35)$$

The system (4.5.33) has been studied from a mathematical point of view in [BFZ03]. The Wigner matrix describing thermal equilibrium of the Kane model has been analyzed in Ref. [Bar04a].

The evolution equation for the Wigner matrix in the case of the M-M model (4.4.2) is

$$\left(\frac{\partial}{\partial t} + p \cdot M_c^{-1}\nabla_x + \frac{i}{\hbar}\Theta[V_{cc}]\right) w_{cc} = \Theta[F_-]w_{cv} - \Theta[F_+]w_{vc}$$

$$\left(\frac{\partial}{\partial t} + p \cdot \frac{M_c^{-1} + M_v^{-1}}{2}\nabla_x - \frac{i\hbar}{4}\nabla_x \cdot \frac{M_c^{-1} - M_v^{-1}}{2}\nabla_x + \frac{ip}{\hbar} \cdot \frac{M_c^{-1} - M_v^{-1}}{2}p\right) w_{cv}$$

$$= -\frac{i}{\hbar}\Theta[V_{cv}]w_{cv} + \Theta[F_-]w_{cc} - \Theta[F_+]w_{vv}$$

$$\left(\frac{\partial}{\partial t} + p \cdot \frac{M_c^{-1} + M_v^{-1}}{2}\nabla_x + \frac{i\hbar}{4}\nabla_x \cdot \frac{M_c^{-1} - M_v^{-1}}{2}\nabla_x - \frac{ip}{\hbar} \cdot \frac{M_c^{-1} - M_v^{-1}}{2}p\right) w_{vc}$$

$$= -\frac{i}{\hbar}\Theta[V_{vc}]w_{vc} - \Theta[F_+]w_{cc} + \Theta[F_-]w_{vv}$$

$$\left(\frac{\partial}{\partial t} + p \cdot M_v^{-1}\nabla_x + \frac{i}{\hbar}\Theta[V_{vv}]\right) w_{vv} = -\Theta[F_+]w_{cv} + \Theta[F_-]w_{vc}, \quad (4.5.36)$$

where we set

$$F_{\pm}(x,\xi) = \frac{\nabla V \cdot P}{mE_g}\left(x \pm \frac{\xi}{2}\right),$$

and the symbols $V_{ij}$ are still given by (4.5.34). From equation (4.4.5) we see that $P$ and, consequently, $F_{\pm}$ are purely imaginary, so that the following relations hold:

$$\overline{\Theta[F_\pm]w_{ij}} = -\Theta[F_\mp]w_{ji}, \qquad i,j \in \{c,v\}.$$

In the special case of constant and opposite effective masses,

$$\mathbb{M}_c = m^*I, \qquad \mathbb{M}_v = -m^*I,$$

the above system reduces to

$$\left(\frac{\partial}{\partial t} + \frac{p}{m^*} \cdot \nabla_x + \frac{i}{\hbar}\Theta[V_{cc}]\right)w_{cc} = \Theta[F_-]w_{cv} - \Theta[F_+]w_{vc}$$

$$\left(\frac{\partial}{\partial t} - \frac{i\hbar}{4m^*}\nabla_x^2 + \frac{ip^2}{\hbar m^*} + \frac{i}{\hbar}\Theta[V_{cv}]\right)w_{cv} = \Theta[F_-]w_{cc} - \Theta[F_+]w_{vv}$$

$$\left(\frac{\partial}{\partial t} + \frac{i\hbar}{4m^*}\nabla_x^2 - \frac{ip^2}{\hbar m^*} + \frac{i}{\hbar}\Theta[V_{vc}]\right)w_{vc} = -\Theta[F_+]w_{cc} + \Theta[F_-]w_{vv}$$

$$\left(\frac{\partial}{\partial t} - \frac{p}{m^*} \cdot \nabla_x + \frac{i}{\hbar}\Theta[V_{vv}]\right)w_{vv} = -\Theta[F_+]w_{cv} + \Theta[F_-]w_{vc}, \qquad (4.5.37)$$

(see also Ref. [FM05]). The negative effective mass introduced in this model has the effect of making the Hamiltonian unbounded from below. As is well known, such a Hamiltonian is not very good, especially for statistical purposes (the thermal equilibrium states are ill defined). However, the correct interpretation is that (4.5.37) should be considered as just an approximation of the true dynamics for small values of the momentum $p$.

## 4.6 Hydrodynamic models

It is universally recognized that the hydrodynamic approach presents important properties both from a theoretical and a numerical point of view because it gives an interpretation of the transport phenomenon by macroscopic quantities and it produces many advantages from a computational point of view.

The literature on hydrodynamic models is very broad, in both the classical as well as the semiclassical and quantum frameworks.

Some very interesting results have been achieved that propose quantum hydrodynamic equations that are able to describe the behaviour of nanometric devices like resonant tunneling diodes. Here, we restrict ourselves to describing the hydrodynamic versions of the Kane model and of the M-M model.

Most of the results published in the literature refer to single-band problems. The generalization to multiband models presents several difficulties, such as the definition of the macroscopic quantities with a realistic physical meaning and the difficulty in imposing boundary conditions.

In this review we give an insight into the classical derivation of a two-band quantum fluid. As we have said, the above-mentioned multiband models are based on the single-electron Schrödinger equation, and the resulting equations are essentially linear. By applying the Wentzel–Kramers–Brillouin (WKB) method, it is possible to derive a zero-temperature hydrodynamic version of the Schrödinger two-band models.

When it is desirable to model the dynamics of a family of electrons, the statistical description requires the introduction of a sequence of mixed states, with an attached occupation probability. In this case, the WKB method leads to a sequence of hydrodynamic equations, from which it is possible to derive a set of equations for certain macroscopic averaged quantities. These hydrodynamic equations share a similar structure with the corresponding equations for a single electron, the only difference being the appearance of terms that can be interpreted as thermal tensors, and of additional source terms. These new terms depend on all states, so the system is not closed unless appropriate closure conditions are provided. It is clear that the final hydrodynamic model with temperature is by no means equivalent to the original quantum model. We could say that the nonlinearity of the resulting hydrodynamic model is "genuine" and is the price to pay for keeping only a finite number of equations.

### 4.6.1 The hydrodynamic quantities

In order to obtain hydrodynamic versions of the kinetic models described in the previous sections, one possibility is to follow the general hydrodynamic approach to quantum mechanics due to Madelung [LL77]. This approach consists in writing the wave function in the quasi-classical form $a\exp(\frac{iS}{\epsilon})$, where $a$ is called the amplitude and $S/\epsilon$ the phase. With this approach, the hydrodynamic limit is valid only for pure states, that is to say, it is valid only for a quantum system at zero temperature. In the case of a two-band model, we have

$$\psi_a(x,t) = \sqrt{n_a(x,t)}\exp\left(\frac{iS_a(x,t)}{\epsilon}\right), \quad a = c, v. \qquad (4.6.1)$$

where the squared amplitude has the physical meaning of the probability density of finding the "particle" at some point in space, and the gradient of the phase corresponds to the classical velocity of the "particle."

In the framework of two-band models, the densities

$$n_{ab} = \overline{\psi}_a \psi_b$$

are introduced, where $\psi_a$, with $a = c, v$, is the envelope function for the conduction and the valence band, respectively. When $a = b$, the quantities $n_{ab} = n_a = |\psi_a|^2$ are real and represent the position probability densities of the conduction band and of the valence band electrons, albeit only in an approximate sense, since $\psi_c$ and $\psi_v$ are envelope functions which mix the Bloch states. Nevertheless, $n = \overline{\psi}_c\psi_c + \overline{\psi}_v\psi_v$ is exactly the total electron density in the conduction and in the valence band, and, as expected, it satisfies a continuity equation. When $a \neq b$, the density $\overline{\psi}_a \psi_b$ is a complex quantity, which does not have a precise physical meaning. Despite this, as will become clear in the next section, the complex quantities $\overline{\psi}_a \psi_b$ appear explicitly in the evolution equation for the total density $n$.

It is customary, after (4.6.1), to write the coupling terms in a more convenient way, by introducing the complex quantity

$$n_{cv} = \overline{\psi}_c \psi_v = \sqrt{n_c}\sqrt{n_v}\, e^{i\sigma}, \qquad (4.6.2)$$

where $\sigma$ is the phase difference defined by

$$\sigma = \frac{S_v - S_c}{\epsilon}. \qquad (4.6.3)$$

In this way, in order to study a zero-temperature quantum hydrodynamic model, we need to use only the three quantities $n_c, n_v$ and $\sigma$ to characterize the zero-order moments.

The situation is more involved for the current densities. Analogously to the one-band case, we introduce the quantum mechanical electron current densities

$$J_{ab} = \epsilon \mathrm{Im}\left(\overline{\psi}_a \nabla \psi_b\right). \qquad (4.6.4)$$

It is natural to recover the classical current densities,

$$J_c = \mathrm{Im}\left(\epsilon \overline{\psi}_c \nabla \psi_c\right) = n_c \nabla S_c, \quad J_v = \mathrm{Im}\left(\epsilon \overline{\psi}_v \nabla \psi_v\right) = n_v \nabla S_v, \qquad (4.6.5)$$

whose physical meaning is clear.

The introduction of the complex quantity (4.6.2) allows us to write $\epsilon \overline{\psi}_a \nabla \psi_b$ in (4.6.4) as

$$\epsilon \overline{\psi}_c \nabla \psi_v = n_{cv} u_v, \qquad \epsilon \overline{\psi}_v \nabla \psi_c = \overline{n}_{cv} u_c, \qquad (4.6.6)$$

where the complex velocities $u_c$ and $u_v$ are given by

$$u_c = u_{\mathrm{os},c} + i u_{\mathrm{el},c}, \quad u_v = u_{\mathrm{os},v} + i u_{\mathrm{el},v}, \qquad (4.6.7)$$

where $u_{\mathrm{os},c}$ and $u_{\mathrm{os},v}$ are the osmotic velocity and current velocity and are given by

$$u_{\mathrm{os},a} = \frac{\epsilon \nabla \sqrt{n_a}}{\sqrt{n_a}}, \quad u_{\mathrm{el},a} = \nabla S_a = \frac{J_a}{n_a}, a = c, v. \qquad (4.6.8)$$

In analogy with the single-band case we have defined the osmotic and current velocities as complex quantities which can be expressed solely by means of $n_c, n_v, J_c$ and $J_v$. In addition, the coupling term $n_{cv}$ has been defined by introducing the phase difference $\sigma$. We note that

$$\epsilon \nabla n_{cv} = n_{cv}(\overline{u}_c + u_v). \qquad (4.6.9)$$

Coming back to the choice of the hydrodynamic quantities, we maintain that, for a zero-temperature quantum hydrodynamic system, it is sufficient to take the usual quantities $n_c, n_v, J_c$ and $J_v$, plus the phase difference $\sigma$. This will be confirmed in the next section.

## 4.6.2 Hydrodynamic version of the Kane system

The Kane model was introduced in Section 4.4.1 by using envelope functions. Before introducing the hydrodynamic form, we rewrite it by using dimensionless variables. To this aim dimensionless form, we introduce the rescaled Planck constant $\epsilon = \hbar/\alpha$, where the dimensional parameter $\alpha$ is given by $\alpha = mx_R^2/t_R$, by using $x_R$ and $t_R$ as characteristic (scalar) length and time variables. The band energy can be rescaled by taking new potential units $V_0 = mx_R^2/t_R^2$. A dimensional argument shows that the original coupling coefficient is a reciprocal of a characteristic length, thus the coefficient is scaled by $Kx_R$, componentwise.

Hence, dropping the primes and without changing the name of the variables, we get the following scaled Kane system, which will be the object of our study:

$$i\epsilon\frac{\partial\psi_c}{\partial t} = -\frac{\epsilon^2}{2}\Delta\psi_c + V_c\psi_c - \epsilon^2 K\cdot\nabla\psi_v$$

$$i\epsilon\frac{\partial\psi_v}{\partial t} = -\frac{\epsilon^2}{2}\Delta\psi_v + V_v\psi_v + \epsilon^2 K\cdot\nabla\psi_c\,, \tag{4.6.10}$$

where $K$ is the rescaled coupling interband coefficient, $\epsilon$ is the rescaled Planck constant, $V_c = E_c + V$ and $V_v = E_v + V$. In the Kane model the coupling parameter has to be considered constant. In realistic heterostructure semiconductor devices, the parameter $K$, expressed in terms of the effective electron mass and the energy gap, depends on the layer composition through the spatial coordinates.

Taking into account the wave form (4.6.1) and using the equations of system (4.6.10), the time derivation of $n_a$, $a = c, v$ gives immediately

$$\frac{\partial n_c}{\partial t} + \nabla\cdot J_c = -2K\cdot\mathrm{Im}\left(n_{cv}u_v\right)$$

$$\frac{\partial n_v}{\partial t} + \nabla\cdot J_v = 2K\cdot\mathrm{Im}\left(\overline{n}_{cv}u_c\right)\,, \tag{4.6.11}$$

where (4.6.5) has been used for $J_c$ and $J_v$. We remark that the right-hand side of (4.6.11), containing the terms $n_{cv}u_v$ and $\overline{n}_{cv}u_c$, can be expressed in terms of osmotic and current velocities, and the phase difference $\sigma$.

Adding the equations in (4.6.11) and using the identity $\mathrm{Im}(\epsilon\overline{\psi}_v\nabla\psi_v) - \mathrm{Im}(\epsilon\overline{\psi}_v\nabla\psi_c) = \epsilon\nabla\mathrm{Im}n_{cv}$, we obtain the balance law for the total density

$$\frac{\partial}{\partial t}(n_c + n_v) + \nabla\cdot(J_c + J_v + 2\epsilon K\mathrm{Im}n_{cv}) = 0\,,$$

which is just the quantum counterpart of the classical continuity equation.

The derivation of the equations for the phases $S_c$, $S_v$, and consequently for $J_c$ and $J_v$, is more involved.

Referring the reader to the original paper [AF05] for more details, the equations for the currents take the form

$$\frac{\partial J_c}{\partial t} + \text{div}\,\left(\frac{J_c \otimes J_c}{n_c} + \epsilon^2 \nabla \sqrt{n_c} \otimes \nabla \sqrt{n_c} - \frac{\epsilon^2}{4} \nabla \otimes \nabla n_c\right) + n_c \nabla V_c$$
$$= -\,\epsilon^2 \text{Re}\left[\nabla(\overline{\psi}_c K \cdot \nabla \psi_v) - 2\nabla \overline{\psi}_c K \cdot \nabla \psi_v\right]. \qquad (4.6.12)$$

$$\frac{\partial J_v}{\partial t} + \text{div}\,\left(\frac{J_v \otimes J_v}{n_v} + \epsilon^2 \nabla \sqrt{n_v} \otimes \nabla \sqrt{n_v} - \frac{\epsilon^2}{4} \nabla \otimes \nabla n_v\right) + n_v \nabla V_v$$
$$= -\,\epsilon^2 \text{Re}\left[\nabla(\overline{\psi}_v K \cdot \nabla \psi_c) - 2\nabla \overline{\psi}_v K \cdot \nabla \psi_c\right]. \qquad (4.6.13)$$

The left-hand sides of the equations for the currents can be put in a more familiar form by using the identity

$$\text{div}\,\left(\nabla \sqrt{n_a} \otimes \nabla \sqrt{n_a} - \frac{1}{4} \nabla \otimes \nabla n_a\right) = -\frac{n_a}{2} \nabla \left[\frac{\Delta \sqrt{n_a}}{\sqrt{n_a}}\right], \quad a = c, v.$$

The correction terms

$$\frac{\epsilon^2}{2} \frac{\Delta \sqrt{n_a}}{\sqrt{n_a}}, \quad a = c, v,$$

can be identified with the quantum Bohm potentials for each band, because they can be interpreted as internal self-consistent potentials, in analogy with the single-band case. The right-hand sides can be further expressed in terms of the hydrodynamic quantities, obtaining the final system

$$\frac{\partial J_c}{\partial t} + \text{div}\,\left(\frac{J_c \otimes J_c}{n_c}\right) - n_c \nabla \left(\frac{\epsilon^2 \Delta \sqrt{n_c}}{2\sqrt{n_c}}\right) + n_c \nabla V_c$$
$$= \quad \epsilon \nabla \text{Re}\,(n_{cv} K \cdot u_v) - 2\text{Re}\,(n_{cv} K \cdot u_v \overline{u}_c)\,,$$
$$(4.6.14)$$
$$\frac{\partial J_v}{\partial t} + \text{div}\,\left(\frac{J_v \otimes J_v}{n_v}\right) - n_v \nabla \left(\frac{\epsilon^2 \Delta \sqrt{n_v}}{2\sqrt{n_v}}\right) + n_v \nabla V_v$$
$$= -\epsilon \nabla \text{Re}\,(\overline{n}_{cv} K \cdot u_c) + 2\text{Re}\,(\overline{n}_{cv} K \cdot u_c \overline{u}_v)\,.$$

It is evident that the equations for the conduction and the valence band are coupled. Also, because of the presence of $\sigma$, it is necessary to "close" the system, in order to obtain an extension of the classical Madelung fluid equations to a two-band quantum fluid. In this context, we choose the following constraint:

$$\epsilon \nabla \sigma = \frac{J_v}{n_v} - \frac{J_c}{n_c}. \qquad (4.6.15)$$

Now we are in position to rewrite the hydrodynamic system (4.6.14) as follows:

$$\frac{\partial n_c}{\partial t} + \mathrm{div} J_c = -2K \cdot \mathrm{Im}(n_{cv} u_v),$$

$$\frac{\partial n_v}{\partial t} + \mathrm{div} J_v = 2K \cdot \mathrm{Im}(\overline{n}_{cv} u_c),$$

$$\frac{\partial J_c}{\partial t} + \mathrm{div}\left(\frac{J_c \otimes J_c}{n_c}\right) - n_c \nabla\left(\frac{\epsilon^2 \Delta \sqrt{n_c}}{2\sqrt{n_c}}\right) + n_c \nabla V_c$$
$$= \epsilon \nabla \mathrm{Re}(n_{cv} K \cdot u_v) - 2\mathrm{Re}\left(n_{cv} K \cdot u_v \overline{u}_c\right), \qquad (4.6.16)$$

$$\frac{\partial J_v}{\partial t} + \mathrm{div}\left(\frac{J_v \otimes J_v}{n_v}\right) - n_v \nabla\left(\frac{\epsilon^2 \Delta \sqrt{n_v}}{2\sqrt{n_v}}\right) + n_v \nabla V_v$$
$$= -\epsilon \nabla \mathrm{Re}(\overline{n}_{cv} K \cdot u_c) + 2\mathrm{Re}\left(\overline{n}_{cv} K \cdot u_c \overline{u}_v\right),$$

$$\epsilon \nabla \sigma = \frac{J_v}{n_v} - \frac{J_c}{n_c},$$

where $n_{cv}, u_v$, and $u_c$ are expressed in the terms of the hydrodynamic quantities $n_c, n_v, J_c, J_v$, and $\sigma$ by (4.6.2) and (4.6.7).

### 4.6.3 The nonzero-temperature case

The extension of the previous analysis to an electron ensemble requires a quantum statistical mechanics treatment. According to the general discussion at the beginning of Section 4.5, it is possible to represent an electron ensemble as a mixed quantum mechanical state given by a sequence of pure states $\Psi_k$, with occupation probabilities $\lambda^k \geq 0$, so that $\sum_k \lambda^k = 1$. In the two-band case, each pure state is represented by a couple of envelope functions, $\psi_c^k$ and $\psi_v^k$ and, therefore, we shall extend the definition of the hydrodynamic quantities as a superposition, with weights $\lambda^k$, of the corresponding pure-state quantities. For example, the density will be $n_a = \sum_k \lambda^k \overline{\psi}_a^k \psi_a^k$, for $a = c, v$.

In the sequel we shall work at the formal level, and we refer to the equations found in [AF05]. The $k$th state for the Kane system is described by the solutions of the system

$$i\epsilon \frac{\partial \psi_c^k}{\partial t} = -\frac{\epsilon^2}{2} \Delta \psi_c^k + V_c \psi_c^k - \epsilon^2 K \cdot \nabla \psi_v^k,$$

$$i\epsilon \frac{\partial \psi_v^k}{\partial t} = -\frac{\epsilon^2}{2} \Delta \psi_v^k + V_v \psi_v^k + \epsilon^2 K \cdot \nabla \psi_c^k. \qquad (4.6.17)$$

Using the expressions (4.6.1) for each state $k$ in (4.6.17),

$$\psi_c^k = \sqrt{n_c^k} \exp\left(iS_c^k/\epsilon\right), \qquad \psi_v^k = \sqrt{n_v^k} \exp\left(iS_v^k/\epsilon\right),$$

under the assumption of positivity of the densities $n_c^k$ and $n_v^k$, a hydrodynamic system analogous to (4.6.16) is obtained for each state $k$. The densities and the currents corresponding to the two mixed states for conduction and valence electrons can be defined as

$$n_c = \sum_{k=0}^{\infty} \lambda^k n_c^k, \quad n_v = \sum_{k=0}^{\infty} \lambda^k n_v^k,$$

$$J_c = \sum_{k=0}^{\infty} \lambda^k J_c^k, \quad J_v = \sum_{k=0}^{\infty} \lambda^k J_v^k.$$

We also define

$$\sigma = \sum_{k=0}^{\infty} \lambda^k \sigma^k, \quad n_{cv} = \sqrt{n_c}\sqrt{n_v}\exp(i\sigma),$$

$$u_c = \frac{\epsilon \nabla \sqrt{n_c}}{\sqrt{n_c}} + i\frac{J_c}{n_c}, \quad u_v = \frac{\epsilon \nabla \sqrt{n_v}}{\sqrt{n_v}} + i\frac{J_v}{n_v}.$$

Multiplying (4.6.16) for the state $k$ by $\lambda^k$ and summing over $k$, we find new quantities that must be manipulated with much care. In analogy with the one-band case [GMU95], new terms containing the total temperature $\vartheta_c$ and $\vartheta_v$, for each band, appear in the current equations. The temperature tensors are defined by the sum of osmotic temperature and electron current temperature

$$\vartheta_c = \vartheta_{\mathrm{os},c} + \vartheta_{\mathrm{el},c} \quad \text{and} \quad \vartheta_v = \vartheta_{\mathrm{os},v} + \vartheta_{\mathrm{el},v}$$

given by

$$\vartheta_{\mathrm{os},c} = \sum_{k=0}^{\infty} \lambda^k \frac{n_c^k}{n_c}(u_{\mathrm{os},c}^k - u_{\mathrm{os},c}) \otimes (u_{\mathrm{os},c}^k - u_{\mathrm{os},c}),$$

$$\vartheta_{\mathrm{el},c} = \sum_{k=0}^{\infty} \lambda^k \frac{n_c^k}{n_c}(u_{\mathrm{el},c}^k - u_{\mathrm{el},c}) \otimes (u_{\mathrm{el},c}^k - u_{\mathrm{el},c}).$$

In conclusion our system becomes

$$\frac{\partial n_c}{\partial t} + \mathrm{div}J_c = -2K \cdot \mathrm{Im}[n_{cv}(\alpha u_v + \beta_v)],$$

$$\frac{\partial n_v}{\partial t} + \mathrm{div}J_v = 2K \cdot \mathrm{Im}[\overline{n_{cv}}(\overline{\alpha}\, u_c + \beta_c)],$$

$$\frac{\partial J_c}{\partial t} + \mathrm{div}\left(\frac{J_c \otimes J_c}{n_c} + n_c\vartheta_c\right) - n_c\nabla\left(\frac{\epsilon^2\Delta\sqrt{n_c}}{2\sqrt{n_c}}\right) + n_c\nabla V_c$$

$$= \epsilon K \cdot \nabla\mathrm{Re}\left(n_{cv}(\alpha u_v + \beta_v)\right)$$

$$- 2K \cdot \mathrm{Re}\left(n_{cv}(\alpha u_v \otimes \overline{u_c} + \beta_v \otimes \overline{u_c} + u_v \otimes \overline{\beta_c} + \vartheta_{cv})\right),$$

$$\frac{\partial J_v}{\partial t} + \mathrm{div}\left(\frac{J_v \otimes J_v}{n_v} + n_v\vartheta_v\right) - n_v\nabla\left(\frac{\epsilon^2\Delta\sqrt{n_v}}{2\sqrt{n_v}}\right) + n_v\nabla V_v$$

$$= -\epsilon K \cdot \nabla\mathrm{Re}\left(\overline{n_{cv}}(\overline{\alpha}u_c + \beta_c)\right)$$

$$+ 2K \cdot \mathrm{Re}\left(\overline{n_{cv}}\left(\overline{\alpha u_c} \otimes \overline{u_v} + \beta_c \otimes \overline{u_v} + u_c \otimes \overline{\beta_v} + \vartheta_{vc}\right)\right),$$

$$\epsilon \nabla \sigma - \frac{J_v}{n_v} + \frac{J_c}{n_c} = -\mathrm{Im}\left\{\frac{1}{\alpha}\left(\epsilon \nabla \alpha - \beta_v - \overline{\beta_c}\right)\right\}, \qquad (4.6.18)$$

where the new quantities are defined by

$$\alpha = \sum_{k=0}^{\infty} \lambda^k \frac{n_{cv}^k}{n_{cv}}, \qquad \beta_v = \sum_{k=0}^{\infty} \lambda^k \frac{n_{cv}^k}{n_{cv}}(u_v^k - u_v), \qquad \beta_c = \sum_{k=0}^{\infty} \lambda^k \frac{\overline{n_{cv}^k}}{n_{cv}}(u_c^k - u_c),$$

and, in the expression of the coupling terms between the two bands, there appears a sum of temperature tensors, given by

$$\vartheta_{cv} = \sum_{k=0}^{\infty} \lambda^k \frac{n_{cv}^k}{n_{cv}}(u_v^k - u_v) \otimes (\overline{u_c^k} - \overline{u_c}),$$

$$\vartheta_{vc} = \sum_{k=0}^{\infty} \lambda^k \frac{\overline{n_{cv}^k}}{n_{cv}}(u_c^k - u_c) \otimes (\overline{u_v^k} - \overline{u_v}).$$

Equations (4.6.18) can be considered as a nonzero-temperature quantum fluid model. The quantities $n_{cv}$, $u_c$, and $u_v$, already present in (4.6.16), are expressed in terms of $n_c$, $n_v$, $J_c$, $J_v$, and $\sigma$, while the new quantities $\alpha$, $\beta_c$, and $\beta_v$ satisfy the relation

$$\mathrm{Re}\left\{\frac{1}{\alpha}\left(\epsilon \nabla \alpha - \beta_v - \overline{\beta_c}\right)\right\} = 0$$

and need appropriate closure relations. Moreover, we must assign constitutive relations for the tensor components $\vartheta_c$, $\vartheta_v$, $\vartheta_{cv}$ and $\vartheta_{vc}$; $\vartheta_c$ and $\vartheta_v$ are formally analogous to the temperature tensor of kinetic theory.

A simple class of closure conditions can be obtained by assigning a function $\alpha = \alpha(n_c, n_v, \sigma)$ and taking

$$\beta_c = 2n_c \frac{\partial \bar{\alpha}}{\partial n_c} u_{\mathrm{os},c} - \frac{\partial \bar{\alpha}}{\partial \sigma} u_{\mathrm{el},c}, \qquad \beta_v = 2n_v \frac{\partial \alpha}{\partial n_v} u_{\mathrm{os},v} + \frac{\partial \alpha}{\partial \sigma} u_{\mathrm{el},v}. \qquad (4.6.19)$$

Then, we have

$$\epsilon \nabla \alpha - \beta_v - \overline{\beta_c} = 0,$$

which implies

$$\epsilon \nabla \sigma - \frac{J_v}{n_v} + \frac{J_c}{n_c} = 0.$$

In particular, it is possible to choose

$$\alpha = 1, \qquad \beta_c = \beta_v = 0. \qquad (4.6.20)$$

We still need to consider the temperature tensors $\vartheta_c$, $\vartheta_v$, $\vartheta_{cv}$ and $\vartheta_{vc}$. Heuristically, following the analogy with the single-band fluid-dynamical model [Jun01], the simplest closure relation is

$$\vartheta_c = \frac{1}{n_c} p_c(n_c) I, \quad \vartheta_v = \frac{1}{n_v} p_v(n_v) I, \quad \vartheta_{cv} = \vartheta_{vc} = 0, \qquad (4.6.21)$$

where $I$ is the identity tensor and the functions $p_c$ and $p_v$ can be interpreted as pressures. In this way we obtain the simplest two-band, isentropic, fluid-dynamical model:

$$\frac{\partial n_c}{\partial t} + \operatorname{div} J_c = -2K \cdot \operatorname{Im}(n_{cv} u_v),$$

$$\frac{\partial n_v}{\partial t} + \operatorname{div} J_v = \phantom{-}2K \cdot \operatorname{Im}(\overline{n_{cv}} u_c),$$

$$\frac{\partial J_c}{\partial t} + \operatorname{div}\left(\frac{J_c \otimes J_c}{n_c} + p_c(n_c) I\right) - n_c \nabla\left(\frac{\epsilon^2 \Delta \sqrt{n_c}}{2\sqrt{n_c}}\right) + n_c \nabla V_c$$
$$= \epsilon K \cdot \nabla \operatorname{Re}(n_{cv} u_v) - 2K \cdot \operatorname{Re}(n_{cv} u_v \otimes \overline{u_c}), \qquad (4.6.22)$$

$$\frac{\partial J_v}{\partial t} + \operatorname{div}\left(\frac{J_v \otimes J_v}{n_v} + p_v(n_v) I\right) - n_v \nabla\left(\frac{\epsilon^2 \Delta \sqrt{n_v}}{2\sqrt{n_v}}\right) + n_v \nabla V_v$$
$$= -\epsilon K \cdot \nabla \operatorname{Re}(\overline{n_{cv}} u_c) + 2K \cdot \operatorname{Re}(\overline{n_{cv}} u_c \otimes \overline{u_v}),$$

$$\epsilon \nabla \sigma - \frac{J_v}{n_v} + \frac{J_c}{n_c} = 0.$$

We remark that if the (classical) pressures are linear functions of $n_c$ and $n_v$, equations (4.6.22) reduce to the isothermal case.

### 4.6.4 Hydrodynamic version of the M-M system

The method used in the previous section is also suitable to be applied to the multiband envelope function model introduced by Modugno and Morandi in [MM05] and described in Section 4.4.2. However, as we have remarked at the end of Section 4.5.3, when mixed states become important (namely, for nonzero-temperature models), the M-M model has some undesirable features that make the discussion more complicated, beyond the scope of the present review. For this reason we shall restrict ourselves to the zero-temperature case.

By using dimensionless variables, the system (4.4.8) reads as follows:

$$ie\frac{\partial \psi_c}{\partial t} = -\frac{\epsilon^2}{2} \Delta \psi_c + (E_c + V)\psi_c - \epsilon^2 P \psi_v,$$

$$\phantom{ie\frac{\partial \psi_c}{\partial t} =} \qquad (4.6.23)$$

$$ie\frac{\partial \psi_v}{\partial t} = \phantom{-}\frac{\epsilon^2}{2} \Delta \psi_v + (E_v + V)\psi_v - \epsilon^2 P \psi_c,$$

where $P$ is the rescaled coupling interband coefficient and $\epsilon$ is the rescaled Planck constant.

By using the Madelung form (4.6.1) for the wave functions, and proceeding in the same way as for the Kane model in Section 4.6.2, we obtain the hydrodynamic equations for the two-band M-M model

$$\frac{\partial n_c}{\partial t} + \nabla \cdot J_c = -2P\mathrm{Im}\left(\epsilon\overline{\psi}_c\psi_v\right),$$

$$\frac{\partial n_v}{\partial t} - \nabla \cdot J_v = 2P\mathrm{Im}\left(\epsilon\overline{\psi}_c\psi_v\right).$$

(4.6.24)

By summing the two equations in (4.6.24), we obtain the balance law for the total density,

$$\frac{\partial \rho}{\partial t} + \nabla \cdot J = 0,$$

(4.6.25)

where $\rho = n_c + n_v$ is the total density and $J = J_c - J_v$ is the total current.

We remark that, in contrast with the Kane model, interband current terms do not appear in the conservation of the total density.

Next, the equations for the phases $S_c$, $S_v$, and the currents $J_c$ and $J_v$ are derived. Referring the reader to the paper [AFM05] for the details, here we only write the equations for the currents in the final form

$$\frac{\partial J_c}{\partial t} + \mathrm{div}\left(\frac{J_c \otimes J_c}{n_c}\right) - n_c\nabla\left(\frac{\epsilon^2\Delta\sqrt{n_c}}{2\sqrt{n_c}}\right) + n_c(\nabla E_c + \nabla V)$$
$$= \epsilon^2\nabla PRe n_{cv} + \epsilon P\sqrt{n_c}\sqrt{n_v}(\cos\sigma(u_{\mathrm{os},v} - u_{\mathrm{os},c}) - \sin\sigma(u_{\mathrm{el},c} + u_{\mathrm{el},v})),$$

$$\frac{\partial J_v}{\partial t} - \mathrm{div}\left(\frac{J_v \otimes J_v}{n_v}\right) + n_v\nabla\left(\frac{\epsilon^2\Delta\sqrt{n_v}}{2\sqrt{n_v}}\right) + n_v(\nabla E_v + \nabla V)$$
$$= \epsilon^2\nabla PRe n_{cv} - \epsilon P\sqrt{n_c}\sqrt{n_v}(\cos\sigma(u_{\mathrm{os},v} - u_{\mathrm{os},c}) - \sin\sigma(u_{\mathrm{el},c} + u_{\mathrm{el},v})).$$

(4.6.26)

Also in this case, we have introduced the internal self-consistent potentials for each band (the Bohm potentials) and the osmotic velocities $(u_{\mathrm{os},c}, u_{\mathrm{os},v})$ and current velocities $(u_{\mathrm{el},c}, u_{\mathrm{el},v})$; $\sigma$ is again the phase difference defined by $\sigma = \frac{S_v - S_c}{\epsilon}$.

The systems (4.6.24) and (4.6.26) are not equivalent to the original system (4.6.23), due to the presence of $\sigma$. By using the constraint (4.6.15), we finally obtain the hydrodynamic system

$$\frac{\partial n_c}{\partial t} + \mathrm{div}J_c = -2\epsilon P\mathrm{Im}n_{cv},$$

$$\frac{\partial n_v}{\partial t} - \mathrm{div}J_v = 2\epsilon P\mathrm{Im}n_{cv},$$

$$\frac{\partial J_c}{\partial t} + \text{div}\left(\frac{J_c \otimes J_c}{n_c}\right) - n_c \nabla \left(\frac{\epsilon^2 \Delta \sqrt{n_c}}{2\sqrt{n_c}}\right) + n_c(\nabla E_c + \nabla V)$$
$$= \epsilon^2 \nabla \text{PRe} n_{cv} + \epsilon \text{PRe}\left(n_{cv}(u_v - \overline{u}_c)\right),$$

$$\frac{\partial J_v}{\partial t} - \text{div}\left(\frac{J_v \otimes J_v}{n_v}\right) + n_v \nabla \left(\frac{\epsilon^2 \Delta \sqrt{n_v}}{2\sqrt{n_v}}\right) + n_v(\nabla E_v + \nabla V) \qquad (4.6.27)$$
$$= \epsilon^2 \nabla \text{PRe} n_{cv} - \epsilon \text{PRe}\left(n_{cv}(u_v - \overline{u}_c)\right),$$

$$\epsilon \nabla \sigma = \frac{J_v}{n_v} - \frac{J_c}{n_c},$$

where $n_{cv}$, $u_v$, $u_c$ are expressed in the terms of the hydrodynamic quantities $n_c$, $n_v$, $J_c$, $J_v$, $\sigma$. System (4.6.27) is the extension of the classical Madelung fluid equations to a two-band quantum fluid.

## Acknowledgments

The authors are grateful to Giuseppe Alì, Paolo Bordone, Giovanni Borgioli, Carlo Jacoboni, Chiara Manzini and Omar Morandi for the collaborations which brought many of the results reviewed in this chapter.

This work was performed under the auspices of the *National Group for Mathematical Physics* of the *Istituto Nazionale di Alta Matematica* and was partly supported by the *Italian Ministery of University (MIUR* National Project "Mathematical Problems of Kinetic Theories," Cofin2004).

# References

[AF05]     Alì, G., Frosali, G.: Quantum hydrodynamic models for the two-band Kane system, Nuovo Cimento B, **120**(12), 1279–1298 (2005).

[AFM05]    Alì, G., Frosali, G., and Manzini, C.: On the drift-diffusion model for a two-band quantum fluid at zero-temperature. Ukrainian Math. J., **57**(6), 723–730 (2005).

[AP05]     Allaire, G., Piatnitski, A.: Homogenization of the Schrödinger equation and effective mass theorems. Comm. Math. Phys., **258**(1), 1–22 (2005).

[Bar03a]   Barletti, L.: A mathematical introduction to the Wigner formulation of quantum mechanics. Boll. Unione Mat. Ital., B, **6-B**, 693–716 (2003).

[Bar03b]   Barletti, L.: Wigner envelope functions for electron transport in semi-conductor devices. Transport Theory Statist. Phys., **32**(3&4), 253–277 (2003).

[Bar04a]   Barletti, L.: A "spinorial" Wigner function describing the two-band kp dynamics of electrons in crystals. In: Primicerio, M., Spigler, R., and Valente, V. (eds.) Applied and Industrial Mathematics in Italy. World Scientific, Singapore (2005).

[Bar04b]   Barletti, L.: On the thermal equilibrium of a quantum system described by a two-band Kane Hamiltonian. Nuovo Cimento B, **119**(12), 1125–1140 (2004).

[Bar05]     Barletti, L.: Quantum moment equations for a two-band k.p Hamiltonian. Boll. Unione Mat. Ital., B, **8-B**, 103–121 (2005).

[BD02]      Barletti, L., Demeio, L.: Wigner function approach to multiband transport in semiconductor devices, Proc. VI Congresso SIMAI, Chia Laguna (CA-Italy) May 27–31, 2002. Versione su CD-rom.

[BLP78]     Bensoussan, A., Lions, J.-L., and Papanicolaou, G.: Asymptotic Analysis for Periodic Structures. North-Holland, Amsterdam (1978).

[BJ99]      Bordone, P., Pascoli, M., Brunetti, R., Bertoni, A., and Jacoboni, C.: Quantum transport of electrons in open nanostructures with the Wigner function formalism, Phys. Rev. B, **59**, 3060–3069 (1999).

[BFZ03]     Borgioli, G., Frosali, G., and Zweifel, P. F.: Wigner approach to the two-band Kane model for a tunneling diode. Transport Theory Statist. Phys. **32**(3&4), 347–366 (2003).

[Buo74]     Buot, F. A.: Method for calculating $\mathrm{Tr}H^n$ in solid-state theory. Phys. Rev. B, **10**, 3700–3705 (1974).

[Buo76]     Buot, F. A.: Magnetic susceptibility of interacting free and Bloch electrons. Phys. Rev. B, **14**, 3310–3328 (1976).

[Buo86]     Buot, F. A.: Direct construction of path integrals in the lattice-space multiband dynamics of electrons in a solid. Phys. Rev. A, **33**, 2544–2562 (1986).

[BJ90]      Buot, F. A., Jensen, K. L.: Lattice Weyl-Wigner formulation of exact many-body quantum-transport theory and applications to novel solid-state quantum-based devices. Phys. Rev. B, **42**, 9429–9457 (1990).

[Bur92]     Burt, M. G.: The justification for applying the effective-mass approximation to microstructure. J. Phys: Condens. Matter, **4**, 6651–6690 (1992).

[DR03]      Degond, P., Ringhofer, C.: Quantum moment hdrodynamics and the entropy principle. Journal Stat. Phys., **112**(3), 587–628 (2003).

[DBBBJ02]   Demeio, L., Barletti, L., Bertoni, A., Bordone, P., and Jacoboni, C.: Wigner function approach to multiband transport in semiconductors. Physica B, **314**, 104–107 (2002).

[DBBJ02]    Demeio, L., Barletti, L., Bordone P., and Jacoboni, C.: Wigner function for multiband transport in semiconductors. Transport Theory Statist. Phys., **32**(3&4), 307–325 (2003).

[DBJ03a]    Demeio, L., Bordone, P., and Jacoboni, C.: Numerical and analytical applications of multiband transport in semiconductors. Proc. XXIII Symposium on Rarefied Gas Dynamics, Whistler, BC, Canada, July 20–25, 2002, pp. 92–98 (AIP Conference Proceedings vol. 663, New York, 2003).

[DBJ03b]    Demeio, L., Bordone, P., and Jacoboni, C.: Multi-band, non-parabolic Wigner function approach to electron transport in semiconductors. Internal Report N. 3/2003, Dipartimento di Scienze Matematiche, Università Politecnica delle Marche, April 2003, Transport'Theory Statist. Phys., (to appear).

[Fey72]     Feynman, R. P.: Statistical Mechanics: A Set of Lectures, Addison-Wesley, Reading (1972).

[Fre90]     Frensley, W. R.: Boundary conditions for open quantum systems far from equilibrium. Rev. Mod. Phys., **62**, 745–791 (1990).

[FM05]      Frosali, G., Morandi, O.: A quantum kinetic approach for modeling a
            two-band resonant tunneling diode. Transport Theory Statist. Phys.
            (submitted).

[Gar94]     Gardner, C. L.: The quantum hydrodynamic model for semiconductor
            devices, SIAM J. Appl. Math., **54**, 409–427 (1994).

[GMU95]     Gasser, I., Markowich, P. A., and Unterreiter, A.: Quantum hydrody-
            namics. In Proceedings of the SPARCH GdR Conference, held in St.
            Malo (1995).

[IK86]      Iafrate, G. J., Krieger, J. B.: Quantum transport for Bloch electrons
            for inhomogeneous electric fields. Phys. Rev. B, **40**, 6144–6148 (1989).

[JAC92]     Jacoboni, C.: Comparison between quantum and classical results in
            hot-electron transport, Semicomd. Sci. Technol., **7**, B6-B11 (1992).

[JBBB01]    Jacoboni, C., Brunetti, R., Bordone, P., and Bertoni, A.: Quantum
            transport and its simulation with the Wigner function approach. In
            Brennan, K., Paul Ruden, P. (eds.) Topics in High Field Transport in
            Semiconductors. World Scientific, Singapore (2001), pp. 25–61.

[Jun01]     Jüngel, A.: Quasi-hydrodynamic Semiconductor Equations. Birk-
            häuser, Basel (2001).

[Kan56]     Kane, E. O.: Energy band structure in $p$-type Germanium and Silicon.
            J. Phys. Chem. Solids, **1**, 82–89 (1956).

[Kef03]     Kefi, J.: Analyse mathématique et numérique de modèles quantiques
            pour les semiconducteurs. PhD Thesis, Université Toulouse III - Paul
            Sabatier (2003).

[KKFR89]    Kluksdahl, N. C., Kriman, A. M., Ferry, D. K., and Ringhofer, C.:
            Self-consistent study of the resonant-tunneling diode. Phys. Rev. B,
            **39**(11), 7720–7735 (1989).

[KI87]      Krieger, J. B., Iafrate, G. J.: Quantum transport for bloch electrons
            in a spatially homogeneous electric field. Phys. Rev. B, **35**, 9644–9658
            (1987).

[LL77]      Landau, L. D., Lifshitz, E. M.: Quantum Mechanics: Non-Relativistic
            Theory. Pergamon Press, Oxford (1977).

[LK55]      Luttinger, J. M., Kohn, W.: Motion of electrons and holes in perturbed
            periodic fields. Phys. Rev. II, **97**, 869–882 (1955).

[MMP94]     Markowich, P. A., Mauser, N. J., and Poupaud, F.: A Wigner function
            approach to (semi)classical limits: Electrons in a periodic potential. J.
            Math. Phys., **35**(3), 1066–1094 (1994).

[MRS90]     Markowich, P. A., Ringhofer, Ch. A., and Schmeiser, Ch.: Semiconduc-
            tor Equations. Springer-Verlag, Wien (1990).

[MM05]      Modugno, M., Morandi, O.: A multiband envelope function model for
            quantum transport in a tunneling diode. Phys. Rev. B, **71**, 235331
            (2005).

[PR96]      Poupaud, F., Ringhofer, C.: Semi-classical limits in a crystal with exte-
            rior potentials and effective mass theorems. Comm. Partial Differential
            Equations **21**(11–12), 1897–1918 (1996).

[RS72]      Reed, M., Simon, B.: Methods of Modern Mathematical Physics. I:
            Functional Analysis. Academic Press, New York (1972).

[RBJ91]     Rossi, F., Brunetti, R., and Jacoboni, C.: An introduction to charge
            quantum transport in semiconductors and numerical approaches. In
            Ferry, D. K., Barker, J. R. (eds.) Granular Nanoelectronics. Plenum
            Press, New York (1991), pp. 43–61.

[RBJ92]    Rossi, F., Brunetti, R., and Jacoboni, C.: Quantum Transport, in Shah, J. (ed.) Hot Carriers in Semiconductors Nanostructures: Physics and Applications. Academic Press, San Diego (1992), pp. 153–188.

[SX89]     Sweeney M., Xu, J. M.: Resonant interband tunnel diodes. Appl. Phys. Lett., **54**(6), 546–548 (1989).

[Wan62]    Wannier, G. H.: Dynamics of band electrons in electric and magnetic fields. Rev. Mod. Phys., **34**, 645–655 (1962).

[Wen99]    Wenckebach, W. T.: Essential of Semiconductor Physics. John Wiley & Sons, Chichester (1999).

[Wig32]    Wigner, E.: On the quantum correction for thermodynamic equilibrium. Phys. Rev., **40**, 749–759 (1932).

[YSDX91]   Yang, R. Q., Sweeny, M., Day, D., and Xu, J. M.: Interband tunneling in heterostructure tunnel diodes. IEEE Transactions on Electron Devices, **38**(3), 442–446 (1991).

# 5

# Optimization models for semiconductor dopant profiling

Martin Burger,[1] Michael Hinze[2] and Rene Pinnau[3]

[1] Institut für Industriemathematik, Johannes Kepler Universität, Altenbergerstr.
69, A 4040 Linz, Austria `martin.burger@jku.at`
[2] Institut für Numerische Mathematik, Technische Universität Dresden,
Willersbau, C318, D-01062 Dresden, Germany `hinze@math.tu-dresden.de`
[3] Fachbereich Mathematik, Technische Universität Kaiserslautern, Erwin
Schrödinger Str., D-67663 Kaiserslautern, Germany
`pinnau@mathematik.uni-kl.de`

## 5.1 Introduction

The design of semiconductor devices is an important and challenging task in
modern microelectronics, which is increasingly being carried out via mathematical optimization with models for the device behavior. The design variable
(and correspondingly the unknown in the associated optimization problems)
is the device doping profile, which describes the (charge) density of ion impurities in the device and is therefore modeled as a spatially inhomogeneous
function. The optimization goals are usually related to the device characteristics, in particular to outflow currents on some contacts. This is also the typical
setup we shall confine ourselves to in this chapter, namely to (approximately)
achieve a certain goal related to the outflow current on a contact (e.g., a max-
imization or just an increase of the current), ideally with minimal change of
the doping profile to some given reference state.

In order to solve such optimal design problems it is important to find
suitable models of objective functionals to be minimized, so that a reasonable compromise between conflicting design goals (e.g., maximizing current
and keeping the doping profile close to the reference state) can be achieved.
We shall study and compare two different models that have been proposed
for the optimization (and used for numerical solutions, cf. [HP02a, HP02b,
BP03, HP05, HP06]). In any of the models, the weighting of the different
goals leads to some parameters in the objective functionals, and we shall pay
particular attention to the limiting behavior of minimizers with respect to
these parameters.

We shall start with an overview of models for the simulation and in particular for the optimization of semiconductor devices, which we carry out in
a rather general setup. Then we turn our attention to a simple model case,

namely the unipolar drift-diffusion model, for which a very detailed analysis of the optimization models can be carried out. We shall verify some fundamental properties such as the existence of minimizers and the existence of Lagrange multipliers, before we analyze the regularity of minimizers and the quite challenging problem of uniqueness. Moreover, we also investigate the asymptotic behavior of the minimizers for large and small parameters in the objective functionals. Finally, we discuss the numerical solution of the optimization problems for the particular case of drift-diffusion models, but allowing bipolarity and multiple dimensions, and give some computational results.

## 5.2 Models for optimal dopant profiling

Macroscopic models for semiconductor devices are usually composed of two basic state variables, namely the electric potential $V$ and a set of densities $\rho$ (e.g., electron and hole densities), which satisfy a nonlinear system of the form

$$-\lambda^2 \Delta V = Q(\rho) + C \qquad (5.2.1)$$

$$F(\rho, V) = 0. \qquad (5.2.2)$$

Here $\lambda$ denotes a scaling parameter (called *Debye length*), $Q(\rho)$ is the total charge density generated by $\rho$, $C$ is the doping profile (modeled as a function of space), and $F$ symbolizes nonlinear differential equations for $\rho$ (which also include the electric potential $V$). All equations are to be solved in a domain $\Omega$ modeling the device geometry and with suitable boundary conditions, which we do not further discuss here. For an overview of device models and their asymptotic relations we refer to [JP01, MRS90].

The primary optimization goal can usually be modeled in a straightforward way as a functional of the densities and the voltage, i.e.,

$$R(V, \rho) \to \min_{(V,\rho,C) \text{ satisfying } (5.2.1),(5.2.2)} . \qquad (5.2.3)$$

The functional $R$ could, e.g., be the negative current outflow on a contact (in order to maximize the current (cf. [PSSS98, St00, Stea98]) or the square of the current minus a target current (cf. [HP02a, HP02b]).

As an example we consider the most frequently used case, namely the *bipolar drift-diffusion model*, where $\rho = (n, p)$ with $n$ being the electron and $p$ the hole density. In this case the charge density is simply $Q = p - n$ and the differential operators included in $F$ are given by

$$F(n, p) = \begin{pmatrix} \nabla \cdot (D_n \nabla n - \mu_n n \nabla V) \\ \nabla \cdot (D_p \nabla p + \mu_p p \nabla V) \end{pmatrix}.$$

The current flowing out over a contact $\Gamma \subset \partial \Omega$ is then given by

$$I = \int_\Gamma J \cdot d\nu, \quad J = D_n \partial_x blan - \mu_n n \partial_x blaV - D_p \partial_x blap - \mu_p p \partial_x blaV,$$

and a prototypical optimization problem would be to minimize $-I$ or $|I - I^*|^2$.

It turns out that an optimization problem of the form (5.2.3) is not well posed, i.e., the existence of solutions and the robustness of the problem cannot be guaranteed. In order to achieve these goals, a second term has to be introduced to the objective functional. In [HP02a, HP02b] an optimization of the form

$$G_\alpha(V, \rho, C) := R(V, \rho) + \alpha \|C - C^*\|^2 \rightarrow \min_{(V,\rho,C) \text{ satisfying } (5.2.1),(5.2.2)} \tag{5.2.4}$$

has been proposed. Here $C^*$ is a given prior for the doping profile and $\alpha$ is a positive parameter. If the norm is chosen appropriately it can be shown that a minimizer of (5.2.4) exists for $\alpha > 0$. Moreover, one can formulate first-order optimality conditions, as usual in optimization based on the Lagrangian

$$L(V, \rho, C; p, q) = G_\alpha(V, \rho, C) + \int_\Omega \left(\lambda^2 \partial_x blaV \cdot \partial_x blap - Q(\rho)p - Cp\right) dx$$
$$+ \langle F(\rho, V), q \rangle$$

having zero variations with respect to the primal variables $(V, \rho, C)$ and the dual variables $(p, q)$. The latter just yields the constraints (5.2.1), (5.2.2), which have to be coupled with

$$0 = \frac{\partial L}{\partial V} = \frac{\partial R}{\partial V}(V, \rho) - \lambda^2 \Delta p + \frac{\partial F}{\partial V}(\rho, V)^* q \tag{5.2.5}$$

$$0 = \frac{\partial L}{\partial \rho} = \frac{\partial R}{\partial \rho}(V, \rho) - Q'(\rho)p + \frac{\partial F}{\partial \rho}(\rho, V)^* q \tag{5.2.6}$$

$$0 = \frac{\partial L}{\partial C} = \alpha E^* E(C - C^*) - p. \tag{5.2.7}$$

Here $A^*$ denotes the adjoint of an operator $A$, and $E$ is the embedding operator from the space used for $C$ (with norm $\|.\|$ as used in the functional $G_\alpha$) into $L^2(\Omega)$. Hence, the optimality conditions yield a system of five strongly coupled nonlinear equations (5.2.1), (5.2.2), (5.2.5), (5.2.6), (5.2.12), which is solved by a minimizer of (5.2.4). Both the analysis and the computation of minimizers turn out to be challenging tasks, which we shall investigate in more detail for a very special device model in the next section.

As an alternative to (5.2.4) a different approach has been introduced in [BP03], which is motivated from the structure of the above optimality system. It turns out that the optimality system can be simplified partially if a variable change from the doping profile $C$ to the total charge density

$$W := Q(\rho) + C \tag{5.2.8}$$

is performed. Obviously, if one minimizes with respect to $(V, \rho, W)$ then one can reconstruct $C$ uniquely from this formula. On the other hand, the Poisson equation simplifies to

$$-\lambda^2 \Delta V = W, \tag{5.2.9}$$

i.e., the densities $\rho$ do not appear any more and the coupling between $V$ and $\rho$ becomes one directional only. Since one is using a novel design variable $W$ in this setup it seems natural to adjust the penalizing term to this fact, i.e., to minimize

$$H_\beta(V, \rho, W) := R(V, \rho) + \beta \|W - W^*\|^2 \to \min_{(V, \rho, W) \text{ satisfying } (5.2.9), (5.2.2)}. \tag{5.2.10}$$

The Lagrangian associated to (5.2.10) is given by

$$L(V, \rho, W; p, q) = H_\beta(V, \rho, C) + \int_\Omega \left( \lambda^2 \partial_x bla V \cdot \partial_x bla p - W p \right) dx + \langle F(\rho, V), q \rangle,$$

and the optimality conditions are given by (5.2.9),(5.2.2) together with

$$0 = \frac{\partial L}{\partial V} = \frac{\partial R}{\partial V}(V, \rho) - \lambda^2 \Delta p + \frac{\partial F}{\partial V}(\rho, V)^* q \tag{5.2.11}$$

$$0 = \frac{\partial L}{\partial \rho} = \frac{\partial R}{\partial \rho}(V, \rho) + \frac{\partial F}{\partial \rho}(\rho, V)^* q \tag{5.2.12}$$

$$0 = \frac{\partial L}{\partial C} = \beta E^* E(W - W^*) - p. \tag{5.2.13}$$

The structure of the optimality system for (5.2.10) turns out to be more convenient than the one for (5.2.4). For given design variable $W$ one can subsequently solve (5.2.9) for $V$, (5.2.2) for $\rho$, (5.2.12) for $q$, and (5.2.13) for $p$. For the case of the drift-diffusion model as stated above, this can be realized by solving scalar linear differential equations only instead of nonlinear coupled systems. As a direct consequence, the analysis of the optimality systems and important properties such as the existence of Lagrange multipliers $p$ and $q$ are rather straightforward (see [BP03] for the drift-diffusion model). Moreover, this decoupling of the optimality system can be used to construct efficient numerical methods as we will discuss in Section 5.4.

## 5.3 Optimization of unipolar diodes

In the following we provide a detailed analysis for the optimization of the unipolar drift-diffusion model for diodes. In this situation a spatially one-dimensional analysis can be carried out, with a single density (namely the electron density $n$). For convenience we shall use a scaled version of the model and a standard transformation to the *Slotboom variables* and consider the unknown $u \sim e^{-V} n$ (cf., e.g., [MRS90] for details on scaling and density variables for the drift-diffusion model).

Motivated by the above discussion we consider the optimization problems (note from the model below that the current $J = e^V u_x$ is spatially homogeneous, so that $I = J$)

$$G_\alpha(u, V, C) := \frac{1}{2} \int_0^1 |e^V u_x - J^*|^2 dx + \frac{\alpha}{2} \|C - C^*\|^2 \qquad (5.3.1)$$

and

$$H_\beta(u, V, C) := \frac{1}{2} \int_0^1 |e^V u_x - J^*|^2 dx + \frac{\beta}{2} \|V_{xx} - V_{xx}^*\|^2, \qquad (5.3.2)$$

both of them subject to

$$
\begin{aligned}
\lambda^2 V_{xx} - e^V u &= -C &&\text{in } (0,1) \\
(e^V u_x)_x &= 0 &&\text{in } (0,1) \\
V &= V^* &&\text{in } \{0,1\} \\
V_{xx} &= V_{xx}^* &&\text{in } \{0,1\} \\
u &= u_D &&\text{in } \{0,1\}.
\end{aligned}
\qquad (5.3.3)
$$

In order to keep the notation as unified as possible in this section we shall not use the variable $W$ but directly write the problems in terms of $V$ and its derivatives.

Our aim is to study the parametric behavior of these functionals with respect to the positive real parameters $\alpha$ and $\beta$, respectively. In particular we shall investigate the asymptotic behavior of the minimizers as the parameters tend to zero or infinity, respectively. In the latter case, it seems obvious that the design variables ($C$ and $V$, respectively) converge to their priors, which we will prove with a rate of at least $\alpha^{-1/2}$ and $\beta^{-1/2}$. In the case of parameters tending to zero numerical experiments indicate that the current $J = e^V u_x$ tends to the desired current $J^*$, which we will prove in both cases. For the functional $H_\beta$, we shall even prove that this convergence arises with rate $\sqrt{\beta}$ as $\beta \to 0$.

### 5.3.1 Optimization for positive and finite parameters

In this section we shall investigate the optimization problems for parameter values $\alpha$ and $\beta$ in the open interval $(0, +\infty)$. This provides some basic results for the later asymptotic analysis, but also a variety of interesting results that yield further insight with respect to the parametric behavior of the optimization problems.

**Minimization of $G_\alpha$**

We start with a discussion of basic properties of the optimization problem (5.3.1) subject to (5.3.3), which was originally introduced in [HP02a, HP02b].

**Theorem 1.** *Let $\alpha > 0, J^* \in \mathbb{R}, C^* \in H^1([0,1])$, and*

$$\|C - C^*\|^2 = \int_0^1 \left( |C_x - C_x^*|^2 + |C - C^*|^2 \right) dx.$$

*Moreover let $u^* \in H^1([0,1]), V^* \in H^3([0,1])$ satisfy*

$$
\begin{aligned}
\lambda^2 V_{xx}^* &= e^{V^*} u^* - C^* \quad && in \ (0,1) \\
\left( e^{V^*} u_x^* \right)_x &= 0 && in \ (0,1) \\
u^* &= u_D^* && in \ \{0,1\}.
\end{aligned}
$$

*Then there exists a solution*

$$(\overline{u}, \overline{V}, \overline{C}) \in H^1([0,1]) \times H^2([0,1]) \times H^1([0,1])$$

*of the optimization problem (5.3.1), (5.3.3).*

*Proof.* The existence of a solution $(\overline{u}, \overline{V}, \overline{C}) \in H^1([0,1])^3$ follows from a more general result in [HP02b]. The additional regularity $\overline{V} \in H^2([0,1])$ in this one-dimensional case follows from

$$\overline{V}_{xx} = e^{\overline{V}} \overline{u} - \overline{C} \in L^2([0,1]).$$

$\square$

Besides the existence of a solution, the Karush–Kuhn–Tucker-system and the existence of Lagrange multipliers are of particular interest.

**Proposition 1.** *Under the conditions of Theorem 1, there exist Lagrange multipliers $(\overline{p}, \overline{q}) \in H_0^1([0,1])^2$ such that a stationary point $(\overline{u}, \overline{V}, \overline{C})$ of (5.3.1), (5.3.3) satisfies*

$$
\begin{aligned}
0 &= -\alpha(\overline{C}_{xx} - C_{xx}^*) + \alpha(\overline{C} - C^*) - \overline{p} && in \ (0,1) \\
0 &= \left( e^{\overline{V}} \overline{u}_x - J^* \right) e^{\overline{V}} \overline{u}_x - \lambda^2 \overline{p}_{xx} + e^{\overline{V}} \overline{p} \overline{u} + e^{\overline{V}} \overline{u}_x \overline{q}_x && in \ (0,1) \\
0 &= -\left( e^{\overline{V}} (e^{\overline{V}} \overline{u}_x - J^*) \right)_x + e^{\overline{V}} \overline{p} - (e^{\overline{V}} \overline{q}_x)_x && in \ (0,1) \quad (5.3.4) \\
\overline{p} &= 0 && in \ \{0,1\} \\
\overline{q} &= 0 && in \ \{0,1\}.
\end{aligned}
$$

*Proof.* See [HP02b]. $\square$

In general, one cannot expect the uniqueness of the Lagrange multipliers defined in (5.2.5)–(5.2.7). But for the unipolar diodes considered here, the Lagrange multipliers are unique.

**Theorem 2.** *Under the conditions of Proposition 1, the Lagrange multipliers* $(\bar{p}, \bar{q}) \in H_0^1([0, 1])^2$ *are unique.*

*Proof.* We consider the homogeneous system

$$
\begin{aligned}
0 &= -\lambda^2 \bar{p}_{xx} + e^{\overline{V}} \overline{pu} + e^{\overline{V}} \overline{u}_x \overline{q}_x &&\text{in } (0, 1) \\
0 &= \phantom{-\lambda^2} + e^{\overline{V}} \bar{p} - (e^{\overline{V}} \overline{q}_x)_x &&\text{in } (0, 1)
\end{aligned}
\tag{5.3.5}
$$

with $\bar{p} = \bar{q} = 0$ in $\{0, 1\}$. The second equation can be written as

$$
\bar{p} = \overline{V}_x \overline{q}_x + \overline{q}_{xx}
$$

and plugging this in the first equation we get (with $n = e^{\overline{V}} \overline{u}$)

$$
-\lambda^2 \bar{p}_{xx} + (n \overline{q}_x)_x = 0.
$$

Hence, $-\lambda^2 \bar{p}_x + n \overline{q}_x = k$ is constant. Now, we introduce $\xi := e^{\overline{V}} \overline{q}_x$ and start again from the second equation to get

$$
\xi_{xx} = (e^{\overline{V}} \bar{p})_x = e^{\overline{V}} \bar{p}_x + e^{\overline{V}} \overline{V}_x \bar{p}_x = \frac{1}{\lambda^2}(n\xi - ke^{\overline{V}}) + \overline{V}_x \xi_x
$$

as an equation for $\xi$ supplemented with the boundary data $\xi_x(0) = \xi_x(1) = 0$. From the boundary data for $\bar{q}$ we deduce

$$
\int_0^1 e^{-\overline{V}} \xi \, dx = 0.
$$

Let $\underline{\xi} \le \xi \le \overline{\xi}$ be sharp bounds. Choose a point $x_0 \in [0, 1]$ such that $\xi(x_0) = \overline{\xi}$ and $\xi_x(x_0) = 0$. Then we have

$$
\xi_{xx} = \frac{1}{\lambda^2}(n\overline{\xi} - ke^{\overline{V}}) \le 0
$$

and hence

$$
\overline{\xi} \le \max_x \frac{ke^{\overline{V}}}{n}.
$$

In analogy, one shows that

$$
\underline{\xi} \ge \min_x \frac{ke^{\overline{V}}}{n}.
$$

We deduce that $\xi$ does not change its sign and thus $\int_0^1 e^{-\overline{V}} \xi \, dx = 0$ implies that $\xi \equiv 0$ and then $\bar{q} \equiv 0$ and $\bar{p} \equiv 0$. Hence, the homogeneous problem has only the trivial solution which implies the uniqueness of the Lagrange multipliers. $\qquad\square$

Another typical property of an objective functional like (5.3.1) is that nonsmooth features of the solution $\overline{C}$ correspond to those in the prior $C^*$, or, in other words, $\overline{C} - C^*$ is very smooth. For rather general semiconductor devices optimized with respect to the objective $G_\alpha$, this effect was discussed in a formal way in [HP02b]. In the case of a unipolar diode as considered here, this statement can be made rigorous as follows:

**Theorem 3.** *Under the conditions of Proposition 1, a doping profile $\overline{C}$ corresponding to a stationary point $(\overline{u}, \overline{V}, \overline{C}, \overline{p}, \overline{q})$ solving (5.3.3), (5.3.4) satisfies*

$$\overline{C} - C^* \in H^6([0,1]) \hookrightarrow C^5([0,1]). \tag{5.3.6}$$

*Proof.* First of all, due to Proposition 1 we have

$$(\overline{C} - C^*)_{xx} = (\overline{C} - C^*) + \frac{1}{\alpha}\overline{p} \in L^2(\Omega),$$

from which we may conclude that $\overline{C} - C^* \in H^2([0,1])$. Moreover, (5.3.4) implies (note that $e^{\overline{V}}\overline{u}_x$ is constant)

$$\overline{p}_{xx} = \lambda^{-2}\left[(e^{\overline{V}}\overline{u}_x - J^*)e^{\overline{V}}\overline{u}_x + e^{\overline{V}}\overline{pu} + e^{\overline{V}}\overline{u}_x\overline{q}_x\right] \quad \in L^2([0,1]),$$

$$\overline{q}_{xx} = \overline{p} - \overline{V}_x\left(e^{\overline{V}}\overline{u}_x - J^*\right) \quad\quad\quad\quad \in L^2([0,1]),$$

and thus, $\overline{p} \in H^2([0,1])$ and $\overline{q} \in H^2([0,1])$. Using this result, we deduce from the first line in (5.3.4) that

$$\frac{\partial^3}{\partial x^3}(\overline{C} - C^*) = (\overline{C} - C^*)_x + \alpha^{-1}\overline{p}_x \quad\quad \in L^2([0,1]),$$

$$\frac{\partial^4}{\partial x^4}(\overline{C} - C^*) = (\overline{C} - C^*)_{xx} + \alpha^{-1}\overline{p}_{xx} \quad \in L^2([0,1]),$$

i.e., $\overline{C} - C^* \in H^4([0,1])$.

By a further iteration of this process we obtain that

$$\frac{\partial^j \overline{p}}{\partial x^j} \in L^2([0,1]),$$

$$\frac{\partial^j \overline{q}}{\partial x^j} \in L^2([0,1]),$$

for $j = 3, 4$, and consequently

$$\frac{\partial^{j+2}}{\partial x^{j+2}}(\overline{C} - C^*) = \frac{\partial^j}{\partial x^j}(\overline{C} - C^*) + \alpha^{-1}\frac{\partial^j \overline{p}}{\partial x^j} \in L^2(\Omega),$$

which implies $\overline{C} - C^* \in H^6([0,1]) \hookrightarrow C^5([0,1])$.  $\square$

Note that the above result is obtained by a bootstrapping technique, from which one often derives $C^\infty$-regularity. The reason to stop at the sixth derivative of $\overline{C} - C^*$ is that one cannot proceed further without assuming higher regularity than $C^* \in L^2(\Omega)$. If we want to have a bound on the seventh derivative of $\overline{C} - C^*$, we need bounds on the fifth derivative of $\overline{p}$, the third derivative of $\overline{V}$, and consequently, the first derivative of $\overline{C}$. But $\overline{C} \in H^1([0,1])$ is obtained only for

$$C^* = \overline{C} - (\overline{C} - C^*) \in H^1([0,1]).$$

## Minimization of $H_\beta$

For the optimization problem (5.3.2), (5.3.3) we can prove similar results on the existence of minimizers and Lagrange multipliers as above.

**Theorem 4.** *Let* $\beta > 0$, $J^* \in \mathbb{R}$, *and* $V^* \in H^2([0,1])$. *Moreover, let* $C^* \in L^2([0,1])$ *be defined by*

$$C^* = \lambda^2 V_{xx}^* - e^{V^*} u^*,$$

*where* $u^*$ *is the unique solution of*

$$-(e^{V^*} u_x^*)_x = 0 \quad in \ (0,1),$$

*satisfying* $u^* = u_D$ *in* $\{0,1\}$. *Then, there exists a minimizer*

$$(\overline{u}, \overline{V}, \overline{C}) \in H^1([0,1]) \times H^2([0,1]) \times L^2([0,1])$$

*of* (5.3.2) *subject to* (5.3.3).

*Proof.* See [BP03].                                                    □

**Proposition 2.** *Under the assumptions of Theorem 4, there exists a Lagrange multiplier* $\overline{q} \in H_0^1([0,1])$ *such that a stationary point* $(\overline{u}, \overline{V})$ *of* (5.3.2), (5.3.3) *satisfies*

$$\begin{aligned}
\beta(\overline{V}_{xx} - V_{xx}^*)_{xx} + e^{\overline{V}}\overline{u}_x(e^{\overline{V}}\overline{u}_x - J^*) + e^{\overline{V}}\overline{u}_x\overline{q}_x &= 0 \ in \ (0,1), \\
-e^{\overline{V}}\overline{V}_x(e^{\overline{V}}\overline{u}_x - J^*) - (e^{\overline{V}}\overline{q}_x)_x &= 0 \ in \ (0,1).
\end{aligned}$$

(5.3.7)

*Proof.* See [BP03].                                                    □

The uniqueness of stationary points is a challenging problem, which seems to depend strongly on the problem setup. Here we consider a case related to numerical computations in [HP02b, BP03]), where the optimization was used to amplify an original given current $J^0$ by 50%, i.e., $J^* = \frac{3}{2} J^0$. Since an optimization algorithm will be started with the reference state (and current $J = J^0$) and then increase the current, one can expect that $J^0 \le J \le J^*$ is the relevant situation for the current, and in this range uniqueness can be guaranteed.

**Theorem 5.** *Let, in addition to the above conditions, $u_D(1) - u_D(0) > 0$ and $J^* > 0$. Then there is a unique stationary point $(\overline{u}, \overline{V}, \overline{q})$ of (5.3.2), (5.3.3), (5.3.7) among those functions satisfying $\frac{2}{3}J^* \leq J = e^{\overline{V}}\overline{u}_x \leq J^*$.*

*Proof.* By integrating the second equation of (5.3.7) we obtain (with $J = e^{\overline{V}}\overline{u}_x$)

$$\overline{q}_x = J^* - J + ce^{-\overline{V}}$$

for a constant $c$, which can be determined after integration from 0 to 1 as

$$c = \frac{J - J^*}{\int_0^1 e^{-\overline{V}}\,dx} = \frac{J}{u_D(1) - u_D(0)}(J - J^*).$$

After plugging this into the first equation of (5.3.7), we end up with the equation

$$\beta(\overline{V}_{xx} - V^*_{xx})_{xx} + Jce^{-\overline{V}} = 0.$$

The linearization of this equation is given by

$$\beta\psi_{xxxx} - Jce^{-\overline{V}}\psi + (J'c + Jc')e^{-\overline{V}} = 0, \tag{5.3.8}$$

where $J'$ and $c'$ denote the derivatives of the functionals $J$ and $c$ with respect to $V$ in direction $\psi$, i.e.,

$$J' = \frac{u_D(1) - u_D(0)}{(\int_0^1 e^{-\overline{V}}\,dx)^2}\int_0^1 e^{-\overline{V}}\psi\,dx, \quad c' = \frac{J'(J - J^*) + JJ'}{u_D(1) - u_D(0)}.$$

Note that for $0 \leq \frac{2}{3}J^* \leq J = e^{\overline{V}}\overline{u}_x \leq J^*$ we have

$$J'c + Jc' = J'J\frac{3J - 2J^*}{u_D(1) - u_D(0)} \geq 0, \quad c = \frac{J}{u_D(1) - u_D(0)}(J - J^*) \leq 0.$$

After multiplying (5.3.8) with $\psi$ and integration we obtain

$$\int_0^1 \left(\beta|\psi_{xx}|^2 - Jce^{-\overline{V}}\psi^2\right)\,dx + J\frac{3J - 2J^*}{u_D(1) - u_D(0)}\left(\int_0^1 e^{-\overline{V}}\psi\,dx\right)^2 = 0.$$

Taking into account the signs of all the terms, this implies that $\psi = 0$. Hence, the linearized problem has only the trivial solution, which implies the uniqueness of $\overline{V}$, and consequently of $\overline{u}$ and $\overline{q}$. □

As for the minimization of $R_\alpha$, we can also derive a regularity result for the design variable. Since in this case, the real design variable is $\overline{V} - V^*$ it should not surprise us that high regularity for this function can be obtained.

**Theorem 6.** *Under the assumptions of Theorem 4, a stationary point $(\overline{u}, \overline{V}, \overline{C})$ solving (5.3.3), (5.3.7) satisfies*

$$\overline{V} - V^* \in H^6([0,1]) \hookrightarrow C^5([0,1]),$$
$$\overline{C} - C^* \in H^2([0,1]) \hookrightarrow C^1([0,1]).$$

*Proof.* First of all, the function $w = \overline{V} - V_{xx}^*$ satisfies the Poisson equation

$$-w_x x = \beta^{-1} \left( e^{\overline{V}} \overline{u}_x (e^{\overline{V}} \overline{u}_x - J^*) + e^{\overline{V}} \overline{u}_x \overline{q}_x \right) \in L^2([0,1]),$$

with homogeneous boundary data, and by standard elliptic regularity we may conclude that $w \in H^2([0,1])$. Taking into account that $e^{\overline{V}} \overline{u}_x$ is constant, we obtain that

$$\frac{\partial^{j+2} w}{\partial x^{j+2}} = -e^{\overline{V}} \overline{u}_x \frac{\partial^{j+1} \overline{q}}{\partial x^{j+1}},$$

for $j \geq 0$.

Due to (5.3.7) we have

$$\frac{\partial^2 \overline{q}}{\partial x^2} = -\overline{V}_x \overline{q}_x - \overline{V}_x (e^{\overline{V}} \overline{u}_x - J^*) \qquad \in L^2([0,1]),$$

$$\frac{\partial^3 \overline{q}}{\partial x^3} = -\overline{V}_x \overline{q}_{xx} - \overline{V}_{xx} \overline{q}_x - \overline{V}_{xx} (e^{\overline{V}} \overline{u}_x - J^*) \quad \in L^2([0,1]),$$

and as a consequence we obtain that $\overline{q} \in H^3([0,1])$, $w \in H^4([0,1])$, and hence, $\overline{V} - V^* \in H^6([0,1]) \hookrightarrow C^5([0,1])$.

Finally, from the Poisson equation we deduce

$$\frac{\partial^j}{\partial x^j} (\overline{C} - C^*) = \frac{\partial^j}{\partial x^j} \left( \lambda^2 W - e^{\overline{V}} \overline{u} + e^{V^*} u^* \right) \in L^2([0,1]).$$

for $j = 0, 1, 2$, and thus, $(\overline{C} - C^*) \in H^2([0,1])$. $\qquad\square$

Note that in this case we also obtain $H^6$-regularity of the design variable with a rather weak penalization term on $W$ in the $L^2$-norm. If we were to use the $H^1$-norm of $W$ for the penalty instead, this would even imply $H^8$-regularity of $\overline{V} - V^*$. For the change in the doping profile $\overline{C} - C^*$ we cannot obtain higher regularity than $H^2$, since this would enforce the existence of higher than second derivatives of $\overline{V}$ and $V^*$, or, due to the Poisson equation, the existence of derivatives for $\overline{C}$ and $C^*$, which we do not assume here.

### 5.3.2 Asymptotic behavior

In the following we investigate the two different limits for the parameters $\alpha$ and $\beta$, namely convergence to zero and infinity, respectively

### Large parameters

The limit of the parameters $\alpha$ and $\beta$ tending to infinity is the easier case. It seems obvious that the solutions of the optimization problem (5.3.1) or (5.3.2) subject to (5.3.3) converge to the prior $(u^*, V^*, C^*)$. From the structure of the objective we can prove that this convergence occurs with rate $\mathcal{O}(\alpha^{-1/2})$ and $\mathcal{O}(\beta^{-1/2})$, respectively.

**Theorem 7.** *Let $\alpha > 0$ and $J^* \in \mathbb{R}$. Moreover, denote by $(u^\alpha, V^\alpha, C^\alpha)$ the unique solution of the optimization problems (5.3.1), (5.3.2) for fixed $\alpha$. Then there exists a positive real constant $M$ such that*

$$\|V^\alpha - V^*\|_{H^3} + \|u^\alpha - u^*\|_{H^1} + \|C^\alpha - C^*\|_{H^1} \leq \frac{M}{\sqrt{\alpha}},$$

*for $\alpha$ sufficiently large. In particular, $(u^\alpha, V^\alpha, C^\alpha) \to (u^*, V^*, C^*)$.*

*Proof.* From (5.3.1) we immediately obtain

$$\|e^{V^\alpha} u_x^\alpha - J^*\|_{L^2}^2 + \alpha \|C^\alpha - C^*\|_{H^1}^2 \leq \|e^{V^*} u_x^* - J^*\|^2,$$

and thus,

$$\|C^\alpha - C^*\|_{H^1} \leq \frac{M_0}{\sqrt{\alpha}}.$$

From the well-posedness of the drift-diffusion model for unipolar diodes [MRS90, GS92] we may conclude that

$$\|V^\alpha - V^*\|_{H^1}^2 + \|u^\alpha - u^*\| \leq \gamma \|C^\alpha - C^*\|_{H^1} \leq \gamma \frac{M_0}{\sqrt{\alpha}},$$

for some constant $\gamma > 0$ independent of $\alpha$. Finally, from the Poisson equation we deduce with the above $H^1$-estimates that

$$\frac{\partial^{j+2}}{\partial x^{j+2}} (V^\alpha - V^*)$$
$$= \lambda^{-2} \frac{\partial^j}{\partial x^j} \left( e^{V^\alpha} (u^\alpha - u^*) + (e^{V^\alpha} - e^{V^*}) u^* + (C^\alpha - C^*) \right) \in L^2([0,1]),$$

for $j = 0, 1$, and from the above estimates on the $H^1$-norms we can also conclude that

$$\|V^\alpha - V^*\|_{H^3} \leq \frac{M_1}{\sqrt{\alpha}},$$

for some constant $M_1$, which completes the proof. $\square$

An analogous result holds for the limit $\beta \to \infty$ in the minimization of $H_\beta$.

**Theorem 8.** *Let $\beta > 0$ be sufficiently large, $J^* \in \mathbb{R}$, and denote by $(u^\beta, V^\beta, C^\beta)$ the unique minimizer of (5.3.2), (5.3.3) for fixed $\beta$. Then there exists a constant $M > 0$ such that*

$$\|V^\beta - V^*\|_{H^2} + \|u^\beta - u^*\|_{H^1} + \|C^\beta - C^*\|_{L^2} \leq \frac{M}{\sqrt{\beta}}.$$

*Proof.* As in the proof of Theorem 7 we may deduce that

$$\|e^{V^\beta} u_x^\beta - J^*\|_{L^2}^2 + \beta \|V_{xx}^\beta - V_{xx}^*\|_{L^2}^2 \leq \|e^{V^*} u_x^* - J^*\|^2,$$

which implies by standard reasoning that

$$\|V^\beta - V^*\|_{H^2} \le \frac{M_0}{\sqrt{\beta}},$$

for some constant $M_0$. Since

$$\int_0^1 e^{V^*} |u_x^\beta - u^*|^2 \, dx = \int_0^1 (e^{V^*} - e^{V^\beta}) u_x^\beta (u_x^\beta - u^*) \, dx,$$

we may conclude from the Cauchy–Schwarz inequality and the uniform boundedness of $u_x^\beta$ that

$$\|u^\beta - u^*\|_{H^1} \le \gamma \|V^\beta - V^*\|_{H^2} \le \gamma \frac{M_0}{\sqrt{\beta}},$$

for some constant $\gamma$. Finally, the estimate for $\|C^\beta - C^*\|_{L^2}$ follows from

$$C^\beta - C^* = \lambda^2 (V_{xx}^\beta - V_{xx}^*) - e^{V^*}(u^\beta - u^*) + (e^{V^*} - e^{V^\beta}) u^\beta$$

and a standard Lipschitz estimate. $\qquad\square$

**Small parameters**

We shall now turn our attention to the limit case of $\alpha \to 0$ and $\beta \to 0$, respectively, where we may expect that $e^{\overline{V}} \overline{u} \to J^*$. In order to obtain further insight, we directly start with the limit problem, which is the same for $\alpha = 0$ and $\beta = 0$. One might expect that the limit is determined by the equation

$$e^{V^0} u_x^0 = J^* \tag{5.3.9}$$

subject to (5.3.3), for which one has to expect an infinite number of solutions. Taking into account the fact that we actually want to compute an optimal design that is as close as possible to the original one, a more suitable limit problem for (5.3.1), (5.3.3) is given by

$$\frac{1}{2} \|C - C^*\|_{H^1}^2 \to \min_{(u,V,C)}, \tag{5.3.10}$$

subject to (5.3.3), (5.3.9). In an analogous way we define a limit problem for (5.3.2), (5.3.3) as

$$\frac{1}{2} \|V_{xx} - V_{xx}^*\|_{L^2}^2 \to \min_{(u,V,C)}, \tag{5.3.11}$$

subject to (5.3.3), (5.3.9).

We first make sure that the feasible set of these problems defined by (5.3.3), (5.3.9) is nonempty.

**Lemma 1.** *Let* $V \in H^2([0,1])$ *satisfy*

$$\int_0^1 e^{-V} \, dx = \frac{u_D(1) - u_D(1)}{J^*} \neq 0 \qquad (5.3.12)$$

*as well as the boundary conditions*

$$V = V^*, \quad V_{xx} = V_{xx}^* \quad in \; \{0,1\}.$$

*Then, there exists* $u \in H^1([0,1])$ *and* $C \in L^2([0,1])$ *such that* (5.3.3) *and* (5.3.9) *hold. Moreover,* $C \in H^1([0,1])$ *if* $V \in H^3([0,1])$ *and* $C^* \in H^1([0,1])$.

*Proof.* For $V$ satisfying the above conditions, there exists a unique weak solution $u \in H^1([0,1])$ of the elliptic equation

$$(e^V u_x)_x = 0 \quad in \; (0,1)$$

with boundary values $u = u_D$ in $\{0,1\}$. Moreover, there exists a unique doping profile $C \in L^2([0,1])$ defined via

$$C = C^* - \lambda^2 (V_{xx} - V_{xx}^*) + e^V u - e^{V^*} u^*.$$

Since the derivative of the last two terms involves only first derivatives of the variables, which exist anyway under the above assumptions, we obtain $C \in H^1([0,1])$ if $V \in H^3([0,1])$ and $C^* \in H^1([0,1])$.

Now let $J := e^V u_x$, which is a constant in $(0,1)$. Then, from (5.3.12) we deduce that

$$u_D(1) - u_D(0) = \int_0^1 u_x \, dx = J \int_0^1 e^{-V} \, dx$$
$$= \frac{J}{J^*}(u_D(1) - u_D(0)),$$

and hence, $J = J^*$. From the above construction we observe that (5.3.3) and (5.3.9) hold, which completes the proof. $\qquad \square$

This result indicates that at least for a reasonable range of parameters, it should be possible to find a solution of (5.3.3), (5.3.9). Indeed, we shall show that this range is determined by the signs of $u_D(1) - u_D(0)$ only.

**Theorem 9.** *Let* $u_D(1) - u_D(0) > 0$ *(*$< 0$*, respectively). Then the feasible set determined by* (5.3.3) *and* (5.3.9) *is nonempty for each* $J^* > 0$ *(*$< 0$*).*

*Proof.* Due to Lemma 1, it suffices to find a $V \in H^3([0,1])$ satisfying the boundary conditions and (5.3.12). Since the right-hand side in this relation does not change when we change the sign of both $(u_D(1) - u_D(0))$ and $J^*$, we may restrict our attention to the case of both being positive. Let $\tilde{V}$ be a function in $H^3([0,1])$ that satisfies the boundary conditions in (5.3.3), e.g., a polynomial of order three. Moreover, let $W \in H^3([0,1])$ be a nonnegative function with compact support and $W \equiv 1$ for $x \in [\frac{1}{4}, \frac{3}{4}]$. We define $V_t = \tilde{V} + tW$ for $t \in \mathbb{R}$, and

$$I(t) := \int_0^1 e^{-V_t} \, dx.$$

It is easy to see that $I$ is a continuous function and

$$\lim_{t \to -\infty} I(t) = 0, \qquad \lim_{t \to +\infty} I(t) = \infty,$$

and hence $\mathbb{R}^+$ is included in the range of the function $I$. Consequently, we can obtain a feasible point of (5.3.3), (5.3.9) for each $J^* \in \mathbb{R}^+$ (by choosing $t$ such that $I(t) = J^*$).   $\square$

Due to the above result we shall assume in the following that $(u_D(1) - u_D(0))J^* > 0$, such that there exists a feasible point of the limit problem.

We shall consider the limit problem of minimizing

$$\frac{1}{2} \|V_{xx} - V_{xx}^*\|_{L^2}^2 \to \min_{V \in H^2([0,1])} \tag{5.3.13}$$

subject to (5.3.12) and the boundary conditions in (5.3.3). Since the feasible set is nonempty and weakly closed under the above conditions, we may conclude the existence of a minimizer. Due to standard first-order optimality, the minimizer $\overline{V}$ satisfies

$$\int_0^1 (\overline{V}_{xx} - V_{xx}^*) W_{xx} \, dx = 0$$

for all $W \in H^2([0,1])$ with homogeneous boundary values and

$$\int_0^1 e^{-\overline{V}} W \, dx = 0.$$

For arbitrary $W$ we can find a decomposition of the form $W = \mu W_0 + (W - \mu W_0)$ for a fixed element $W_0$ satisfying

$$\int_0^1 e^{-\overline{V}} W_0 \, dx \neq 0$$

and

$$\mu = \frac{\int_0^1 e^{-\overline{V}} W \, dx}{\int_0^1 e^{-\overline{V}} W_0 \, dx}.$$

Thus, we have

$$\int_0^1 (\overline{V}_{xx} - V_{xx}^*) W_{xx} \, dx - p_0 \int_0^1 e^{\overline{V}} W \, dx = 0,$$

for arbitrary $W \in H^2([0,1])$ with homogeneous boundary values, and the Lagrange multiplier

$$p_0 = \frac{\int_0^1 e^{-\overline{V}} W \, dx}{\int_0^1 (\overline{V}_{xx} - V_{xx}^*)(W_0)_{xx} \, dx}.$$

That means, $\overline{V}$ is a weak solution of the fourth-order equation

$$(V_{xx} - V_{xx}^*)_{xx} = p_0 \, e^{-V}.$$

As a consequence, we can derive the following result on the existence of Lagrange multipliers for the limit problem.

**Proposition 3.** *Let* $(u_D(1) - u_D(0))J^* > 0$, *and let* $(\overline{V}, \overline{u}, \overline{C})$ *be a minimizer of* (5.3.13) *subject to* (5.3.3), (5.3.9). *Then there exist Lagrangian variables*

$$(\overline{p}, \overline{q}, \overline{r}) \in H_0^1([0,1]) \times H^1([0,1]) \times L^2([0,1]),$$

*such that*

$$\mathcal{L}'(\overline{u}, \overline{V}, \overline{C}; \overline{p}, \overline{q}, \overline{r}) = 0,$$

*for the Lagrangian*

$$\mathcal{L}(u, V, C; p, q, r)$$
$$= \frac{1}{2}\|V_{xx} - V_{xx}^*\|_{L^2}^2 + \int_0^1 \left(e^V(u_x p_x + u_x q) - J^* q + \lambda^2 V_x r_x e^V ur - Cr\right) dx.$$

*Moreover,* $\overline{p} = \overline{r} = 0$.

*Proof.* Let $\overline{q} = -\frac{p_0}{J^*} e^{-\overline{V}}$, with $p_0$ and $\overline{V}$ as above, and let $\overline{p} = \overline{r} = 0$. Then we obtain

$$\frac{\partial}{\partial u}\mathcal{L}(\overline{u}, \overline{V}, \overline{C}; \overline{p}, \overline{q}, \overline{r}) = -(e^{\overline{V}}\overline{p}_x + e^{\overline{V}}\overline{q})_x + e^{\overline{V}}\overline{r} = \left(\frac{p_0}{J^*}\right)_x = 0$$

$$\frac{\partial}{\partial V}\mathcal{L}(\overline{u}, \overline{V}, \overline{C}; \overline{p}, \overline{q}, \overline{r}) = (\overline{V}_{xx} - V_{xx}^*)_{xx} + e^{\overline{V}}\overline{u}_x\overline{q} + e^{\overline{V}}\overline{u}_x\overline{p}_x - \lambda^2\overline{r}_{xx} + e^{\overline{V}}\overline{u}r$$
$$= (\overline{V}_{xx} - V_{xx}^*)_{xx} - p_0 e^{-\overline{V}} = 0$$

$$\frac{\partial}{\partial C}\mathcal{L}(\overline{u}, \overline{V}, \overline{C}; \overline{p}, \overline{q}, \overline{r}) = -\overline{r} = 0.$$

Since $(\overline{u}, \overline{V}, \overline{C})$ satisfies the constraints (5.3.3) and (5.3.9) the derivatives with respect to the Lagrangian variables vanish, too, and thus

$$\mathcal{L}'(\overline{u}, \overline{V}, \overline{V}; \overline{p}, \overline{q}, \overline{r}) = 0.$$

$\square$

The existence of Lagrange multipliers for the limit problem allows us to derive a quantitative convergence result for $\beta \to 0$.

**Theorem 10.** *Let $(\beta_k)$ be a sequence of positive numbers converging to zero, and let $(u_k, V_k, C_k)$ be a sequence of minimizers of (5.3.2), (5.3.3). Then there exists a subsequence converging to a minimizer $(\overline{u}, \overline{V}, \overline{V})$ of (5.3.13), (5.3.3). Moreover, each such subsequence (without restriction of generality $(u_k, V_k, C_k)$ itself) satisfies*

$$\|V_k - \overline{V}\|_{H^2} + \|u_k - \overline{u}\|_{H^1} + \|C_k - C\|_{L^2} \le m\sqrt{\beta_k}, \qquad (5.3.14)$$

*for some constant $m \in \mathbb{R}^+$. Moreover, the fitting term satisfies*

$$\|e^{V_k}(u_k)_x - J^*\|_{L^2} \le m_0 \beta_k$$

*for some constant $m_0 \in \mathbb{R}^+$.*

*Proof.* Due to the existence of Lagrangian variables $(\overline{p}, \overline{q}, \overline{r})$ we obtain that

$$\mathcal{L}(\overline{u}, \overline{V}, \overline{C}; \overline{p}, \overline{q}, \overline{r}) \le \mathcal{L}(u_k, V_k, C_k; \overline{p}, \overline{q}, \overline{r})$$

and because $\overline{p} = \overline{r} = 0$, this implies

$$\frac{1}{2}\|\overline{V}_{xx} - V_{xx}^*\|_{L^2}^2 \le \frac{1}{2}\|(V_k)_{xx} - V_{xx}^*\|_{L^2}^2 + \int_0^1 \overline{q}(e^{V_k}(u_k)_x - J^*)\, dx$$

$$\le \frac{1}{2}\|(V_k)_{xx} - V_{xx}^*\|_{L^2}^2 + \|\overline{q}\|_{L^2}\|e^{V_k}(u_k)_x - J^*\|_{L^2}.$$

On the other hand, since $(V_k, u_k, C_k)$ is a minimizer of (5.3.2), (5.3.3) with $\beta = \beta_k$, we have

$$\frac{1}{2}\|(V_k)_{xx} - V_{xx}^*\|_{L^2}^2 + \frac{1}{\beta_k}\|e^{V_k}(u_k)_x - J^*\|_{L^2}^2 \le \frac{1}{2}\|\overline{V}_{xx} - V_{xx}^*\|_{L^2}^2.$$

By combining these estimates, we may conclude that

$$\|e^{V_k}(u_k)_x - J^*\|_{L^2} \le \beta_k\|\overline{q}\|_{L^2},$$

and subsequently

$$\left| \|\overline{V}_{xx} - V_{xx}^*\|_{L^2}^2 - \|(V_k)_{xx} - V_{xx}^*\|_{L^2}^2 \right| \le 2\beta_k\|\overline{q}\|_{L^2}.$$

Thus,

$$\|\overline{V}_{xx} - (V_k)_{xx}\|_{L^2}^2 = \|\overline{V}_{xx} - V_{xx}^*\|_{L^2}^2 - \|(V_k)_{xx} - V_{xx}^*\|_{L^2}^2$$

$$-2\int_0^1 (\overline{V}_{xx} - (V_k)_{xx})(\overline{V}_{xx} - V_{xx}^*)\, dx$$

$$\le 2\beta_k\|\overline{q}\|_{L^2} - 2\int_0^1 e^{\overline{V}}(\overline{V} - V_k)\overline{u}_x\overline{q}\, dx$$

$$= 2\beta_k\|\overline{q}\|_{L^2} + 2p_0\int_0^1 e^{-\overline{V}}(\overline{V} - V_k)\, dx = 2\beta_k\|\overline{q}\|_{L^2},$$

where we have used $\bar{q} = -\frac{p_0}{J^*}e^{-\bar{V}}$ and $J^* = e^{\bar{V}}\bar{u}_x$ in the last identity. From the first-order optimality for the limit problem the second term on the right-hand side vanishes and hence,

$$\|\bar{V}_{xx} - (V_k)_{xx}\|_{L^2} \leq \sqrt{2\|\bar{q}\|_{L^2}\beta_k}.$$

The estimate (5.3.14) follows from Poincaré inequalities and standard stability estimates for the equations in (5.3.3). □

## 5.4 Numerical solution of the optimization problems

Numerical algorithms for the solution of (5.2.4) are either based on a steepest descent approach or on the solution of the first-order optimality condition given by (5.2.1), (5.2.2) and (5.2.5)–(5.2.7) via Newton's method (cf. [HP05, HP06]). The same approaches might be used for the minimization problem (5.2.10), but due to the special structure of the first-order optimality condition one might use a variant of the well-known Gummel iteration instead (cf. [BP03]). All three approaches are discussed in the following, and numerical examples for the optimal dopant profiling of an unsymmetric np-diode are presented, where the underlying model equations are given by the stationary bipolar drift-diffusion equations without generation-recombination terms (cf. [MRS90]). Stated on the interval $\Omega = (0, 1)$ the scaled bipolar model reads

$$J_n = (n_x - n\,V_x)\,, \quad J_p = -\,(p_x + p\,V_x)\,, \tag{5.4.1}$$

$$\partial_x J_n = 0, \quad \partial_x J_p = 0, \tag{5.4.2}$$

$$-\lambda^2 V_{xx} = C - n + p. \tag{5.4.3}$$

Here, we have only considered regimes in which we can assume the Einstein relations

$$D_n = U_T\,\mu_n, \quad D_p = U_T\,\mu_p,$$

where $U_T = k_B\,T/q$ is the thermal voltage of the device and $T$ denotes its temperature and $k_B$ the Boltzmann constant.

This system is supplemented with the following boundary conditions:

$$n = n_D, \quad p = p_D, \quad V = V_D \quad \text{in } \{0, 1\}\,,$$

where $n_D, p_D, V_D$ are the $H^1(0, 1)$-extensions of

$$n_D = \frac{C + \sqrt{C^2 + 4\,\delta^4}}{2}, \quad p_D = \frac{-C + \sqrt{C^2 + 4\,\delta^4}}{2}, \tag{5.4.4}$$

$$V_D = -\log\left(\frac{n_D}{\delta^2}\right) + U, \tag{5.4.5}$$

where $\delta^2 = n_i/C_m$ denotes the scaled intrinsic density.

### 5.4.1 Gradient-based methods

The formulation of a gradient-based steepest descent method for the optimization problem (5.2.4) relies on the introduction of the *reduced cost functional*
$\hat{G}_\alpha(C) := G_\alpha(V(C), \rho(C), C)$, where $(V(C), \rho(C))$ is the solution of (5.2.1),
(5.2.2). Clearly, this is only possible if the nonlinear system admits a unique
solution. For the unipolar diodes in one spatial dimension, as discussed in
Section 3, this holds due to a result in [GS92] for the stationary drift-diffusion
model. In general, we can expect the uniqueness only near to the thermal
equilibrium state, i.e., for small applied biasing voltages.

The gradient algorithm for the reduced cost functional $\hat{G}_\alpha$ reads (cf.
[HP05]):

1. Choose an admissible $C_0$.
2. For $k = 1, 2, \ldots$ compute

$$C_k = C_{k-1} - \delta_k \hat{G}'_\alpha(C_{k-1}).$$

The step size $\delta_k$ is computed by an exact one-dimensional line search

$$\delta_k = \operatorname{argmin}_\delta \hat{G}_\alpha \left( C_{k-1} - \delta \hat{G}'_\alpha(C_{k-1}) \right),$$

and the algorithm is terminated if the relative error $\|\hat{G}'_\alpha(C_k)\|/\|\hat{G}'_\alpha(C_0)\|$ is
less than a specified error tolerance.

The evaluation of $\hat{G}'_\alpha$ requires the solution of the nonlinear state system
(5.2.1), (5.2.2) as well as a solution of the linear adjoint system (5.2.5), (5.2.6).
Hence, each gradient step yields a feasible point. Compared with black-box
optimization we have the advantage that this algorithm is independent of the
number of discrete design variables given by a suitable discretization of $C$.

Nevertheless, one has to admit that the main part of the numerical work is
hidden in the line search, since each evaluation of the reduced cost functional
also requires the solution of the nonlinear state system. Instead of the exact
line search one could use here, e.g., Armijo's rule, which will still give sufficient
decrease of the cost functional to ensure convergence.

To get an impression of the performance of the algorithm we present in
Figure 5.1 the optimized doping profiles for an unsymmetric np-diode where
the observation is given by

$$R(J_n \cdot \nu|_\Gamma, J_p \cdot \nu|_\Gamma) = \frac{1}{2} \left| \int_\Gamma J_n \cdot \nu \, ds - I_n^* \right|^2 + \frac{1}{2} \left| \int_\Gamma J_p \cdot \nu \, ds - I_p^* \right|^2$$

and the state system is given by the standard drift-diffusion model without
generation-recombination terms. This allows us to adjust the electron and hole
current separately. In particular, we present the optimized doping profiles for
different choices of $I_n^*, I_p^*$, i.e., we are seeking an amplification of either the
hole current ($I_n^* = J_n^*, I_p^* = 1.5 \cdot J_p^*$) or of the electron current ($I_n^* = 1.5 \cdot J_n^*$,

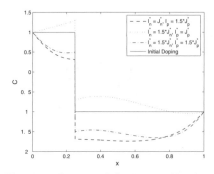

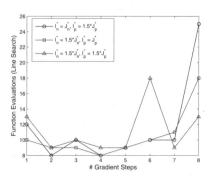

**Fig. 5.1.** Optimized doping profiles for different design goals.

**Fig. 5.2.** Number of evaluations of $\hat{G}_\alpha$ per gradient step.

$I_p^* = J_p^*$) or of both of them ($I_n^* = 1.5 \cdot J_n^*$, $I_p^* = 1.5 \cdot J_p^*$) by 50%. To get an impression of the overall performance of the method we also have to consider the nonlinear solves needed for the exact one-dimensional line search. These are presented in Figure 5.2, and one realizes that this is indeed the numerically most expensive part.

### 5.4.2 Newton methods

Newton-type methods for the solution of the optimization problem (5.2.4) are mainly based on the direct solution of the first-order optimality condition given by (5.2.1), (5.2.2) and (5.2.5)–(5.2.7) or, with the help of the Lagrangian,

$$L'(V, \rho, C; p, q) = 0.$$

This coupled nonlinear system is then solved iteratively using its Jacobian $L''(V, \rho, V; p, q)$, which formally reads

$$L''(V, \rho, C; p, q) =$$
$$\begin{bmatrix} G_{\alpha,xx}(x, C) + \langle e_{xx}(x, C)(\cdot, \cdot), (p, q)\rangle & 0 & e_x(x, C)^* \\ 0 & G_{\alpha,CC}(x, C) & e_C(x, C)^* \\ e_x(x, C) & e_C(x, C) & 0 \end{bmatrix},$$

where we used for brevity the notation $x = (V, \rho)$ and the operator $e$ is defined via $e(x, C) = \big(F(v, \rho), -\lambda^2 \Delta V - Q(\rho) - C\big)$. Further, for notational convenience we define the state-control pair $y \overset{\text{def}}{=} (x, C)$.

If the state system admits a unique solution, we can again introduce the reduced cost functional $\hat{G}_\alpha(C) \overset{\text{def}}{=} G_\alpha(x(C), C)$, where $x(C)$ is determined by $e(x(C), C) = 0$. The derivative of the reduced cost functional is given by

$$\hat{G}'_\alpha(C) = G_{\alpha,C}(y(C)) + e_C^*(y(C))(p, q), \tag{5.4.6}$$

where $(p, q)$ solves the adjoint equations (5.2.5), (5.2.6). We recall that unique solvability of $e(x, C) = 0$ is ensured for devices operated near thermal equilibrium, i.e., for devices with small applied biasing voltages or for the one-dimensional unipolar diode.

Now we derive Newton's method for the solution of

$$\hat{G}'_\alpha(C) = 0,$$

which has the advantage that we have at each iteration level a feasible solution for the state equation. We introduce the operator

$$T(y) \stackrel{\text{def}}{=} \begin{bmatrix} -e_x^{-1}(y)e_C(y) \\ Id_C \end{bmatrix}.$$

Then, for given Lagrange multipliers $(p, q)$ the *reduced Hessian* is defined by

$$H(y; p, q) \stackrel{\text{def}}{=} T^*(y)L_{yy}(y; p, q)T(y), \tag{5.4.7}$$

and it holds that

$$H(y) = G_{\alpha,CC}(y)$$
$$+ e_C^*(y)e_x^{-*}(y) \cdot \{G_{\alpha,xx}(y)(\cdot, \cdot) + \langle e_{xx}(y)(\cdot, \cdot), (p, q)\rangle\} e_x^{-1}(y)e_C(y).$$

Then, the Newton algorithm reads as follows:

Let an admissible $C_0$ be given.
(i) Set $k = 0$ and $C^0 = C_0$.
(ii) Do while the stopping criterion is violated
    (1) Set $y^k = (x(C^k), C^k)$ and $(p^k, q^k) = -e_x^{-*}(y^k)G_{\alpha,x}(y^k)$
    (2) Solve $H(y^k; p^k, q^k)\delta C^k = -\hat{G}'_\alpha(C^k)$
    (3) Set $C^{k+1} = C^k + \delta C^k$, $k = k + 1$
(iii) $C^* \stackrel{\text{def}}{=} C^k, y^* \stackrel{\text{def}}{=} y^k$, STOP.

*Remark 1.* We note that due to the structure of the reduced Hessian the Newton system in step (ii)(2) has to be solved iteratively using, e.g., a conjugate gradient method. Let us refer to this as the inner iteration. To provide the right-hand side in (ii)(2) one has to solve the nonlinear state system (5.2.1), (5.2.2) for $x^k = (V^k, \rho^k)$, and one needs to solve the adjoint system (5.2.5), (5.2.6) for $(p^k, q^k)$. These are all ingredients for the calculation of $\hat{G}'_\alpha$ from (5.4.6).

Every application of $H(y^k; p^k, q^k)$ in the $j$th inner iteration amounts to two linear solves, namely

$$v_j^k = e_x^{-1}(y^k)e_C(y^k)\delta C_j^k$$

and

$$w_j^k = e_C^*(y^k)e_x^{-*}(y^k)\left\{G_{\alpha,xx}(y^k) + \langle e_{xx}(y^k)(v_j^k, v_j^k), (p^k, q^k)\rangle\right\}.$$

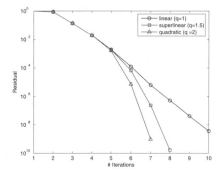
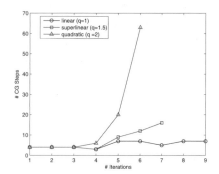

**Fig. 5.3.** Dependence of the residual on $q$.

**Fig. 5.4.** Dependence of the CG iteration on $q$.

Let us come back to our previous numerical example to get more insight into the behavior of the algorithm. We tried to achieve an increase of the electron and hole current by 50% for the unsymmetric np-diode via optimal dopant profiling (for details we refer to [HP06]).

The conjugate gradient algorithm in the inner loop was terminated when the norm of the gradient became sufficiently small; to be more precise, in the $j$th conjugate gradient step for the computation of the update in the Newton step $k$ we stop if the residual $r_j^k$ satisfies

$$\frac{\|r_j^k\|}{\|\hat{G}'_\alpha(C^0)\|} \leq \min\left\{\left(\frac{\|\hat{G}'_\alpha(C^k)\|}{\|\hat{G}'_\alpha(C^0)\|}\right)^q, p\frac{\|\hat{G}'_\alpha(C^k)\|}{\|\hat{G}'_\alpha(C^0)\|}\right\} \quad \text{or} \quad j \geq 100. \quad (5.4.8)$$

Note that $q \in (1, 2)$ determines the convergence order of the outer Newton algorithm. The value of $p \in (0, 1)$ is important for the first step of Newton's method, as for $k = 0$ the norm quotients are all 1; for later steps, the influence of $q$ becomes increasingly dominant. In Figure 5.3 the decrease of the residual is depicted for different values of $q = 1, 1.5$, or 2. As predicted by the general theory [Kel95] one gets linear, superlinear, and quadratic convergence. Clearly, the parameter $q$ strongly influences the number of conjugate gradient steps, which can be seen from Figure 5.4. While in the linear case ($q = 1$) we have an almost constant amount of CG steps in each each iteration, we get, as expected, a drastic increase towards the end of the iteration for the quadratic case ($q = 2$). Hence, the overall numerical effort in terms of CG steps is, despite the quadratic convergence, much larger compared to the relaxed stopping criterion, which only yields linear convergence!

### 5.4.3 Gummel iterations

Finally, we turn our attention to the minimization problem (5.2.10). Clearly, here one can also employ the previously discussed methods. But to exploit the special structure of the optimality system it is favorable to use a different iterative method in the spirit of the well-known Gummel iteration for the solution of the nonlinear state system [BP03]. Using a lower triangular approximation of the optimality system, we start with a potential $V$, and subsequently the continuity equation (5.2.2) with the given potential $V$ for $\rho$. With given potential $V$ and given $\rho$, we solve the adjoint equations (5.2.11), (5.2.12) to obtain the Lagrangian variables $p$ and $q$. Finally, we can perform a gradient step with respect to the design variable $V$ using the optimality equation (5.2.13). Due to the simple structure of this equation, it seems reasonable to discretize the Laplace term in an implicit way and, thus, to solve

$$-\beta(V_{xx} - V_{xx}^*)_{xx} + \tau(V_{xx} - V_{xx}^*) = \tau(V_{xx}^{old} - V_{xx}^*) - e^V u_x p_x + e^{-V} v_x q_x, \tag{5.4.9}$$

for an appropriately chosen damping parameter $\tau$. All together, we can write this iteration in the following form:

1. Choose an admissible $V^0$.
2. For $k = 1, 2, \ldots$ solve consecutively

$$\partial_x(e^{V^k} \partial_x u^k) = 0$$

$$\partial_x(e^{-V^k} \partial_x v^k) = 0$$

$$\partial_x(e^{V^k} \partial_x p^k) = 0$$

$$\partial_x(e^{-V^k} \partial_x q^k) = 0$$

$$-\beta(V_{xx}^k - V_{xx}^*)_{xx} + \tau(V_{xx}^k - V_{xx}^*) = \tau(V_{xx}^{k-1} - V_{xx}^*) - e^{V^k} u_x^k p_x^k + e^{-V^k} v_x^k q_x^k.$$

The corresponding value of the doping profile can be computed independently by

$$C^k - C^* = -\lambda^2(V_{xx}^k - V_{xx}^*)_{xx} + n^k - n^* - p^k + p^*, \tag{5.4.10}$$

where $n^k = e^{V^k} u^k$ and $p^k = e^{-V^k} v^k$.

Finally, we apply this algorithm to our numerical test case, where we want to have an increase of the overall current by 50%. But now, we want to optimize a symmetric np-diode.

The optimal doping profile is depicted in Figure 5.5. But more interesting is the evolution of the cost function, which is shown in Figure 5.6. The objective functional and the observation are reduced in a few iterations. Even if more iterations are necessary, one needs to appreciate this Gummel-like algorithm, since its overall performance is again independent of the number of discrete design parameters and, due to its iterative structure, it is a straightforward extension of the Gummel method. Hence, it is easy to incorporate into existing device simulation codes. Moreover, the numerical performance of this optimization algorithm is optimal, since we need in fact only two Gummel iterations for the solution of the minimization problem, i.e., the numerical

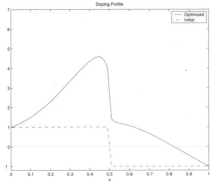

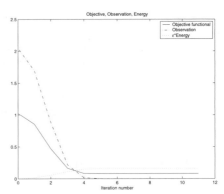

**Fig. 5.5.** Optimized doping profile.

**Fig. 5.6.** Evolution of the cost functional.

complexity is double compared with a forward solve for the nonlinear state system. Note that, on the other hand, here the algorithm is not feasible on each iteration level.

### Acknowledgments

This work was partially supported by the Johann Radon Institute for Computational and Applied Mathematics (Austrian Academy of Sciences ÖAW), by the Austrian National Science Foundation FWF through project SFB F 013/08, by the Sonderforschungsbereich 609 "Elektromagnetische Strömungsbeeinflussung in Matallurgie, Kristallzüchtung und Elektrochemie," by the Sonderforschungsbereich 557 "Beeinflussung komplexer turbulenter Scherströmungen," both sponsored by the Deutsche Forschungsgemeinschaft, and by the Forschungsschwerpunkt "Mathematik & Praxis" (TU Kaiserslautern).

## References

[BP03]   M. Burger, R. Pinnau, Fast optimal design of semiconductor devices, SIAM J. Appl. Math. **64**, 108–126 (2003).

[GS92]   H. Gajewski, J. Sommrey, On the uniqueness of solutions of van Roosbroeck equations, ZAMM **72**, 151–153 (1992).

[HP02a]  M. Hinze, R. Pinnau, Optimal control of the drift-diffusion model for semiconductor devices, in: K.H. Hoffmann, G. Lengering, J. Sprekels, and F. Trottzch (eds.): *Optimal Control of Complex Structures,* Birkhäuser, Basel (2002), 95–106.

[HP02b]  M. Hinze, R. Pinnau, An optimal control approach to semiconductor design, Math. Mod. Meth. Appl. Sci. **12**, 89–107 (2002).

[HP05]   M. Hinze, R. Pinnau, Mathematical tools in optimal semiconductor design, to appear in TTSP (2005).

[HP06]   M. Hinze, R. Pinnau, A second order approach to optimal semiconductor design, to appear in JOTA (2006).

[JP01]   A. Jüngel, Y.J. Peng, A model hierarchy for semiconductors and plasmas, Nonlin. Anal. **47** (2001), 1821–1832.

[Kel95]  C.T. Kelley. *Iterative Methods for Linear and Nonlinear Equations*. SIAM, Philadelphia, 1995.

[MRS90]  P.A. Markowich, C.A. Ringhofer, and C. Schmeiser, *Semiconductor Equations*, Springer, Wien, New York, 1990.

[PSSS98] R. Plasun, M. Stockinger, R. Strasser, and S. Selberherr, *Simulation based optimization environment and its application to semiconductor devices*, in: Proceedings IASTED Intl. Conf. on Applied Modelling and Simulation, 1998, pp. 313–316.

[St00]   M. Stockinger, *Optimization of ultra-low-power CMOS transistors*, PhD Thesis, Technical University Vienna, 2000.

[Stea98] M. Stockinger, R. Strasser, R. Plasun, A. Wild, and S. Selberherr, *A Qualitative Study on Optimized MOSFET Doping Profiles*, in: Proceedings SISPAD 98 Conf., Leuven, 1998, pp. 77–80.

# 6

# Inverse problems for semiconductors: models and methods

A. Leitão,[1] P.A. Markowich[2] and J.P. Zubelli[3]

[1] Department of Mathematics, Federal University of St. Catarina, P.O. Box 476, 88040-900 Florianopolis, Brazil `aleitao@mtm.ufsc.br`
[2] Department of Mathematics, University of Vienna, Boltzmanngasse 9, A-1090 Vienna, Austria `peter.markowich@univie.ac.at`
[3] IMPA, Estr. D. Castorina 110, 22460-320 Rio de Janeiro, Brazil `zubelli@impa.br`

## 6.1 Introduction

### The mathematical model

The starting point of the mathematical model discussed in this chapter is the system of *drift diffusion equations* (see (6.2.1a)–(6.2.1f) below). This system of equations, derived more than fifty years ago [vRo50], is the most widely used to describe semiconductor devices. For the current state of technology, this system represents an accurate compromise between efficient numerical solvability of the mathematical model and realistic description of the underlying physics [Mar86, MRS90, Sel84].

The name *drift diffusion equations* of semiconductors originates from the type of dependence of the current densities on the carrier densities and the electric field. The current densities are the sums of drift terms and diffusion terms. It is worth mentioning that, with the increased miniaturization of semiconductor devices, one comes closer and closer to the limits of validity of the drift diffusion equation. This is due to the fact that in ever-smaller devices the assumption that the free carriers can be modeled as a continuum becomes invalid. On the other hand, the drift diffusion equations are derived by a scaling limit process, where the mean free path of a particle tends to zero.

### The inverse problems

This chapter is devoted to the investigation of inverse problems related to *drift diffusion equations* modeling semiconductor devices. In this context we analyze several inverse problems related to the identification of doping profiles. In all these inverse problems the parameter to be identified corresponds to what is called the *doping profile*. Such a profile enters as a functional parameter in a system of partial differential equations (PDEs). However, the reconstruction problems are related to data generated by different types of measurement techniques.

Identification problems for semiconductor devices, although of increasing technological importance, seem to be poorly understood so far. In the inverse problem literature there has been increasing interest in the identification of a position dependent function $C = C(x)$ representing the doping profile, i.e., the density difference of ionized donors and acceptors. These are the *inverse doping profile problems*. See, for example, [BELM04, BEMP01, BEM02, LMZ05, FI92, FIR02, FI94, BFI93] and references therein.

In some cases, e.g., the p-n diode, it may be assumed that the function $C$ is piecewise constant over the device. In this case, the problem reduces to identifying the curves (or surfaces) between the subdomains where doping is constant. Particularly important are the curves separating subdomains where the doping profile assumes constant values of different signs. These curves are called *p-n junctions* (see Section 6.2 for details). In the *ion implantation* technique, the most important technique for manufacturing silicon devices, only a rough estimate of the doping profile can be obtained by process modelling (see, e.g., [Sel84]). An efficient alternative to determine the real doping profile is the use of reconstruction methods from indirect data.

Another relevant inverse problem concerns identifying the transistor contact resistivity of planar electronic devices, such as MOSFETs (metal oxide semiconductor field-effect transistors); see [FC92]. It is shown that a one-point boundary measurement of the potential is sufficient to identify the resistivity from a one-parameter monotone family, and such identification is both stable and continuously dependent on the parameter. Because of the device miniaturization, it is impossible to measure the contact resistivity in a direct way to satisfactory accuracy. There are extensive experimental and simulation studies for the determination of contact resistivity by certain accessible boundary measurements.

Yet another inverse problem is that of determining the contact resistivity of a semiconductor device from a single voltage measurement [BF91]. It can be modeled as an inverse problem for the elliptic differential equation $\Delta V - p\chi(S)u = 0$ in $\Omega \subset \mathbb{R}^2$, $\partial V/\partial n = g \geq 0$ but $g \not\equiv 0$ on $\partial\Omega$, where $V(x)$ is the measured voltage, and $S \subset \Omega$ and $p > 0$ are unknown. In [BF91], the authors consider the identification of $p$ when the contact location $S$ is also known.

**Outline of the chapter**

In Section 6.2 we introduce and discuss relevant properties of the main mathematical models: the (transient and stationary) systems of drift diffusion equations. In Section 6.3 we derive, from the drift diffusion equations, some special stationary and transient models, which will serve as mathematical background to the formulation the inverse doping profile problems. In Section 6.4 we formulate several inverse problems, which relates to specific measurement procedures for the voltage-current map (namely *pointwise measurements of the current density* and *current flow measurements through a contact*) as well as to specific model idealizations. In Section 6.5 we present a short description of techniques from the theory of inverse problems that is used to handle the doping profile identification problem described in the other sections. In Section 6.6 we present numerical experiments for some models concerning the inverse doping profile problem for the stationary linearized unipolar and bipolar cases.

## 6.2 Drift diffusion equations

### 6.2.1 The transient model

The basic semiconductor device equations in the *transient case* consist of the Poisson equation (6.2.1a), the continuity equations for electrons (6.2.1b) and holes (6.2.1c) and the current relations for electrons (6.2.1d) and holes (6.2.1e). For some applications, in order to account for thermal effects in semiconductor devices, it is also necessary to add to this system the heat flow equation (6.2.1f).

$$\operatorname{div}(\epsilon \nabla V) = q(n - p - C) \qquad \text{in } \Omega \times (0, T) \qquad (6.2.1a)$$
$$\operatorname{div} J_n = q(\partial_t n + R) \qquad \text{in } \Omega \times (0, T) \qquad (6.2.1b)$$
$$\operatorname{div} J_p = q(-\partial_t p - R) \qquad \text{in } \Omega \times (0, T) \qquad (6.2.1c)$$
$$J_n = q(D_n \nabla n - \mu_n n \nabla V) \quad \text{in } \Omega \times (0, T) \qquad (6.2.1d)$$
$$J_p = q(-D_p \nabla p - \mu_p p \nabla V) \text{ in } \Omega \times (0, T) \qquad (6.2.1e)$$
$$\rho\, c\, \partial_t T - H = \operatorname{div} k(T) \nabla T \qquad \text{in } \Omega \times (0, T). \qquad (6.2.1f)$$

This system is defined in $\Omega \times (0, T)$, where $\Omega \subset \mathbb{R}^d$ ($d = 1, 2, 3$) is a domain representing the semiconductor device. Here $V$ denotes the electrostatic potential ($-\nabla V$ is the electric field $E$), $n$ and $p$ are the concentration of free carriers of negative charge (electrons) and positive charge (holes) respectively and $J_n$ and $J_p$ are the densities of the electron and the hole current respectively. $D_n$ and $D_p$ are the diffusion coefficients for electrons and holes respectively. $\mu_n$ and $\mu_p$ denote the mobilities of electrons and holes respectively. The positive constants $\epsilon$ and $q$ denote the permittivity coefficient (for silicon) and the elementary charge.

The function $R$ has the form $R = \mathcal{R}(n, p, x)(np - n_i^2)$ and denotes the *recombination-generation rate* ($n_i$ is the intrinsic carrier density). The *bandgap* is relatively large for semiconductors (the gap between the valence and conduction bands), and a significant amount of energy is necessary to transfer electrons from the valence and to the conduction band. This process is called generation of electron-hole pairs. The reverse process corresponds to the transfer of a conduction electron into the lower energy valence band. This process is called recombination of electron-hole pairs. In our model these phenomena are described by the recombination-generation rate $R$. Frequently adopted in the literature are the Shockley–Read–Hall model ($\mathcal{R}_{SRH}$) and the Auger model ($\mathcal{R}_{AU}$). They are defined by

$$\mathcal{R}_{SRH} \stackrel{\text{def}}{=} \frac{1}{\tau_p(n+n_i) + \tau_p(p+n_i)} \,, \quad \mathcal{R}_{AU} \stackrel{\text{def}}{=} (C_n n + C_p p)\,,$$

where $C_n$, $C_p$, $\tau_n$ and $\tau_p$ are positive constants whose physical values are listed in Table A.1 of the Appendix.

The function $\mathcal{T}$ represents the temperature and the constants $\rho$ and $c$ denote the specific mass density and specific heat of the material respectively. Furthermore, $k(\mathcal{T})$ and $H$ denote the thermal conductivity and the locally generated heat. Equation (6.2.1f) was presented here only for the sake of completeness of the model and shall not be considered in the subsequent development.

The function $C(x)$ models a preconcentration of ions in the crystal, so $C(x) = C_+(x) - C_-(x)$ holds, where $C_+$ and $C_-$ are concentrations of negative and positive ions respectively. In those subregions of $\Omega$ for which the preconcentration of negative ions predominates (p-regions), we have $C(x) < 0$. Analogously, we define the n-regions, where $C(x) > 0$ holds. The boundaries between the p-regions and n-regions (where $C$ changes sign) are called *p-n junctions*.

In the sequel we turn our attention to the boundary conditions. We assume the boundary $\partial\Omega$ of $\Omega$ to be divided into two nonempty disjoint parts: $\partial\Omega = \overline{\partial\Omega_N} \cup \overline{\partial\Omega_D}$. The Dirichlet part of the boundary $\partial\Omega_D$ models the ohmic contacts, where the potential $V$ as well as the concentrations $n$ and $p$ are prescribed. The Neumann part $\partial\Omega_N$ of the boundary corresponds to insulating surfaces, thus a zero current flow and a zero electric field in the normal direction are prescribed. The Neumann boundary conditions for system (6.2.1a)–(6.2.1e) read

$$\frac{\partial V}{\partial \nu}(x,t) = \frac{\partial n}{\partial \nu}(x,t) = \frac{\partial p}{\partial \nu}(x,t) = 0\,, \ \partial\Omega_N \times [0,T]\,. \tag{6.2.2}$$

Moreover, at $\partial\Omega_D \times [0,T]$, the following Dirichlet boundary conditions are imposed:

$$V(x,t) = V_D(x,t) = U(x,t) + V_{\text{bi}}(x) = U(x,t) + U_T \ln(n_D(x)/n_i)$$

$$n(x,t) = n_D(x) = \frac{1}{2}\left(C(x) + \sqrt{C(x)^2 + 4n_i^2}\right) \tag{6.2.3}$$

$$p(x,t) = p_D(x) = \frac{1}{2}\left(-C(x) + \sqrt{C(x)^2 + 4n_i^2}\right)\,.$$

Here, the function $U(x,t)$ denotes the applied potential and $U_T$ is the thermal voltage. We shall consider the simple situation $\partial\Omega_D = \Gamma_0 \cup \Gamma_1$, which occurs, e.g., in a diode. The disjoint boundary parts $\Gamma_i$, $i = 0, 1$, correspond to distinct contacts. Differences in $U(x)$ between different segments of $\partial\Omega_D$ correspond to the applied bias between these two contacts. The constant $U_T$ represents the thermal voltage. Moreover the initial conditions $n(x, 0) \geq 0$, $p(x, 0) \geq 0$ have to be imposed.

We conclude this subsection by discussing the solution theory for the transient drift diffusion system (6.2.1a)–(6.2.1e), (6.2.2), (6.2.3).[4] Twenty years ago, the existence and uniqueness of global in time solutions for the transient drift diffusion equations were demonstrated by Gajewski in [Gaj85]. Under the assumption that the doping profile satisfies $C \in L^r(\Omega)$ for $d \leq r \leq 6$, it is shown that

$$(V - V_D, n - n_D, p - p_D) \in W \stackrel{\text{def}}{=}$$
$$\{C([0, T]; H_0^2(\Omega)) \cap L^2([0, T]; W_0^{2,r}(\Omega)) \cap H^1([0, T]; \widetilde{W})\} \times \widetilde{\widetilde{W}} \times \widetilde{\widetilde{W}},$$
$$(6.2.4)$$

where $\widetilde{W} \stackrel{\text{def}}{=} \{w \in H^1(\Omega); \ w|_{\partial\Omega_D} = 0\}$ and

$$\widetilde{\widetilde{W}} \stackrel{\text{def}}{=} C([0, T]; L^2(\Omega)) \cap L^2([0, T]; \widetilde{W}) \cap H^1([0, T]; \widetilde{W}^*).$$

In the special one-dimensional case $\Omega = (0, L)$, the following stronger result is proved.

**Lemma 1.** *Let the doping profile satisfy $C \in L^r(\Omega)$, for $d \leq r \leq 6$. If the mobilities $\mu_n$ and $\mu_p$ are in $L^\infty(\Omega)$, then every solution $(V, n, p)$ of the transient drift diffusion equations (6.2.1a)–(6.2.1e), (6.2.2) and (6.2.3) satisfies (6.2.4). Moreover,*

$$(V, n, p) \in C([0, T]; H^2(\Omega)) \cap C([0, T]; W^{1,\infty}(\Omega))^2.$$

### 6.2.2 The stationary model

In this subsection we turn our attention to the stationary drift diffusion equations. We neglect the thermal effects and assume further $\frac{\partial n}{\partial t} = \frac{\partial p}{\partial t} = 0$. Thus, the *stationary drift diffusion model* is derived from (6.2.1a)–(6.2.1e) in a straightforward way. Next, motivated by the Einstein relations $D_n = U_T\mu_n$ and $D_p = U_T\mu_p$ (a standard assumption about the mobilities and diffusion coefficients), one introduces the *Slotboom variables* $u$ and $v$. They are related to the original $n$ and $p$ variables by the formula

$$n(x) = n_i \exp\left(\frac{V(x)}{U_T}\right) u(x), \quad p(x) = n_i \exp\left(\frac{-V(x)}{U_T}\right) v(x). \quad (6.2.5)$$

---

[4]In order to simplify the model, we neglect thermal effects.

For convenience, we rescale the potential and the mobilities, i.e., $V(x) \leftarrow V(x)/U_T$, $\mu_n \leftarrow qU_T\mu_n$, $\mu_p \leftarrow qU_T\mu_p$. Obviously, the current relations now read $J_n = \mu_n n_i\, e^V \nabla u$, $J_p = -\mu_p n_i\, e^{-V} \nabla v$.

Now we can write the stationary drift diffusion equations in the form

$$\lambda^2\, \Delta V = \delta^2\left(e^V u - e^{-V} v\right) - C(x) \quad \text{in } \Omega \tag{6.2.6a}$$

$$\operatorname{div} J_n = \delta^4\, Q(V, u, v, x)\,(uv - 1) \quad \text{in } \Omega \tag{6.2.6b}$$

$$\operatorname{div} J_p = -\delta^4\, Q(V, u, v, x)\,(uv - 1) \text{ in } \Omega \tag{6.2.6c}$$

$$V = V_D = U + V_{\mathrm{bi}} \qquad \text{on } \partial\Omega_D \tag{6.2.6d}$$

$$u = u_D = e^{-U} \qquad \text{on } \partial\Omega_D \tag{6.2.6e}$$

$$v = v_D = e^{U} \qquad \text{on } \partial\Omega_D \tag{6.2.6f}$$

$$\nabla V \cdot \nu = J_n \cdot \nu = J_p \cdot \nu = 0 \qquad \text{on } \partial\Omega_{\mathrm{N}}, \tag{6.2.6g}$$

where $\lambda^2 \overset{\text{def}}{=} \epsilon/(qU_T)$ is the Debye length of the device, $\delta^2 \overset{\text{def}}{=} n_i$ and the function $Q$ is defined implicitly by the relation $Q(V, u, v, x) = \mathcal{R}(n, p, x)$.[5]

One should notice that, due to the thermal equilibrium assumption, it follows $np = n_i^2$, and the assumption of vanishing space charge density gives $n - p - C = 0$, for $x \in \partial\Omega_D$. This fact motivates the boundary conditions on the Dirichlet part of the boundary.

It is worth mentioning that, in a realistic model, the mobilities $\mu_n$ and $\mu_p$ usually depend on the electric field strength $|\nabla V|$. In what follows, we assume that $\mu_n$ and $\mu_p$ are positive constants. This assumption simplifies the subsequent analysis, allowing us to concentrate on the inverse doping problems. As a matter of fact, this dependence could be incorporated in the model without changing the results described in the sequel.

Next, we describe some existence and uniqueness results for the stationary drift diffusion equations. We start by presenting a classical existence result.

**Lemma 2. [MRS90, Theorem 3.3.16]** *Let $\kappa > 1$ be a constant satisfying $\kappa^{-1} \leq u_D(x)$, $v_D(x) \leq \kappa$, $x \in \partial\Omega_D$, and let $-\infty < C_m \leq C_M < +\infty$. Then for any $C \in \{L^\infty(\Omega); \ C_m \leq C(x) \leq C_M, \ x \in \Omega\}$, the system (6.2.6a)–(6.2.6g) admits a weak solution $(V, u, v) \in (H^1(\Omega) \cap L^\infty(\Omega))^3$.*

Under stronger assumptions on the boundary parts $\partial\Omega_D$, $\partial\Omega_N$ as well as on the boundary conditions $V_D$, $u_D$, $v_D$, it is even possible to show $H^2$-regularity for a solution $(V, u, v)$ of system (6.2.6a)–(6.2.6g). For details on this result we refer the reader to [MRS90, Theorem 3.3.1].

Regarding the uniqueness of solutions of system (6.2.6a)–(6.2.6g), some results can be obtained if the applied voltage is small (in the norm of $L^\infty(\partial\Omega_D) \cap H^{3/2}(\partial\Omega_D)$).

**Lemma 3. [BEMP01, Theorem 2.4]** *Let the applied voltage $U$ be such that $\|U\|_{L^\infty(\partial\Omega_D)} + \|U\|_{H^{3/2}(\partial\Omega_D)}$ is sufficiently small. Then, system (6.2.6a)–(6.2.6g) has a unique solution $(V, u, v) \in (H^1(\Omega) \cap L^\infty(\Omega))^3$.*

---

[5]Notice that the applied potential also has to be rescaled: $U(x) \leftarrow U(x)/U_T$.

Since the existence and uniqueness of solutions for system (6.2.6a)–(6.2.6g) can only be guaranteed for small applied voltages, it is reasonable to consider, instead of this system, its linearized version around the equilibrium point $U \equiv 0$. We shall return to this point in the next section, where the voltage-current map is introduced.

## 6.3 Special models

In the following subsections we assume several different simplifications of the drift diffusion models introduced in Section 6.2 and derive some special cases which will serve as underlying models for the inverse problems investigated in Section 6.4.

### 6.3.1 The linearized stationary drift diffusion equations (close to equilibrium)

We begin this subsection by introducing the *thermal equilibrium* assumption for the stationary drift diffusion equations. This is a preliminary step for deriving a linearized system of stationary drift diffusion equations (close to equilibrium).

The thermal equilibrium assumption refers to the condition in which the semiconductor is not subject to external excitations, except for a uniform temperature, i.e., no voltages or electric fields are applied. We note that, under the thermal equilibrium assumption, all externally applied potentials to the semiconductor contacts are zero (i.e., $U(x) = 0$). Moreover, the thermal generation is perfectly balanced by recombination (i.e., $\mathcal{R} = 0$).

If the applied voltage satisfies $U = 0$, one immediately sees that the solution of system (6.2.6a)–(6.2.6g) simplifies to $(V, u, v) = (V^0, 1, 1)$, where $V^0$ solves

$$\lambda^2 \, \Delta V^0 = e^{V^0} - e^{-V^0} - C(x) \text{ in } \Omega \tag{6.3.1a}$$

$$V^0 = V_{\mathrm{bi}}(x) \qquad \text{on } \partial\Omega_D \tag{6.3.1b}$$

$$\nabla V^0 \cdot \nu = 0 \qquad \text{on } \partial\Omega_N . \tag{6.3.1c}$$

For some of the models discussed below, we will be interested in the linearized drift diffusion system at equilibrium. Keeping this in mind, we compute the Gateaux derivative of the solution of system (6.2.6a)–(6.2.6g) with respect to the voltage $U$ at the point $U \equiv 0$ in the direction $h$. This directional derivative is given by the solution $(\hat{V}, \hat{u}, \hat{v})$ of

$$\lambda^2 \, \Delta \hat{V} = e^{V^0} \hat{u} + e^{-V^0} \hat{v} + (e^{V^0} + e^{-V^0}) \hat{V} \text{ in } \Omega \tag{6.3.2a}$$

$$\mathrm{div} \left( \mu_n e^{V^0} \nabla \hat{u} \right) = Q_0(V^0, x)(\hat{u} + \hat{v}) \qquad \text{in } \Omega \tag{6.3.2b}$$

$$\mathrm{div} \left( \mu_p e^{-V^0} \nabla \hat{v} \right) = Q_0(V^0, x)(\hat{u} + \hat{v}) \qquad \text{in } \Omega \tag{6.3.2c}$$

$$\hat{V} = h \qquad\qquad \text{on } \partial\Omega_D \quad (6.3.2\text{d})$$

$$\hat{u} = -h \qquad\qquad \text{on } \partial\Omega_D \quad (6.3.2\text{e})$$

$$\hat{v} = h \qquad\qquad \text{on } \partial\Omega_D \quad (6.3.2\text{f})$$

$$\nabla V^0 \cdot \nu = \nabla\hat{u} \cdot \nu = \nabla\hat{v} \cdot \nu = 0 \qquad\qquad \text{on } \partial\Omega_N , (6.3.2\text{g})$$

where the function $Q_0$ satisfies $Q_0(V^0, x) = Q(V^0, 1, 1, x)$.

### 6.3.2 Linearized stationary bipolar case (close to equilibrium)

In this subsection we present a special case, which plays a key role in modeling inverse doping problems related to *current flow* measurements.

The discussion is motivated by the *stationary voltage-current (V-C) map*

$$\Sigma_C : H^{3/2}(\partial\Omega_D) \to \mathbb{R}$$
$$U \mapsto \int_{\Gamma_1} (J_n + J_p) \cdot \nu \, ds.$$

Here $(V, u, v)$ is the solution of (6.2.6) for an applied voltage $U$. This operator models practical experiments where *voltage-current data* are available, i.e., measurements of the averaged outflow current density on $\Gamma_1 \subset \partial\Omega_D$.

The *linearized stationary bipolar case (close to equilibrium)* corresponds to the model obtained from the drift diffusion equations (6.2.6) by linearizing the V-C map at $U \equiv 0$. This simplification is motivated by the fact that, due to hysteresis effects for large applied voltage, the V-C map can only be defined as a single-valued function in a neighborhood of $U = 0$. Moreover, the following simplifying assumptions are also taken into account:

*A1)* The electron mobility $\mu_n$ and hole mobility $\mu_p$ are constant;
*A2)* No recombination-generation rate is present, i.e., $\mathcal{R} = 0$ (or $Q_0 = 0$).

An immediate consequence of our assumptions is the fact that the Poisson equation and the continuity equations decouple. Indeed, from (6.3.2) we see that the Gateaux derivative of the V-C map $\Sigma_C$ at the point $U = 0$ in the direction $h \in H^{3/2}(\partial\Omega_D)$ is given by the expression

$$\Sigma_C'(0)h = \int_{\Gamma_1} \left( \mu_n e^{V_{\mathrm{bi}}} \hat{u}_\nu - \mu_p e^{-V_{\mathrm{bi}}} \hat{v}_\nu \right) ds, \qquad (6.3.3)$$

where $(\hat{u}, \hat{v})$ solve

$$\mathrm{div}\,(\mu_n e^{V^0} \nabla\hat{u}) = 0 \qquad\qquad \text{in } \Omega \qquad (6.3.4\text{a})$$

$$\mathrm{div}\,(\mu_p e^{-V^0} \nabla\hat{v}) = 0 \qquad\qquad \text{in } \Omega \qquad (6.3.4\text{b})$$

$$\hat{u} = -h \qquad\qquad \text{on } \partial\Omega_D \qquad (6.3.4\text{c})$$

$$\hat{v} = h \qquad\qquad \text{on } \partial\Omega_D \qquad (6.3.4\text{d})$$

$$\nabla \hat{u} \cdot \nu = \nabla \hat{v} \cdot \nu = 0 \text{ on } \partial \Omega_N \qquad (6.3.4e)$$

and $V^0$ is the solution of the equilibrium problem (6.3.1); (see Lemma 4 for details).

Notice that the solution of the Poisson equation can be computed a priori, since it does not depend on $h$. The linear operator $\Sigma'_C(0)$ is continuous. Actually, we can prove more: since $(u, v)$ depend continuously in $H^2(\Omega)^2$ on the boundary data $h$ in $H^{3/2}(\partial \Omega_D)$, it follows from the boundedness and compactness of the trace operator $\gamma : H^2(\Omega) \to H^{1/2}(\Gamma_1)$ that $\Sigma'_C(0)$ is a compact operator. The operator $\Sigma'_C(0)$ maps the Dirichlet data for $(\hat{u}, \hat{v})$ to a weighted sum of their Neumann data and can be compared with the DtN map in electrical impedance tomography (EIT). See [Bor02, BU02, Nac96].

### 6.3.3 Linearized stationary unipolar case (close to equilibrium)

The linearized unipolar case (close to equilibrium) corresponds to the model obtained from the unipolar drift diffusion equations by linearizing the V-C map at $U \equiv 0$. In addition to *A1)* and *A2)*, we further assume:

*A3)* The concentration of holes satisfies $p = 0$ (or, equivalently, $v = 0$ in $\Omega$).

Under those assumptions, the Gateaux derivative of the V-C map $\Sigma_C$ at the point $U = 0$ in the direction $h$ is given by

$$\Sigma'_C(0)h = \int_{\Gamma_1} \mu_n e^{V_{\text{bi}}} \hat{u}_\nu \, ds,$$

where $\hat{u}$ solves

$$\text{div}\left(\mu_n e^{V^0} \nabla \hat{u}\right) = 0 \qquad \text{in } \Omega \qquad (6.3.5a)$$
$$\hat{u} = -h(x) \text{ on } \Omega_D \qquad (6.3.5b)$$
$$\nabla \hat{u} \cdot \nu = 0 \qquad \text{on } \Omega_N \qquad (6.3.5c)$$

and $V^0$ is the solution of the equilibrium problem (6.3.1), with (6.3.1a) replaced by

$$\lambda^2 \Delta V^0 = e^{V^0} - C(x) \text{ in } \Omega. \qquad (6.3.1a')$$

### 6.3.4 Linearized transient bipolar case (close to equilibrium)

In this subsection we introduce a transient case, which is the time-dependent counterpart of the bipolar model discussed in Subsection 6.3.2. It will serve as the background for the formulation of inverse doping problems related to transient current flow measurements.

As in Subsection 6.3.2, we begin the discussion by introducing the *transient voltage-current map*. For an applied time dependent voltage $U(x, t)$, the transient V-C map is given by

$$\Sigma_{t,C} : L^2([0,T]; H^{3/2}(\partial\Omega_D)) \to L^2(0,T)$$
$$U(\cdot,t) \mapsto \int_{\Gamma_1} [J_n(\cdot,t) + J_p(\cdot,t)] \cdot \nu \, ds. \qquad (6.3.6)$$

Here $(V,n,p)$ is the solution of (6.2.1), (6.2.2), (6.2.3) for an applied voltage $U$.[6] This operator models practical experiments where time dependent *voltage-current data* are available. In [BEM02] it is shown that the nonlinear operator $\Sigma_{t,C}$ is well defined, continuous and Fréchet differentiable. In the sequel we derive the Gateaux derivative of $\Sigma_{t,C}$ in equilibrium.

As in the stationary cases, we shall consider the transient drift diffusion equations under the *thermal equilibrium* assumption. One immediately observes that, for zero applied voltage $U(\cdot,t) = 0$, the solution $(V^0, n^0, p^0)$ of (6.2.1), (6.2.2), (6.2.3) is constant in time, being the counterpart (in the $n$, $p$ variables) of the solution triplet $(V^0, 1, 1)$ in the Slotboom variables (see (6.2.5)).

Here again, we assume $A1)$, $A2)$ of Subsection 6.3.2. Then, arguing as in Subsection 6.3.1, it follows that the Gateaux derivative of the transient V-C map $\Sigma_{t,C}$ at the point $U = 0$ in the direction $h(\cdot,t) \in L^2([0,T]; H^{3/2}(\partial\Omega_D))$ is given by

$$\Sigma'_{t,C}(0)h = \int_{\Gamma_1} \left[ \mu_n(\hat{n}_\nu - \hat{n}V^0_\nu - n^0\hat{V}_\nu) - \mu_p(\hat{p}_\nu + \hat{p}V^0_\nu + p^0\hat{V}_\nu) \right] ds, \quad (6.3.7)$$

where $(\hat{V}, \hat{n}, \hat{p})$ solve

$$\lambda^2\hat{V} = \hat{n} - \hat{p} \qquad\qquad\qquad \text{in } \Omega \times (0,T) \qquad (6.3.8a)$$
$$\partial_t\hat{n} = \text{div}(\mu_n[\nabla\hat{n} - \hat{n}\nabla V^0 - n^0\nabla\hat{V}]) \text{ in } \Omega \times (0,T) \qquad (6.3.8b)$$
$$\partial_t\hat{p} = \text{div}(\mu_p[\nabla\hat{p} + \hat{p}\nabla V^0 + p^0\nabla\hat{V}]) \text{ in } \Omega \times (0,T) \qquad (6.3.8c)$$
$$\hat{V} = h \qquad\qquad\qquad\qquad \text{on } \partial\Omega_D \times (0,T) \qquad (6.3.8d)$$
$$\hat{n} = \hat{p} = 0 \qquad\qquad\qquad \text{on } \partial\Omega_D \times (0,T) \qquad (6.3.8e)$$
$$\nabla\hat{V} \cdot \nu = \nabla\hat{n} \cdot \nu = \nabla\hat{p} \cdot \nu = 0 \qquad \text{on } \partial\Omega_N \times (0,T). \quad (6.3.8f)$$

Notice that, unlike the stationary case, the Poisson equation (6.3.8a) and the continuity equations (6.3.8b), (6.3.8c) do not decouple.

## 6.4 Inverse problems for semiconductors

In practical experiments there are different types of measurement techniques, such as

- *Laser beam induced current* (LBIC) measurements;
- *Capacitance* measurements;

---

[6]Once more we do not consider equation (6.2.1f).

- *Current flow* measurements.

We refer to [FI92, FI94, FIR02] for the first type and to [BELM04, BEMP01, BEM02, LMZ05] for the last two types. These measurement techniques are related to different types of data and lead to different inverse problems for reconstructing the doping profile. They are called the *inverse doping profile problems*. In the following subsections we address inverse problems related to each one of these measurement techniques.

### 6.4.1 The stationary V-C map

We begin this subsection by verifying that the V-C map $\Sigma_C$, introduced in Subsection 6.3.2, is well defined in a suitable neighborhood of $U = 0$.

**Lemma 4. [BEMP01, Proposition 3.1]** *For each applied voltage $U \in B_r(0) \subset H^{3/2}(\partial\Omega_D)$ with $r > 0$ sufficiently small, the current $J \cdot \nu \in H^{1/2}(\Gamma_1)$ is uniquely defined. Furthermore, $\Sigma_C : H^{3/2}(\partial\Omega_D) \to H^{1/2}(\Gamma_1)$ is continuous and is continuously differentiable in $B_r(0)$. Moreover, its derivative in the direction $h \in H^{3/2}(\partial\Omega_D)$ is given by the operator $\Sigma_C'(0)$ defined in (6.3.3).*

As a matter of fact, we can actually prove that, since $(\hat{u}, \hat{v})$ in (6.3.4) depend continuously (in $H^2(\Omega)^2$) on the boundary data $U \in H^{3/2}(\partial\Omega_D)$, it follows from the boundedness and compactness of the trace operator $\gamma : H^2(\Omega) \to H^{1/2}(\Gamma_1)$ that $\Sigma_C'(0)$ is a bounded and compact operator. The operator $\Sigma_C'(0)$ in (6.3.3) maps the Dirichlet data for $(\hat{u}, \hat{v})$ to a weighted sum of their Neumann data and the related inverse problem can be compared with the identification problem in *electrical impedance tomography* (EIT).

Lemma 4 establishes a basic property to consider the inverse problem of reconstructing the doping profile $C$ from the V-C map. In the sequel we shall consider two possible inverse problems for the V-C map.

#### Current flow measurements through a contact

In the first inverse problem we assume that, for each $C$, the output is given by $\Sigma_C'(0)U_j$ for some $U_j$. A realistic experiment corresponds to measuring, for given $\{U_j\}_{j=1}^N$, with $\|U_j\|$ small, the outputs

$$\left\{ \Sigma_C'(0)U_j \mid j = 1, \dots, N \right\}$$

(recall that $\Sigma_C(0) = (V^0, 1, 1)$). In practice, the functions $U_j$ are chosen to be piecewise constant on the contact $\Gamma_1$ and to vanish on $\Gamma_0$. From the definition of $\Sigma_C'(0)$ we deduce the following abstract formulation of the inverse doping profile problem for the V-C map:

$$F(C) = Y, \tag{6.4.1}$$

where

1) $\{U_j\}_{j=1}^N \subset H^{3/2}(\partial\Omega_D)$ are fixed voltage profiles satisfying $U_j|_{\Gamma_1} = 0$;

2) Parameter: $C = C(x) \in L^2(\Omega) =: \mathcal{X}$;

3) Output: $Y = \{\Sigma'_C(0)U_j\}_{j=1}^N \in \mathbb{R}^N =: \mathcal{Y}$;

4) Parameter-to-output map: $F : \mathcal{X} \to \mathcal{Y}$.

The domain of definition of the operator $F$ is

$$D(F) \overset{\text{def}}{=} \{C \in L^\infty(\Omega); \ C_m \leq C(x) \leq C_M, \ \text{a.e. in } \Omega\},$$

where $C_m$ and $C_M$ are suitable positive constants.

   This approach is motivated by the fact that, in practical applications, the V-C map can only be defined in a neighborhood of $U = 0$ (due to hysteresis effects for large applied voltages). The inverse problem described above corresponds to the problem of identifying the doping profile $C$ from the linearized V-C map at $U = 0$. See the unipolar and bipolar cases in Subsections 6.3.2 and 6.3.3.

   The nonlinear parameter-to-output operator $F$ is well defined and Fréchet differentiable in its domain of definition $D(F)$. This assertion follows from standard regularity results in PDE theory [BELM04, Propositions 2.2 and 2.3].

   Note that the solution of the Poisson equation can be computed *a priori*. The remaining problem (coupled system (6.3.4) for $(\hat{u}, \hat{v})$) is quite similar to the problem of EIT (see [Bor02, Isa98]). In this inverse problem the aim is to identify the conductivity $q = q(x)$ in the equation

$$-\text{div}\,(q\nabla u) \ = \ f \ \text{in } \Omega$$

from measurements of the *Dirichlet-to-Neumann map*, which maps the applied voltage $u|_{\partial\Omega}$ to the electrical flux $qu_\nu|_{\partial\Omega}$. The map $\Sigma'_C(0)$ sends the Dirichlet data for $\hat{u}$ and $\hat{v}$ to the weighted sum of their Neumann data. It can be seen as the counterpart of EIT for common conducting materials.

## Pointwise measurements of the current density

In the sequel, we investigate a different formulation of the same inverse problem related to the V-C map considered above. Unlike the previous discussion, we shall assume that the V-C operator maps the Dirichlet data for $\hat{u}$ and $\hat{v}$ in (6.3.4) to the sum of their Neumann data, i.e.,

$$\Sigma_C : H^{3/2}(\partial\Omega_D) \to H^{1/2}(\Gamma_1)$$
$$U \mapsto (J_n + J_p) \cdot \nu|_{\Gamma_1}$$

where functions $V$, $u$, $v$, $J_n$, $J_p$ and $U$ have the same meaning as in Subsection 6.3.2. We immediately observe that the Gateaux derivative of the V-C map $\Sigma_C$ at the point $U = 0$ in the direction $h \in H^{3/2}(\partial\Omega_D)$ is given by

$$\Sigma'_C(0)h = \left(\mu_n \, e^{V_{\text{bi}}}\hat{u}_\nu - \mu_p \, e^{-V_{\text{bi}}}\hat{v}_\nu\right)|_{\Gamma_1}, \tag{6.4.2}$$

where $(\hat{u}, \hat{v})$ solve system (6.3.4). Notice that, for each voltage profile $U$, the V-C map associates a scalar valued function defined on $\Gamma_1$. In this case, the outputs $\Sigma'_C(0)U_j$ give much more information about the parameter $C$ than in the case of current flow measurements.

Again we can derive an abstract formulation of type (6.4.1) for the inverse doping profile problem for the V-C map with pointwise measurements of the current density. The only difference to the framework described in the previous paragraph concerns the definition of the Hilbert space $Y$, which is now defined by

3') Output: $Y = \left\{ \Sigma'_C(0)U_j \right\}_{j=1}^N \in L^2(\Gamma_1)^N =: \mathcal{Y}.$

The domain of definition of the operator $F$ remains unaltered.

In Section 6.6 we shall consider three numerical implementations concerning inverse doping problems for the V-C map described above, namely:

(i) The stationary linearized unipolar model (close to equilibrium) with current flow measurements through a contact;

(ii) The stationary linearized unipolar model (close to equilibrium) with pointwise measurements of the current density.

(iii) The stationary linearized bipolar model (close to equilibrium) with pointwise measurements of the current density.

### 6.4.2 The transient V-C map

In the sequel we shall consider inverse problems for the map $\Sigma_{t,C}$ in (6.3.6). As already observed in Subsection 6.3.4, this V-C map is well defined, continuous and Fréchet differentiable, its Gateaux derivative in equilibrium $\Sigma'_{t,C}(0)$ being defined by (6.3.7).

As in Subsection 6.4.1 we investigate two possible inverse doping problems for the *linearized transient bipolar case (close to equilibrium)*.

### Transient current flow measurements through a contact

Here we assume that, for each $C$, the output corresponds to $\Sigma'_{t,C}(0)U_j$ for some prescribed $U_j(\cdot, t)$. The experiment corresponds to measuring, for given $\{U_j(\cdot, t)\}_{j=1}^N$, with $\|U_j\|$ small in $L^2([0,T]; H^{3/2}(\partial\Omega_D))$, the (averaged) currents

$$\left\{ \Sigma'_{t,C}(0)U_j(\cdot, t) \mid j = 1, \ldots, N \right\}.$$

The profile of the voltages $U_j$ is chosen analogously to that of Subsection 6.4.1. Notice that in a realistic transient experiment, the amplitude of functions $U_j(\cdot, t)$ may vary with the time, e.g.,

$$U_j(x, t) \stackrel{\text{def}}{=} \begin{cases} 1 + t, & |x - x_j| \leq h \\ 0, & \text{elsewhere} \end{cases},$$

where $\Gamma_0 = (0,1) \times \{0\} \subset \mathbb{R}^2$ and $0 < x_1 < x_2 < \cdots < x_N < 1$ and $h$ is small enough (compare with Subsection 6.6.1).

The inverse doping profile problem for the V-C map $\Sigma'_{t,C}(0)$ can be formulated in the abstract form (6.4.1), where

$1_t$) $\{U_j(\cdot,t)\}_{j=1}^N \subset L^2([0,T]; H^{3/2}(\partial\Omega_D))$ are fixed voltage profiles satisfying $U_j(\cdot,t)|_{\Gamma_1} = 0$, $t \geq 0$;

$2_t$) Parameter: $C = C(x) \in L^2(\Omega) =: \mathcal{X}$;

$3_t$) Output: $Y = \{\Sigma'_{t,C}(0)U_j\}_{j=1}^N \in L^2(0,T)^N =: \mathcal{Y}$;

$4_t$) Parameter-to-output map: $F : \mathcal{X} \to \mathcal{Y}$.

The domain of definition of the operator $F$ is

$$D(F) \overset{\text{def}}{=} \{C \in L^\infty(\Omega); \, C_m \leq C(x) \leq C_M, \text{ a.e. in } \Omega\},$$

where $C_m$ and $C_M$ are suitable positive constants.

From our knowledge about the operator $\Sigma'_{t,C}(0)$ we conclude that the parameter-to-output operator $F$ is well defined and continuous. Moreover, for one-dimensional domains $\Omega$ it is shown in [BEM02] that $F$ is weakly sequentially closed.

### Transient pointwise measurements of the current density

In the previous discussion, we considered $\Sigma_{t,C}$ to be defined by (6.3.6). Now, we shall assume that, for every time instant $t \geq 0$, current measurements are available at every point of the segment $\Gamma_1$. This assumption corresponds to the following definition of the V-C map:

$$\Sigma_{t,C} : L^2([0,T]; H^{3/2}(\partial\Omega_D)) \to L^2([0,T]; H^{1/2}(\Gamma_1)).$$
$$U(\cdot,t) \mapsto (J_n(\cdot,t) + J_p(\cdot,t)) \cdot \nu|_{\Gamma_1},$$

where $(V,n,p)$ is the solution of (6.2.1), (6.2.2), (6.2.3) for an applied voltage $U(\cdot,t)$. We immediately observe that the Gateaux derivative of $\Sigma_{t,C}$ at the point $U(\cdot,t) = 0$ in the direction $h \in L^2([0,T]; H^{3/2}(\partial\Omega_D))$ is given by

$$\Sigma'_{t,C}(0)h = [\mu_n(\hat{n}_\nu - \hat{n}V_\nu^0 - n^0\hat{V}_\nu) - \mu_p(\hat{p}_\nu + \hat{p}V_\nu^0 + p^0\hat{V}_\nu)]|_{\Gamma_1},$$

where $(\hat{V}, \hat{n}, \hat{p})$ solve (6.3.8).

The inverse doping profile problem for this V-C map can again be written in the abstract form $F(C) = Y$. The corresponding framework is now described by $1_t$), $2_t$), $4_t$) and

$3'_t$) Output: $Y = \{\Sigma'_{t,C}(0)U_j\}_{j=1}^N \in L^2([0,T]; H^{1/2}(\Gamma_1))^N =: \mathcal{Y}$;

As in the inverse problem of the previous paragraph, the parameter-to-output map $F$ is well defined and continuous. If the domain $\Omega$ is one dimensional, the results in [BEM02] can be adapted in a straightforward way and we can conclude that $F$ is weakly sequentially closed.

## 6.5 Background on inverse problems and level set methods

In what follows we present some of the tools from the theory of inverse problems that are needed as background to understand the approach we use here. These tools include some classical material, for example, singular value decomposition, regularization and Landweber's method, which are treated in Sections 6.5.1, 6.5.2 and 6.5.3, as well as more recent developments such as the use of level set methods for handling inverse problems. The latter is treated in Section 6.5.4.

### 6.5.1 Singular value decomposition

We briefly review *singular value decomposition (SVD)*. This result has a number of important applications in numerical analysis, inverse problems, and numerical linear algebra.

Let $\mathcal{L}(\mathcal{H}, \mathcal{K})$ denote the space of bounded linear operators from $\mathcal{H}$ to $\mathcal{K}$, where $\mathcal{H}$ and $\mathcal{K}$ are Hilbert spaces. We endow $\mathcal{L}(\mathcal{H}, \mathcal{K})$ with the *uniform operator topology* defined by the norm

$$||T||_{\mathcal{H},\mathcal{K}} \stackrel{\text{def}}{=} \sup_{f \neq 0} \frac{||Tf||_{\mathcal{K}}}{||f||_{\mathcal{H}}} .$$

Whenever no confusion may arise we shall drop the $\mathcal{H}, \mathcal{K}$ subscript in $||T||_{\mathcal{H},\mathcal{K}}$.

We recall the concept of compact operator.

**Definition 1.** Let $\mathcal{H}$ and $\mathcal{K}$ be Banach spaces and $T : \mathcal{H} \to \mathcal{K}$ a linear operator. $T$ is called *compact* (or *completely continuous*) if it maps the unit ball $B_{\mathcal{H}}$ on a pre-compact set, i.e., $T(B_{\mathcal{H}})$ has compact closure. The set of compact operators from $\mathcal{H}$ to $\mathcal{K}$ is denoted by $\kappa(\mathcal{H}, \mathcal{K})$.

It can be easily shown that this definition implies that $T$ is a bounded operator. Furthermore, the set of compact operators is closed under limits in the uniform operator topology of $\mathcal{L}(\mathcal{H}, \mathcal{K})$. Typical examples of compact operators are operators of finite-dimensional range. Another important class of compact operators is given by integral operators. We use the convention that our *complex* Hilbert space inner products are *linear* w.r.t. the first entry and *anti-linear* w.r.t. to the second one. The following result is instrumental in understanding the structure of compact operators.

**Theorem 1.** Let $A \in \kappa(\mathcal{H}, \mathcal{K})$, then we can write

$$A = \sum_{n=1}^{r} \sigma_n \left( \cdot \mid \psi_n \right) \phi_n, \tag{6.5.1}$$

where $r \in \mathbb{N} \cup \{\infty\}$ and $\sigma_1 \geq \sigma_2 \geq \cdots \geq \sigma_r > 0$ ; the sets $\{\phi_n\}_{n=1}^{r}$ and $\{\psi_n\}_{n=1}^{r}$ are orthonormal sets (not necessarily complete) in $\mathcal{K}$ and $\mathcal{H}$, respectively.

As a direct consequence of SVD we get that the equation $Af = g$ has a solution $f$ if, and only if,

1. the vector $g \in \ker A^{*\perp}$, and
2. the sum

$$\sum_{n=1}^{r} \frac{1}{\sigma_n^2} |(g \mid \phi_n)|^2 < \infty .$$

In this case the solution of $x$ will be given by

$$f = \sum_{n=1}^{r} \frac{1}{\sigma_n} (g \mid \phi_n) \psi_n .$$

In the finite-dimensional case, the transformation $A$ will be invertible if, and only if, the dimensions of $\mathcal{H}$ and $\mathcal{K}$ equal $r$. In other words, the eigenvalues of $A^*A$ and $AA^*$ are all nonzero. The presence of singular values close to zero indicates that the solution of the problem $Af = g$ will be doomed to numerical instability. The condition number of a matrix $A \in \mathbb{C}^{n \times n}$ is the ratio $\sigma_1/\sigma_r$, if $r = n$, and $\infty$ otherwise. See [GV89] for more information on numerical aspects related to conditioning.

In the infinite-dimensional case, if $r = \infty$ then the sequence $\{\sigma_n\}$ must necessarily converge to zero. This shows the inherent instability of solving equations of the form $Af = g$ when $A$ is a compact operator in an infinite-dimensional Hilbert space.

### 6.5.2 Regularization

A mathematical problem defined in the form of an equation $F(u) = g$, where $F$ is an operator between two Hilbert spaces $\mathcal{H}$ and $\mathcal{K}$, is said to be well posed (in the sense of Hadamard) if for every $g \in \mathcal{K}$ the solution $u \in \mathcal{H}$ exists, is unique and depends continuously on $g$.

Let $A : \mathcal{H} \to \mathcal{K}$ be a compact linear operator. We now analyze the question of solving a linear equation of the form

$$Af = g .$$

There are three things that can go wrong:

1. The equation may not be solvable (i.e., $g \notin \mathrm{Ran}(A)$).
2. The solution may not be unique (i.e., $A$ is not $1-1$).
3. The solution may not depend continuously on the data (i.e., $A^{-1}$ is not continuous).

The concept of pseudo-inverse (or generalized inverse) $A^\dagger$ is used to handle cases 1 and 2 above. We define

$$A^\dagger g = \sum_{n=1}^{r} \frac{1}{\sigma_n} (g|\phi_n) \psi_n , \ g \in D(A^\dagger) ,$$

where

$$D(A^\dagger) \overset{\text{def}}{=} \left\{ g \in \mathcal{K} \,\middle|\, \sum_{n=1}^{r} \frac{1}{\sigma_n^2} |(g \mid \phi_n)|^2 < \infty \right\} .$$

Note that $A^\dagger g$ is the unique solution of $Af = g$ in $(\ker A)^\perp$. To tackle the problem of discontinuous $A^{-1}$, one needs to introduce the notion of *regularization*.

Let us consider a family of continuous operators $T_\alpha : \mathcal{K} \to \mathcal{H}$ such that

$$\lim_{\alpha \downarrow 0} T_\alpha g = A^\dagger g , \; g \in D(A^\dagger) . \tag{6.5.2}$$

Note that if $A^\dagger$ is not bounded, then $||T_\alpha|| \to \infty$ when $\alpha \downarrow 0$. Let us solve $Af = g$ approximately in the sense that $g^\epsilon \in \mathcal{K}$ is an approximation to $g$ such that $||g - g^\epsilon|| \leq \epsilon$. Consider $\alpha(\epsilon)$ such that $\alpha(\epsilon) \downarrow 0$ and $||T_{\alpha(\epsilon)}||\epsilon \to 0$. Thus,

$$||T_{\alpha(\epsilon)} g^\epsilon - A^\dagger g|| \leq ||T_{\alpha(\epsilon)} (g^\epsilon - g)|| + ||T_{\alpha(\epsilon)} g - A^\dagger g||$$
$$\leq ||T_{\alpha(\epsilon)}||\epsilon + ||T_{\alpha(\epsilon)} g - A^\dagger g|| \to 0.$$

So, $T_{\alpha(\epsilon)} g^\epsilon$ is close to $A^\dagger g$ provided $g^\epsilon$ is close to $g$.

The following three techniques are used for the regularization of ill-posed problems.

- Truncated SVD:

$$T_\alpha = \sum_{\sigma_k \geq \alpha} \frac{1}{\sigma_k} (\cdot | \phi_k) \psi_k .$$

- Tikhonov–Phillips regularization:

$$T_\alpha = (A^* A + \alpha I)^{-1} A^* ,$$

which is associated to minimizing the quadratic form

$$||Af - g^\epsilon||^2 + \alpha ||f||^2 .$$

More generally, it is associated to minimizing expressions of the form

$$||Af - g^\epsilon||^2 + \alpha q(f - f_0) ,$$

where $q$ is some penalty function designed to keep $f$ close to a prior $f_0$.
- Early stop of an iterative method: Assume that

$$f^{k+1} = B_k f^k + C_k g^\epsilon$$

is an iterative method to solve $Af = g$ with $B_k$ and $C_k$ bounded and $\lim_{k \to \infty} f^k = A^\dagger g$. For $\alpha > 0$, let the value of $k(\alpha)$ be such that $k(\alpha) \to \infty$ when $\alpha \to 0$. Then, under suitable conditions on $B_k$ and $C_k$, we have that $T_\alpha f \overset{\text{def}}{=} f^{k(\alpha)}$ is a regularization of the problem.

In the case of infinite-dimensional problems for compact operator equations of the form $Af = g$, it is natural to analyze how ill posed the problem is by looking at the rate of convergence of the singular values of $A$ to zero. Problems for which the rate of decay of $\sigma_n$ is not faster than that of a polynomial are considered tractable. Problems for which the decay is exponential or faster are considered to be severely ill posed. More details on regularization theory can be found, for example, in [EHN96, EKN89, ES00, TA77].

### 6.5.3 Landweber–Kaczmarz

We first consider Landweber's iteration to solve a nonlinear problem of the form $F(\gamma) = g$ where $F : D(F) \subset \mathcal{H} \to \mathcal{K}$ is a Fréchet differentiable map between the Hilbert spaces $\mathcal{H}$ and $\mathcal{K}$. The iteration is defined by

$$\gamma_{k+1} = \gamma_k - F'(\gamma_k)^*(F(\gamma_k) - g) \, , \qquad (6.5.3)$$

where $\gamma_0 \in D(F)$.

Under suitable conditions, the method converges [BL05]. Furthermore, this iteration is known to generate a regularization method for the inverse problem if we apply the early stopping technique mentioned above. Landweber's iteration has been the subject of intense study both for theoretical as well as for practical applications. See, for example, [EHN96, ES00, HNS95].

We now describe the Landweber–Kaczmarz method for the doping profile identification problem of Section 6.6.1. The notation and definitions of the different operators follow that of Section 6.6.

- Parameter space: $\mathcal{H} \overset{\text{def}}{=} L^2(\Omega)$;
- Input (fixed): $U_j \in H^{3/2}(\partial\Omega_D)$, with $U_j|_{\Gamma_1} = 0$, $1 \leq j \leq N$;
- Output (data): $Y = \{\Lambda_\gamma(U_j)\}_{j=1}^N \in \mathcal{K} \overset{\text{def}}{=} [L^2(\Gamma_1)]^N$;
- Parameter to output map: $F : D(F) \subset \mathcal{H} \to \mathcal{K}$
$$\gamma(x) \mapsto \{\Lambda_\gamma(U_j)\}_{j=1}^N,$$

where the domain of definition of the operator $F$ is

$$D(F) \overset{\text{def}}{=} \{\gamma \in L^2(\Omega); \gamma_+ \geq \gamma(x) \geq \gamma_- > 0, \text{ a.e. in } \Omega\} \, .$$

Here $\gamma_-$ and $\gamma_+$ are appropriate positive constants. We shall denote the noisy data by $Y^\delta$ and assume that the data error is bounded by $\|Y - Y^\delta\| \leq \delta$. Thus, we are able to represent the inverse doping problem in the form of finding

$$F(\gamma) = Y^\delta. \qquad (6.5.4)$$

It can be shown that [BEMP01] if we let the voltages $\{U_j\}_{j=1}^N$ be chosen in the neighborhood of $U \equiv 0$, then the parameter-to-output map $F$ defined above is well defined and Fréchet differentiable on $D(F)$. Thus, the *Landweber iteration* [DES98, EHN96, ES00, HNS95] becomes

$$\gamma_{k+1}^{\delta} = \gamma_k^{\delta} - F'(\gamma_k^{\delta})^* \left( F(\gamma_k^{\delta}) - Y^{\delta} \right).$$

One possible variation of the Landweber iteration consists in coupling it with the Kaczmarz strategy of considering an inner iteration where, at each step, one takes into account one component of the measurement vector only. A detailed analysis of the Landweber–Kaczmarz method can be found in [KS02].

In the specific example mentioned above for equation (6.5.4), we consider the components of the parameter-to-output map: $F = \{\mathcal{F}_j\}_{j=1}^N$, where

$$\mathcal{F}_j : L^2(\Omega) \supset D(F) \ni \gamma \mapsto \Lambda_\gamma(U_j) \in L^2(\Gamma_1).$$

Now, setting $Y_j^{\delta} \stackrel{\text{def}}{=} \mathcal{F}_j(\gamma^{\delta})$, $1 \leq j \leq N$, the Landweber–Kaczmarz iteration can be written in the form

$$\gamma_{k+1}^{\delta} = \gamma_k^{\delta} - \mathcal{F}_k'(\gamma_k^{\delta})^* \left( \mathcal{F}_k(\gamma_k^{\delta}) - Y_k^{\delta} \right), \tag{6.5.5}$$

for $k = 1, 2, \ldots$, where we adopted the notation

$$\mathcal{F}_k \stackrel{\text{def}}{=} \mathcal{F}_j, \quad Y_k^{\delta} \stackrel{\text{def}}{=} Y_j^{\delta}, \quad \text{whenever} \quad k = iN + j, \quad \text{and} \quad \begin{cases} i = 0, 1, \ldots \\ j = 1, \ldots, N \end{cases}.$$

Notice that each step of the Landweber–Kaczmarz method consists of one Landweber iterative step with respect to the $j$th component of the residual in (6.5.4). These Landweber steps are performed in a cyclic way, using the components of the residual $\mathcal{F}_j(\gamma) - Y_j^{\delta}$, $1 \leq j \leq N$, one at a time.

In the next section we shall describe how level set methods can be applied to tackle inverse problems. A comparison between the Landweber–Kaczmarz method and a level set approach to the doping profile identification problem was developed in [LMZ05]. The preliminary conclusion obtained therein is that in general the level set method performed better than its Landweber–Kaczmarz counterpart.

## 6.5.4 Level set methods in inverse problems

The level set methodology has established itself as a promising alternative for the solution of several inverse problems that involve boundaries or obstacles. The original formulation of level sets, as applied to curve and surface motion, is due to Osher and Sethian [OS88]. The use of such methods in obstacle inverse problems is due to Santosa [San95]. Burger [Bur01] presented a rigorous mathematical treatment for level set methods in inverse problems. See also [LS03, FSL05] for a constrained optimization treatment of the method. In what follows we focus on the application of level set methods in inverse problems.

Let $\Omega \subset \mathbb{R}^n$ be a given set and $F : \mathcal{H} \to \mathcal{K}$ a Fréchet differentiable operator. The problem consists of finding $D \subset \text{int}(\Omega)$ in the equation

$$F(u) = g, \tag{6.5.6}$$

where

$$u = \begin{cases} u_{\text{int}}, x \in D \\ u_{\text{ext}}, x \in \Omega \setminus D \end{cases}.$$

We now consider the boundary of the region $D$ in $\Omega$ as described by $\partial D = \{x \in \Omega | \phi(x) = 0\}$. The function $\phi$ shall be referred to as the *level set function*. The level set function evolves according to a parameter $t$ in such a way that

$$\partial D_t \overset{\text{def}}{=} \{x \in \Omega \mid \phi_t(x) = 0\} \longrightarrow D \text{ as } t \to \infty.$$

There are several possible dynamics for the evolution of $\phi_t$ with $t$. See [BL05] for discussion and motivation, as well as [Bur01, LS03]. One possibility is to use the dynamics introduced in [LS03, FSL05]. According to this approach, one represents a zero level set by an $H^1$-function $\phi : \Omega \to \mathbb{R}$, in such a way that $\phi(x) > 0$ if $\gamma(x) = u_{\text{ext}}$ and $\phi(x) < 0$ if $\gamma(x) = u_{\text{int}}$. Starting from some initial guess $\phi_0 \in H^1(\Omega)$, one solves the Hamilton–Jacobi equation

$$\frac{\partial \phi}{\partial t} + V \nabla \phi = 0, \tag{6.5.7}$$

where $V = v \frac{-\nabla \phi}{|\nabla \phi|^2}$ and the *velocity* $v$ solves

$$\begin{cases} \alpha(\Delta - I)v = \frac{\delta(\phi(t))}{|\nabla \phi(t)|} \left[ F'(\chi(t))^*(F(\chi(t)) - Y^\delta) - \alpha \nabla \cdot \left( \frac{\nabla P(\phi)}{|\nabla P(\phi)|} \right) \right], \text{ in } \Omega \\ \frac{\partial v}{\partial \nu} = 0, \text{ on } \partial \Omega \end{cases}. \tag{6.5.8}$$

Here, $\alpha > 0$ is a regularization parameter and $\chi = \chi(x, t)$ is the projection of the level set function $\phi(x, t)$ defined by

$$\chi(x, t) = P(\phi(x, t)) \overset{\text{def}}{=} \begin{cases} u_{\text{ext}}, \text{ if } \phi(x, t) > 0 \\ u_{\text{int}}, \text{ if } \phi(x, t) < 0 \end{cases}.$$

The above dynamics leads, for the problem under consideration, to the following algorithm.

**Algorithm:**

1. Evaluate the residual $r_k \overset{\text{def}}{=} F(P(\phi_k)) - Y^\delta$;
2. Evaluate $v_k \overset{\text{def}}{=} F'(P(\phi_k))^*(r_k)$;
3. Evaluate $w_k \in H^1(\Omega)$, satisfying

$$\alpha(I - \Delta)w_k = -P'(\phi_k)v_k + \alpha P'(\phi_k)\nabla \cdot \left( \frac{\nabla P(\phi_k)}{|\nabla P(\phi_k)|} \right), \text{ in } \Omega;$$

$$\left. \frac{\partial v_k}{\partial \nu} \right|_{\partial \Omega} = 0.$$

4. Update the level set function $\phi_{k+1} = \phi_k + \frac{1}{\alpha} v_k$.

In practical implementations, instead of $P$, we use a smooth version $P_\varepsilon$.

## 6.6 Some numerical experiments

In this section we apply numerical methods to solve inverse doping profile problems related to the V-C map. In the first two subsections, we address the linearized unipolar case (close to equilibrium). In Subsection 6.6.1 pointwise measurements of the current density are considered, and in Subsection 6.6.2 current flow measurements through a contact are used as data. In the last subsection we present some numerical results for the linearized bipolar case (close to equilibrium).

### 6.6.1 Stationary linearized unipolar model: pointwise measurements of the current density

In this specific model, due to the assumptions $p = 0$ and $Q = 0$, the Poisson equation and the continuity equation for the electron density decouple. Therefore, we have to identify $C = C(x)$ from measurements of the current density $\mu_n e^{V_{\mathrm{bi}}} \hat{u}_\nu|_{\Gamma_1}$, where $(V^0, \hat{u})$ solve, for each applied voltage $U$, the system

$$\begin{cases} \lambda^2 \,\Delta V^0 = e^{V^0} - C(x) \text{ in } \Omega \\ \quad\quad V^0 = V_{\mathrm{bi}}(x) \quad\quad \text{on } \partial\Omega_D \\ \quad\nabla V^0 \cdot \nu = 0 \quad\quad \text{on } \partial\Omega_N \end{cases} \quad \begin{cases} \mathrm{div}\,(\mu_n e^{V^0} \nabla \hat{u}) = 0 \quad\quad \text{in } \Omega \\ \quad\quad\quad\quad \hat{u} = U(x) \text{ on } \partial\Omega_D \\ \quad\quad\nabla \hat{u} \cdot \nu = 0 \quad\quad \text{on } \partial\Omega_N \,. \end{cases}$$

Notice that we split the problem in two parts: First we define the function $\gamma(x) \stackrel{\mathrm{def}}{=} \mu_n e^{V^0(x)}$, $x \in \Omega$, and solve the parameter identification problem

$$\begin{cases} \mathrm{div}\,(\gamma \nabla \hat{u}) = 0 \quad\quad \text{in } \Omega \\ \quad\quad\quad \hat{u} = U(x) \text{ on } \Omega_D \\ \quad\nabla \hat{u} \cdot \nu = 0 \quad\quad \text{on } \Omega_N \,, \end{cases} \tag{6.6.1}$$

for $\gamma$ from measurements of $\gamma \hat{u}_\nu|_{\Gamma_1}$. The second step consists in the determination of $C$ in

$$C(x) = \mu_n^{-1} \gamma(x) - \lambda^2 \,\Delta(\ln \mu_n^{-1} \gamma(x)), \; x \in \Omega \,.$$

The evaluation of $C$ from $\gamma$ is a mildly ill-posed problem and can be explicitly performed in a routine way. We shall focus on the problem of identifying the function parameter $\gamma$ in (6.6.1). Therefore, the inverse doping profile problem in the linearized unipolar model for pointwise measurements of the current density reduces to the identification of the parameter $\gamma$ in (6.6.1) from measurements of the Dirichlet-to-Neumann (DtN) map

$$\begin{aligned} \Lambda_\gamma : H^{3/2}(\partial\Omega_D) &\to H^{1/2}(\Gamma_1) \,. \\ U &\mapsto \gamma \, \hat{u}_\nu|_{\Gamma_1} \,. \end{aligned}$$

If we take into account the restrictions imposed by the practical experiments described in Subsection 6.4.1, it follows that:

(i) The voltage profiles $U \in H^{3/2}(\partial \Omega_D)$ must satisfy $U|_{\Gamma_1} = 0$;

(ii) The identification of $\gamma$ has to be performed from a finite number of measurements, i.e., from the data

$$\left\{ (U_j, \Lambda_\gamma(U_j)) \right\}_{j=1}^{N} \in \left[ H^{3/2}(\partial \Omega_D) \times H^{1/2}(\Gamma_1) \right]^{N}. \qquad (6.6.2)$$

For the concrete numerical tests presented in this chapter, we apply an iterative method of level set type to solve problem (6.4.1) See [LMZ05] for details. The domain $\Omega \subset \mathbb{R}^2$ is the unit square, and the boundary parts are defined as follows:

$$\Gamma_1 \overset{\text{def}}{=} \{(x,1);\ x \in (0,1)\}, \quad \Gamma_0 \overset{\text{def}}{=} \{(x,0);\ x \in (0,1)\},$$

$$\partial \Omega_N \overset{\text{def}}{=} \{(0,y);\ y \in (0,1)\} \cup \{(1,y);\ y \in (0,1)\}.$$

The fixed inputs $U_j$ are chosen to be piecewise constant functions supported in $\Gamma_0$:

$$U_j(x) \overset{\text{def}}{=} \begin{cases} 1, & |x - x_j| \le h \\ 0, & \text{else} \end{cases},$$

where the points $x_j$ are equally spaced in the interval $(0,1)$. The doping profiles to be reconstructed are shown in Figure 6.1. In these pictures, as well as in the forthcoming ones, $\Gamma_1$ is the lower left edge and $\Gamma_0$ is the top right edge (the origin corresponds to the upper right corner).

For the experiments concerning pointwise measurements of the current density, we assume that only one measurement is available, i.e., $N = 1$ in (6.6.2).

The first numerical experiment is shown in Figure 6.2. Here exact data is used for the reconstruction of the p-n junction in Figure 6.1 (b). The pictures correspond to plots of the iteration error after 5, 10 and 100 steps of the level set method.

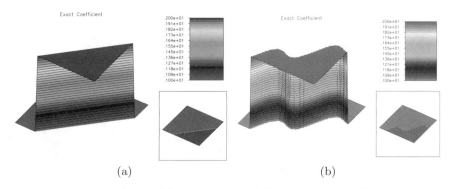

(a)                              (b)

**Fig. 6.1.** Pictures (a) and (b) show the two different doping profiles to be reconstructed in the numerical experiments.

The second experiment (see Figure 6.3) concerns the reconstruction of the p-n junction in Figure 6.1 (a). In this experiment the data is contaminated with 10% random noise. The pictures correspond to plots of the iteration error after 10, 100 and 400 steps of the level set method.

### 6.6.2 Stationary linearized unipolar model: current flow measurements through a contact

In what follows we consider the same unipolar model as in Subsection 6.6.1. Again we shall focus on the identification problem related to (6.6.1). However, the coefficient $\gamma$ has to be identified from measurements of the current flow through the contact $\Gamma_1$, i.e., from

$$\int_{\Gamma_1} \gamma \hat{u}_\nu \, ds \,,$$

where $\hat{u}$ solves (6.6.1) for prescribed inputs $U \in H^{3/2}(\partial\Omega_D)$.

An immediate remark is that the amount of available data is much larger for pointwise measurements of the current density than for flow measurements through a contact. Notice that the inverse doping profile problem in the linearized unipolar model for measurements of the current flow through the contact $\Gamma_1$ reduces to the identification of the parameter $\gamma$ in (6.6.1) from measurements of the (averaged) DtN map

$$\widetilde{\Lambda}_\gamma : H^{3/2}(\partial\Omega_D) \to \mathbb{R}$$
$$U \mapsto \int_{\Gamma_1} \gamma \, \hat{u}_\nu \, ds.$$

As in the previous subsection, we take into account the restrictions imposed by practical experiments, which lead to the following assumptions:

(i) The voltage profile $U \in H^{3/2}(\partial\Omega_D)$ must satisfy $U|_{\Gamma_1} = 0$;

(ii) The identification of $\gamma$ has to be performed from a finite number of measurements, i.e., from the data

$$\left\{ (U_j, \widetilde{\Lambda}_\gamma(U_j)) \right\}_{j=1}^N \in \left[ H^{3/2}(\partial\Omega_D) \times \mathbb{R} \right]^N. \tag{6.6.3}$$

The subsequent numerical tests were performed using the same iterative method of level set type as in the previous subsection. The domain $\Omega \subset \mathbb{R}^2$ as well as the boundary parts $\Gamma_0$, $\Gamma_1$ and $\partial\Omega_N$ are defined as before.

For the experiments concerning current flow measurements through the contact $\Gamma_1$, we assume that several measurements are available, i.e., $N \gg 1$ in (6.6.3).

The first numerical experiment is shown in Figure 6.4. Here exact data is used for the reconstruction of the p-n junction in Figure 6.1 (a). The picture on the left-hand side shows the error for the initial guess of the iterative

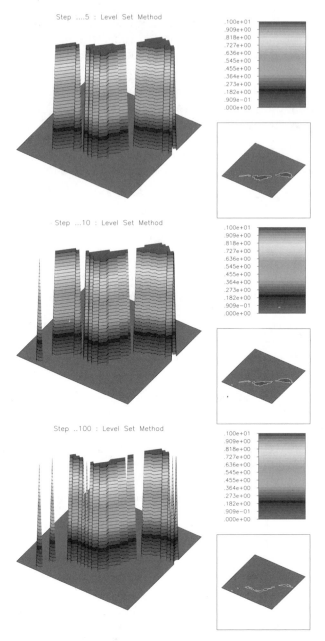

**Fig. 6.2.** First experiment for the unipolar model with pointwise measurements of the current density: Reconstruction of the p-n junction in Figure 6.1 (b). Evolution of the iteration error for exact data and one measurement of the DtN map $\Lambda_\gamma$ (i.e., $N = 1$ in (6.6.2)).

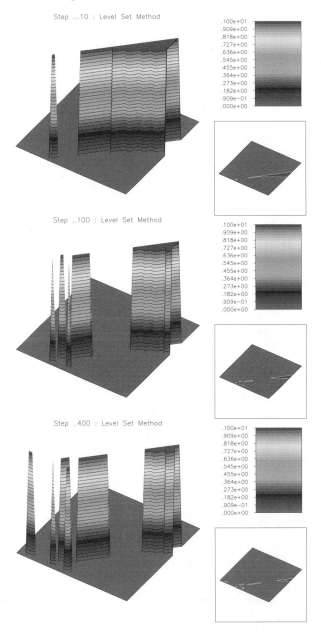

**Fig. 6.3.** Second experiment for the unipolar model with pointwise measurements of the current density: Reconstruction of the p-n junction in Figure 6.1 (a). Only one measurement of the DtN map $\Lambda_\gamma$ is available (i.e., $N = 1$ in (6.6.2)). Evolution of the iteration error for the level set method and data contaminated with 10% random noise.

method.[7] The other two pictures correspond to plots of the iteration error after 50 and 250 steps of the level set method respectively.

The second experiment (see Figure 6.5) concerns the reconstruction of the p-n junction in Figure 6.1 (b). The available data is contaminated with 1% random noise. The pictures correspond to plots of the iteration error after 100, 2000 and 3000 steps of the level set method.

### 6.6.3 Stationary linearized bipolar model: pointwise measurements of the current density

We now consider the bipolar model introduced in Subsection 6.3.2. As in the unipolar model, it follows from the assumption $Q = 0$ that the Poisson equation (6.3.1a) and the continuity equations (6.3.4a), (6.3.4b) decouple. The inverse doping profile problem corresponds to the identification of $C = C(x)$ from pointwise measurements of the total current density $J = J_n + J_p$, namely

$$(\mu_n e^{V_{bi}} \hat{u}_\nu - \mu_p e^{-V_{bi}} \hat{v}_\nu)|_{\Gamma_1} .$$

Compare with the Gateaux derivative of the V-C map $\Sigma_C$ at the point $U = 0$ in (6.3.3). Here $(V^0, \hat{u}, \hat{v})$ solve, for each applied voltage $U$, the system (6.3.1), (6.3.4) (with $h$ substituted by $U$).

As in the unipolar case, we can split the inverse problem in two parts: First we define the function $\gamma(x) \stackrel{\text{def}}{=} e^{V^0(x)}$, $x \in \Omega$, and solve the parameter identification problem

$$\begin{cases} \operatorname{div}(\mu_n \gamma \nabla \hat{u}) = 0 & \text{in } \Omega \\ \hat{u} = -U(x) & \text{on } \partial\Omega_D \\ \nabla \hat{u} \cdot \nu = 0 & \text{on } \partial\Omega_N \end{cases} \qquad \begin{cases} \operatorname{div}(\mu_p \gamma^{-1} \nabla \hat{v}) = 0 & \text{in } \Omega \\ \hat{v} = U(x) & \text{on } \partial\Omega_D \\ \nabla \hat{v} \cdot \nu = 0 & \text{on } \partial\Omega_N \end{cases}$$

$$(6.6.4)$$

for $\gamma$, from measurements of $(\mu_n \gamma \hat{u}_\nu - \mu_p \gamma^{-1} \hat{v}_\nu)|_{\Gamma_1}$. The second step consists in the determination of $C$ in

$$C(x) = \gamma(x) - \gamma^{-1}(x) - \lambda^2 \Delta(\ln \gamma(x)), \quad x \in \Omega .$$

Analogously to the unipolar case, the evaluation of $C$ from $\gamma$ can be performed in a stable way. Therefore, we shall focus on the problem of identifying the function parameter $\gamma$ in (6.6.4). Notice that the inverse doping profile problem in the linearized bipolar model for pointwise measurements of the current density reduces to the identification of the parameter $\gamma$ in (6.6.4) from measurements of the DtN map

$$\Phi_\gamma : H^{3/2}(\partial\Omega_D) \to H^{1/2}(\Gamma_1) .$$
$$U \mapsto (\mu_n \gamma \hat{u}_\nu - \mu_p \gamma^{-1} \hat{v}_\nu)|_{\Gamma_1} .$$

---

[7]In all numerical experiments presented in this chapter we used the same initial guess for the iterative methods. We observed that the choice of the initial guess does not significantly influence the overall performance of the iterative method.

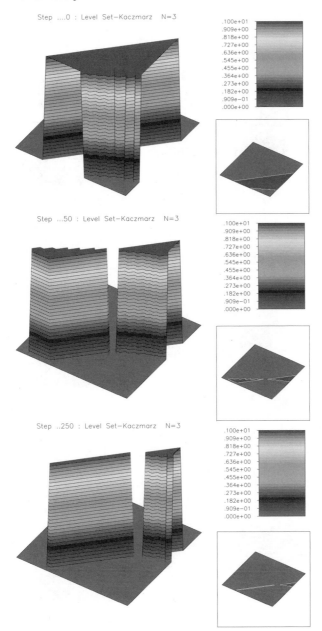

**Fig. 6.4.** First experiment for the unipolar model with current flow measurements through the contact $\Gamma_1$: Reconstruction of the p-n junction in Figure 6.1 (a). Three measurements of the DtN map $\widetilde{\Lambda}_\gamma$ are used in the reconstruction (i.e., $N = 3$ in (6.6.3)). Evolution of the iteration error for exact data.

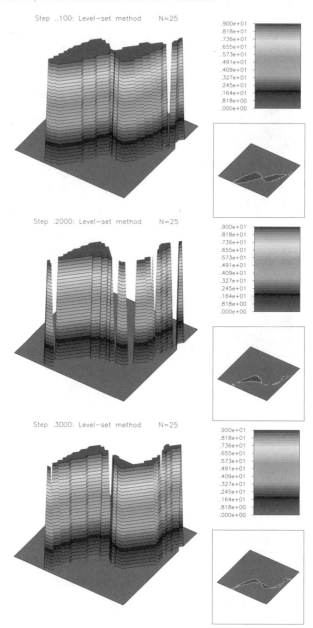

**Fig. 6.5.** Second experiment for the unipolar model with current flow measurements through the contact $\Gamma_1$: Reconstruction of the p-n junction in Figure 6.1 (b). The data consists of 25 measurements of the DtN map $\widetilde{\Lambda}_\gamma$ (i.e., $N = 25$ in (6.6.3)). Plots of the iteration error after 100, 2000 and 3000 steps. Data with 1% random noise.

As before, we take into account the restrictions imposed by the practical experiments:

(i) The voltage profiles $U \in H^{3/2}(\partial\Omega_D)$ must satisfy $U|_{\Gamma_1} = 0$;

(ii) The identification of $\gamma$ has to be performed from a finite number of measurements, i.e., from the data

$$\{(U_j, \Phi_\gamma(U_j))\}_{j=1}^N \in \left[H^{3/2}(\partial\Omega_D) \times H^{1/2}(\Gamma_1)\right]^N. \qquad (6.6.5)$$

In Figure 6.6 we present a numerical experiment for the bipolar model with pointwise measurements of the current density. Here exact data is used for the reconstruction of the p-n junction in Figure 6.1 (b). The pictures show plots of the iteration error after 1, 10 and 100 steps of the level set method respectively.

### 6.6.4 Remarks and conclusions

The best numerical results are obtained for the experiments concerning the linearized unipolar case with pointwise measurements of the current density. In this model, a single measurement of the DtN map $\Lambda_\gamma$, i.e., $N = 1$ in (6.6.2), contains enough information about the structure of the doping profile and suffices to obtain a very precise reconstruction of the p-n junction. This is the case even for highly oscillating p-n junctions as shown in Figure 6.2 and also in the presence of noise (see Figure 6.3). We observed that the iteration is extremely robust with respect to the choice of the initial guess and also with respect to high levels of noise.

Concerning the linearized unipolar model with current flow measurements through the contact $\Gamma_1$, our experiments show that the (averaged) DtN map $\widetilde{\Lambda}_\gamma$ furnishes much less information about the solution structure than the map $\Lambda_\gamma$. Depending on the complexity of the p-n junction, more measurements of $\widetilde{\Lambda}_\gamma$ may be needed in order to obtain an acceptable reconstruction. The experiments show that a single measurement ($N = 1$ in (6.6.3)) is not enough to identify the doping profile in Figure 6.1 (a). Moreover, although three measurements have shown to be enough to reconstruct this p-n junction (see Figure 6.4), this is not the case for the p-n junction in Figure 6.1 (b). For this second and more complex junction, we first obtained more accurate reconstructions with $N = 19$ in (6.6.3). The quality of the reconstruction obtained for $N = 25$ is already very high (see Figure 6.5) and does not qualitatively improve for larger values of $N$ (we experimented up to $N = 49$).

Note that the number of iterative steps required by the level set algorithm to reach the stopping criteria for the inverse problem related to the map $\widetilde{\Lambda}_\gamma$ is greater than that for the operator $\Lambda_\gamma$. This is again explained by the fact that the range of $\widetilde{\Lambda}_\gamma$ lies in $\mathbb{R}$, while the range of $\Lambda_\gamma$ lies in $H^{1/2}(\Gamma_1)$.

Concerning the experiments for the linearized bipolar model with pointwise measurements of the current density, the quality of the results is comparable to those in Subsection 6.6.1 and, as in that subsection, a single measurement of the operator $\Phi_\gamma$ ($N = 1$ in (6.6.5)) suffices to precisely reconstruct the

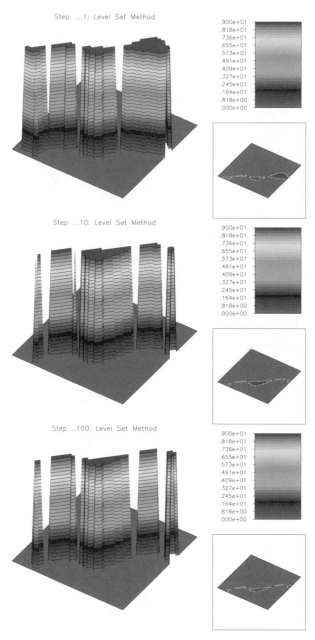

**Fig. 6.6.** Experiment for the bipolar model with pointwise measurements of the current density: Reconstruction of the p-n junction in Figure 6.1 (b). Evolution of the iteration error for exact data and one measurement of the DtN map $\Phi_\gamma$ (i.e., $N = 1$ in (6.6.5)).

p-n junction. We observed, however, that convergence of the iterative method is more sensitive to the choice of the initial condition than in Subsection 6.6.1.

## Appendix

Properties of silicon at room temperature:

**Table A.1.** Typical values of the main constants in the model.

| Parameter | Typical value |
|:---:|:---:|
| $\epsilon_s$ | 11.9 $\epsilon_0$ |
| $\mu_n$ | $\approx 1500$ cm$^2$ V$^{-1}$ s$^{-1}$ |
| $\mu_p$ | $\approx 450$ cm$^2$ V$^{-1}$ s$^{-1}$ |
| $C_n$ | $2.8 \times 10^{-31}$ cm$^6$/s |
| $C_p$ | $9.9 \times 10^{-32}$ cm$^6$/s |
| $\tau_n$ | $10^{-6}$ s |
| $\tau_p$ | $10^{-5}$ s |

Relevant physical constants:

Permittivity of vacuum: $\epsilon_0 = 8.85 \times 10^{-14}$ As V$^{-1}$ cm$^{-1}$;
Elementary charge: $q = 1.6 \times 10^{-19}$ As.

## Acknowledgments

A.L. acknowledges support from the Brazilian National Research Council CNPq, under project grants 305823/03-5 and 478099/04-5. P.A.M. acknowledges support from the Austrian National Science Foundation FWF through his Wittgenstein Award 2000. J.P.Z. acknowledges financial support from CNPq through grants 302161/2003-1 and 474085/2003-1.

## References

[APL05]  Astala, K., Päivärinta, L., Lassas, M.: Calderón's inverse problem for anisotropic conductivity in the plane. Comm. Partial Differential Equations **30**, 207–224 (2005).

[BL05]     Baumeister, J., Leitão, A.: Topics in inverse problems. Mathematical Publications of IMPA. ISBN: 85-244-0224-5 25th Brazilian Mathematics Colloquium. Rio de Janeiro, Brazil (2005).

[Bor02]     Borcea, L.: Electrical impedance tomography. Inverse Problems **18**, R99–R136 (2002).

[BU02]     Bukhgeim, A., Uhlmann, G.: Recovering a potential from partial Cauchy data. Comm. Partial Differential Equations **27**, 653–668 (2002).

[BELM04]     Burger, M., Engl, H.W., Leitão, A., Markowich, P.A.: On inverse problems for semiconductor equations. Milan Journal of Mathematics **72**, 273–314 (2004).

[BEMP01]     Burger, M., Engl, H.W., Markowich, P.A., Pietra, P.: Identification of doping profiles in semiconductor devices. Inverse Problems **17**, 1765–1795 (2001).

[BEM02]     Burger, M., Engl, H.W., Markowich, P.A.: Inverse doping problems for semiconductor devices. In: Chan, T.F., Huang, Y., Tang, T., Xu, J.A, Ying, L.A. (eds) Recent Progress in Computational and Applied PDEs. Kluwer/Plenum, New York, 39–53 (2002).

[BF91]     Busenberg, S., Fang, W.: Identification of semiconductor contact resistivity. Quart. Appl. Math. **49**, 639–649 (1991).

[BFI93]     Busenberg, S., Fang, W., Ito, K.: Modeling and analysis of laser-beam-inducted current images in semiconductors. SIAM J. Appl. Math. **53**, 187–204 (1993).

[Bur01]     Burger, M.: A level set method for inverse problems. Inverse Problems **17**, 1327–1355 (2001).

[DES98]     Deuflhard, P., Engl, H.W., Scherzer, O.: A convergence analysis of iterative methods for the solution of nonlinear ill-posed problems under affinely invariant conditions. Inverse Problems **14**, 1081–1106 (1998).

[EHN96]     Engl, H.W., Hanke, M., Neubauer, A.: Regularization of Inverse Problems. Kluwer Academic Publishers, Dordrecht (1996).

[EKN89]     Engl, H.W., Kunisch, K., Neubauer, A.: Convergence rates for Tikhonov regularization of nonlinear ill-posed problems. Inverse Problems **5**, 523–540 (1989).

[ES00]     Engl, H.W., Scherzer, O.: Convergence rates results for iterative methods for solving nonlinear ill-posed problems. In: Colton, D., Engl, H.W., Louis, A.K., McLaughlin, J.R., Rundell, W. (eds), Surveys on Solution Methods for Inverse Problems. Springer, Vienna, 7–34 (2000).

[FC92]     Fang, W., Cumberbatch, E.: Inverse problems for metal oxide semiconductor field-effect transistor contact resistivity. SIAM J. Appl. Math. **52**, 699–709 (1992).

[FI92]     Fang, W., Ito, K.: Identifiability of semiconductor defects from LBIC images. SIAM J. Appl. Math. **52**, 1611–1626 (1992).

[FI94]     Fang, W., Ito, K.: Reconstruction of semiconductor doping profile from laser-beam-induced current image. SIAM J. Appl. Math. **54**, 1067–1082 (1994).

[FIR02]     Fang, W., Ito, K., Redfern, D.A.: Parameter identification for semiconductor diodes by LBIC imaging. SIAM J. Appl. Math. **62**, 2149–2174 (2002).

[FSL05]     Frühauf, F., Scherzer, O., Leitão, A.: Analysis of regularization methods for the solution of ill–posed problems involving discontinuous operators. SIAM J Numerical Analysis **43**, 767–786 (2005).

[Gaj85]   Gajewski, H.: On existence, uniqueness and asymptotic behavior of so-
          lutions of the basic equations for carrier transport in semiconductors. Z.
          Angew. Math. Mech. **65**, 101–108 (1985).
[GV89]    Golub, G.H., Van Loan, C.F.: Matrix Computations (Second ed.) Johns
          Hopkins University Press, Baltimore, MD (1989).
[Gri85]   Grisvard, P.: Elliptic Problems in Nonsmooth Domains. Pittman Pub-
          lishing, London (1985).
[HNS95]   Hanke, M., Neubauer, A., Scherzer, O.: A convergence analysis of the
          Landweber iteration for nonlinear ill-posed problems. Numer. Math. **72**,
          21–37 (1995).
[Isa98]   Isakov, V.: Inverse problems for partial differential equations. Applied
          Mathematical Sciences, Springer, New York (1998).
[KS02]    Kowar, R., Scherzer, O.: Convergence analysis of a Landweber–Kaczmarz
          method for solving nonlinear ill-posed problems. In: Kabanikhin, S.I.,
          Romanov, V.G. (eds), Ill-Posed and Inverse Problems. VSP, Boston, 253–
          270 (2002).
[LS03]    Leitão, A., Scherzer, O.: On the relation between constraint regulariza-
          tion, level sets, and shape optimization. Inverse Problems **19**, L1–L11
          (2003).
[LMZ05]   Leitão, A., Markowich, P.A., Zubelli, J.P.: On inverse doping profile prob-
          lems for the stationary voltage-current map. Inverse Problems, **22**, no.
          3, 1071–1088 (2006).
[Mar86]   Markowich, P.A.: The Stationary Semiconductor Device Equations.
          Springer, Vienna (1986).
[MRS90]   Markowich, P.A., Ringhofer, C.A., Schmeiser, C.: Semiconductor Equa-
          tions. Springer, Vienna (1990).
[Nac96]   Nachman, A.I.: Global uniqueness for a two-dimensional inverse bound-
          ary value problem. Ann. of Math. **143**, 71–96 (1996).
[OS88]    Osher, S., Sethian, J.: Fronts propagation with curvature dependent
          speed: Algorithms based on Hamilton–Jacobi formulation. J. of Com-
          putational Physics., **56**, 12–49 (1988).
[San95]   Santosa, F.: A level-set approach for inverse problems involving obstacles,
          ESAIM Contrôle Optim. Calc. Var., **1**, 17–33 (1995/96).
[Sch91]   Scherzer, O.: Tikhonov regularization of nonlinear ill-posed problems
          with applications to parameter identification in partial differential equa-
          tions. PhD Thesis Johannes-Kepler-Universität, Linz (1991).
[Sel84]   Selberherr, S.: Analysis and Simulation of Semiconductor Devices.
          Springer, New York (1984).
[TA77]    Tikhonov, A.N., Arsenin, V.Y.: Solutions of Ill-posed Problems. John
          Wiley & Sons, New York (1977).
[vRo50]   van Roosbroeck, W.R.: Theory of flow of electrons and holes in germa-
          nium and other semiconductors. Bell Syst. Tech. J. **29**, 560–607 (1950).

# 7

# Deterministic kinetic solvers for charged particle transport in semiconductor devices

M.J. Cáceres,[1] J.A. Carrillo,[2] I.M. Gamba,[3] A. Majorana[4] and C.-W. Shu[5]

[1] Departamento de Matematica Aplicada, Universidad de Granada, 18071 Granada, Spain. caceresg@ugr.es

[2] ICREA (Institució Catalana de Recerca i Estudis Avançats) and Departament de Matemàtiques, Universitat Autònoma de Barcelona, E-08193 Bellaterra, Spain. carrillo@mat.uab.es

[3] Department of Mathematics and ICES, The University of Texas at Austin, USA gamba@math.utexas.edu

[4] Dipartimento di Matematica e Informatica, Università di Catania, Catania, Italy majorana@dmi.unict.it

[5] Division of Applied Mathematics, Brown University, Providence, RI 02912, USA shu@dam.brown.edu

**Keywords:** Weighted essentially non-oscillatory (WENO) schemes; Boltzmann transport equation (BTE); semiconductor device simulation; metal semiconductor field effect transistor (MESFET); metal oxide semiconductor field effect transistor (MOSFET); direct simulation Monte Carlo (DSMC).

## 7.1 Introduction

Statistical models [F91, L00, MRS90, To93] are used to describe electron transport in semiconductors at a mesoscopic level. The basic model is given by the Boltzmann transport equation (BTE) for semiconductors in the semiclassical approximation:

$$\frac{\partial f}{\partial t} + \frac{1}{\hbar}\nabla_{\mathbf{k}}\varepsilon \cdot \nabla_{\mathbf{x}}f - \frac{\mathfrak{e}}{\hbar}\mathbf{E}\cdot\nabla_{\mathbf{k}}f = Q(f), \qquad (7.1.1)$$

where $f$ represents the electron probability density function (pdf) in phase space $\mathbf{k}$ at the physical location $\mathbf{x}$ and time $t$. $\hbar$ and $\mathfrak{e}$ are physical constants; the Planck constant divided by $2\pi$ and the positive electric charge, respectively. The energy-band function $\varepsilon$ is given by the Kane non-parabolic band model, which is a non-negative continuous function of the form

$$\varepsilon(\mathbf{k}) = \frac{1}{1 + \sqrt{1 + 2\dfrac{\alpha}{m^*}\hbar^2|\mathbf{k}|^2}}\frac{\hbar^2}{m^*}|\mathbf{k}|^2, \qquad (7.1.2)$$

where $m^*$ is the effective mass and $\alpha$ is the non-parabolicity factor. In this way we observe that setting $\alpha = 0$ in Equation (7.1.2) the model is reduced to the widely used parabolic approximation.

The electric field $\mathbf{E}$ is self-consistently computed by the Poisson equation:

$$\nabla_{\mathbf{x}} \left[ \epsilon_r(\mathbf{x}) \nabla_{\mathbf{x}} V \right] = \frac{\mathfrak{e}}{\epsilon_v} \left[ \rho(t, \mathbf{x}) - N_D(\mathbf{x}) \right], \qquad (7.1.3)$$

$$\mathbf{E} = -\nabla_{\mathbf{x}} V, \qquad (7.1.4)$$

since we take into account the electrostatics produced by the electrons and the dopants in the semiconductor. We represent the dielectric constant in a vacuum by $\epsilon_v$ and $\epsilon_r(\mathbf{x})$ labels the relative dielectric function depending on the material, $\rho(t, \mathbf{x}) = \int_{\mathbb{R}^3} f(t, \mathbf{x}, \mathbf{k}) \, d\mathbf{k}$ is the electron density, $N_D(\mathbf{x})$ is the doping and $V$ is the electric potential. Equations (7.1.1), (7.1.3), (7.1.4) give the Boltzmann–Poisson system for electron transport in semiconductors.

The right-hand side of Equation (7.1.1) models the interaction of electrons with lattice vibrations of the crystal and can be written as

$$Q(f)(t, \mathbf{x}, \mathbf{k}) = \int_{\mathbb{R}^3} \left[ S(\mathbf{k}', \mathbf{k}) f(t, \mathbf{x}, \mathbf{k}) a - S(\mathbf{k}, \mathbf{k}') f(t, \mathbf{x}, \mathbf{k}) \right] d\mathbf{k}', \qquad (7.1.5)$$

where $S(\mathbf{k}, \mathbf{k}')$ is the transition probability from state $\mathbf{k}$ to $\mathbf{k}'$ per unit of time for each scattering mechanism. Therefore, the collision term $Q(f)$ depends on the device semiconductor material. For Si-based technology, the scattering phenomena taken into account are the acoustic phonon and optical non-polar phonon, while for a GaAs device the scattering mechanisms are impurities, and the acoustic phonon, non-polar optical phonon and polar optical phonon. Here, we will just give the expression of some of them relevant to the discussions below. For instance, the transition probability for randomly placed impurities reads

$$S^{(imp)}(\mathbf{k}, \mathbf{k}') = \frac{K_{imp}}{(|\mathbf{k} - \mathbf{k}'|^2 + \beta^2)^2} \, \delta(\varepsilon' - \varepsilon),$$

with

$$K_{imp} = \frac{N_I Z^2 \mathfrak{e}^4}{4\pi^2 \hbar \epsilon_v^2 \epsilon_r^2}, \qquad \beta = \sqrt{\frac{\mathfrak{e}^2 N_I}{\epsilon_v \epsilon_r k_B T_L}},$$

where $\beta$ is the inverse of the Debye length, $N_I$ and $Z\mathfrak{e}$ are the impurities concentration and its charge, respectively, $k_B$ is the Boltzmann constant and $T_L$ is the lattice temperature. We refer to Table 7.1 for a complete list of the physical parameters.

The scattering with crystal vibrations in the acoustic mode is taken into account in the elastic approximation and is given by

$$S^{(ac)}(\mathbf{k}, \mathbf{k}') = K_{ac} \, \delta(\varepsilon' - \varepsilon),$$

with

**Table 7.1.** Parameters for Si and GaAs.

| | | |
|---|---|---|
| $m$ | electron mass | $9.1095 \cdot 10^{-31}$ Kg |
| $m_*^*$ | effective electron mass in Si | $0.32\ m$ |
| $m_\Gamma^*$ | effective electron mass in the $\Gamma$-valley (GaAs) | $0.067\ m$ |
| $m_L^*$ | effective electron mass in the $L$-valley (GaAs) | $0.35\ m$ |
| $\rho_0$ | density lattice (Si) | $2330$ Kg/m$^3$ |
| | density lattice (GaAs) | $5360$ Kg/m$^3$ |
| $v_s$ | longitudinal sound speed (Si) | $9040$ m/s |
| | longitudinal sound speed (GaAs) | $5240$ m/s |
| $\alpha$ | non-parabolicity factor in Si | $0.5$ eV$^{-1}$ |
| $\alpha_\Gamma$ | non-parabolicity factor in the $\Gamma$-valley (GaAs) | $0.611$ eV$^{-1}$ |
| $\alpha_L$ | non-parabolicity factor in the $L$-valley (GaAs) | $0.242$ eV$^{-1}$ |
| $\epsilon_r$ | relative dielectric constant (Si) | $11.7$ |
| | relative dielectric constant (GaAs) | $12.90$ |
| $\epsilon_\infty$ | relative dielectric constant at optical frequency (GaAs) | $10.92$ |
| $\epsilon_v$ | vacuum dielectric constant | $8.85419 \cdot 10^{-12}$ F/m |
| $\Xi_d$ | acoustic-phonon deformation potential (Si) | $9$ eV |
| | acoustic-phonon deformation potential (GaAs) | $7$ eV |
| $D_t K$ | non-polar optical phonon deformation potential (Si) | $11.4 \cdot 10^{10}$ eV/m |
| | non-polar optical phonon deformation potential (GaAs) | $10^{11}$, eV/m |
| $\hbar\omega_{np}$ | non-polar optical phonon energy (Si) | $0.063$ eV |
| | non-polar optical phonon energy (GaAs) | $0.032$ eV |
| $\hbar\omega_p$ | polar optical phonon energy (GaAs) | $0.032$ eV |
| $\varepsilon_{0\Gamma}$ | $\Gamma$-valley bottom energy (GaAs) | $0$ eV |
| $\varepsilon_{0L}$ | $L$-valley bottom energy (GaAs) | $0.32$ eV |
| $Z_{\Gamma L}$ | degeneracy from $\Gamma$ to $L$ valleys (GaAs) | $4$ |
| $Z_{L\Gamma}$ | degeneracy form $L$ to $\Gamma$ valley (GaAs) | $1$ |
| $Z_{LL}$ | degeneracy from $L$ to $L$ valleys (GaAs) | $3$ |
| $n_i$ | intrinsic concentration (GaAs) | $1.79 \cdot 10^6$, cm$^{-3}$ |
| $N_I$ | impurity concentration (GaAs) | $10^{14}$ cm$^{-3}$ |
| $Z\mathfrak{e} = \mathfrak{e}$ | elementary charge | $1.6021 \cdot 10^{-19}$ C |

$$K_{ac} = \frac{k_B T_L \Xi_d^2}{4\pi^2 \hbar \rho_o v_s^2}.$$

Polar optical phonon scattering has a transition probability

$$S^{(p)}(\mathbf{k}, \mathbf{k}') = K_p \frac{G(\mathbf{k}, \mathbf{k}')}{|\mathbf{k} - \mathbf{k}'|^2} \left[ n_p \delta(\varepsilon' - \varepsilon - \hbar\omega_p) + (1 + n_p)\delta(\varepsilon' - \varepsilon + \hbar\omega_p) \right],$$

with

$$K_p = \frac{\hbar\omega_p \mathbf{e}^2}{8\pi^2 \hbar \epsilon_v} \left( \frac{1}{\epsilon_\infty} - \frac{1}{\epsilon_r} \right), \qquad n_p = \frac{1}{\exp(\hbar\omega_p/(k_B T_L)) - 1},$$

$\hbar\omega_p$ the polar optical phonon energy, $n_p$ its occupation number and the overlapping factor

$$G(\mathbf{k}, \mathbf{k}') = [aa' + cc' \cos(\mathbf{k}, \mathbf{k}')]^2,$$

with

$$a = \sqrt{\frac{1 + \alpha\varepsilon}{1 + 2\,\alpha\varepsilon}}, \qquad a' = \sqrt{\frac{1 + \alpha\varepsilon'}{1 + 2\,\alpha\varepsilon'}},$$

$$c = \sqrt{\frac{\alpha\varepsilon}{1 + 2\,\alpha\varepsilon}}, \qquad c' = \sqrt{\frac{\alpha\varepsilon'}{1 + 2\,\alpha\varepsilon'}}.$$

Before we describe the last scattering, non-polar optical phonon, we recall the band structure of the GaAs material. On the lowest energy conduction band, the most important information comes from the absolute and local minima since they typically concentrate most of the electrons. These minima are called valleys and we will consider them as different populations in our model. In GaAs the most interesting part of the valley structure is given by an absolute minimum at the center of the Brillouin zone called the $\Gamma$-valley and four local minima through the crystal orientation called the $L$-valleys which are considered equivalent [To93]. In this way, the transport of electrons in each valley is studied by a linear Boltzmann equation (7.1.1). Therefore, for a GaAs device the Boltzmann–Poisson system (7.1.1), (7.1.3), (7.1.4) is augmented since we need to use two Boltzmann equations, one for each valley. The Poisson equation couples both kinetic equations since the total density of electrons consists in a suitable average of the density of electrons in each valley. Moreover, electrons can jump to different valleys; this fact is reflected in the collision term $Q$ by means of the inter-valley scattering mechanism. Non-polar optical phonon scattering is considered in the inelastic approximation, both intra- and inter-valley, and thus, we consider emission and absorption of phonons of energy $\hbar\omega_{np}$ ($A$, $B$ denote either different or equal valleys):

$$S^{(np)}(\mathbf{k}_A, \mathbf{k}_B) = K_{np} \left[ n_{np}\delta(\varepsilon_B - \varepsilon_A - \hbar\omega_{np}^+) + (1 + n_{np})\delta(\varepsilon_B - \varepsilon_A + \hbar\omega_{np}^-) \right],$$

with

$$K_{np} = \frac{(D_t K)^2}{8\pi^2 \rho_0 \omega_{np}}, \qquad n_{np} = \frac{1}{\exp(\hbar\omega_{np}/(k_B T_L)) - 1}$$

$$\hbar\omega_{np}^+ = \hbar\omega_{np} + \Delta_{AB}, \qquad \hbar\omega_{np}^- = \hbar\omega_{np} - \Delta_{AB}.$$

The occupation number $n_{np}$ is given by the equilibrium Bose–Einstein distribution. GaAs solid-state properties imply that non-polar optical phonon scattering contributes to the inter-valley scattering but only to the intra-valley scattering for the $L$-valley. $\Delta_{AB} = \varepsilon_{0A} - \varepsilon_{0B}$ is the energy gap between the minima of the two valleys.

In order to develop a competitive numerical scheme the equations are simplified by changing variables to suitable energy-type variables in which the scattering operators become either simple evaluations or linear integral operators with singular kernels. This change of variables was introduced first in [FO93] in the parabolic band approximation and in [MP01] for general dispersion relations for the energy formula (7.1.2). This change of variables and a suitable adimensionalization were performed for Si and GaAs devices [CGMS03-2, CCM06].

The main numerical tools to simulate the Boltzmann–Poisson system for semiconductor devices have practically been based on the direct simulation Monte Carlo (DSMC) method. Since the dimension of the Boltzmann–Poisson system is 7, numerical simulations are heavily costly. To overcome this problem several approximated systems and corresponding deterministic numerical methods have been proposed in the literature: spherical harmonics approximation, hydrodynamic models, diffusion limits leading to drift-diffusion equations, Child–Langmuir asymptotics, and others. The literature of the different approximation systems is considerably large, therefore we just refer to [AH02, AMR03, BDMS01, CC95, D04, JS94, MRS90, Sel84] and the references therein. We do not claim that the list is at all exhaustive.

The earliest work in the direct deterministic simulation of Boltzmann–Poisson systems started with particle methods in [DDM90]. We also mention the works of C. Ringhofer, see [R00, R03] and the references therein; he has proposed a general approach by means of series expansion in momentum vectors to solve the stationary and time dependent BTE. This strategy leads to a hyperbolic system for the coefficients of the expansion that depend both on time and position. Moreover, the proposed scheme can deal with regimes close to drift-diffusion equations in a natural way. Recent developments are the subject of [JPMRB06].

Starting in [CGJS00, CGS00], direct finite difference methods based on high-order WENO (weighted essentially non-oscillatory) approximations were used to compute the advection part of the transport equation for the one-dimensional (1D) relaxation-time kinetic system for semiconductors. In [CGS00] the considered model distinguishes the materials only with some representative parameters: mobility, mass and relaxation time. The collision operator (relaxation) does not take into account the different scattering mechanisms concerning the material of the device. However, the simplicity of this operator allows us to validate the use of a WENO scheme to simulate the spatial and velocity derivatives of the equation. Later, using the change of variables cited above [MP01] the real collision operators were analyzed, considering all possible scattering mechanisms in Si [CGMS02, CGMS03-2] and GaAs [CCM06], and thus even extended to 2D spatial semiconductors [CGMS03-3, CGMS06].

Recently, these developments have been coupled to alternative ideas producing other deterministic solvers which are applied to different semiconductor materials. A brief introduction to these schemes is given in Section 7.5. Here, we wish to cite the recent and relevant papers of a research group in Graz. In [AS05, AS05-2, DS04, DGS05, ES05, GS05, GM06, GS06] the numerical technique is used to solve the Boltzmann–Poisson system in 1D or 2D cases for semiconductor devices based on Si, GaAs or AlGaN/GaN material. Moreover, they also consider [ASK04, GS04, GSb05] a model consisting of a couple of kinetic equations: one for the free electron and the other for the phonons, which are no longer constant. This model must be considered when hot phonon phenomena occur. The book by M. Galler [Ga05] furnishes both an introduction to semiconductor physics and a complete description of these numerical schemes; many examples of applications are also given.

We discuss in Section 7.2 the 1D case analyzing the results from the relaxation case to the full collision operator for GaAs devices. In Section 7.3, previous results are extended to dimension 2 where new difficulties arise to implement the boundary conditions. We conclude this section with simulations for double gate MOSFETs. Section 7.4 is devoted to summarizing the works developed using alternative ideas in energy and angles variables by multigroup techniques borrowed from neutron transport numerical schemes. In Section 7.5, we briefly state the goals achieved in this research and we comment on possible new directions in this field.

## 7.2 1D deterministic device simulations

WENO schemes [JS96, Sh98] were used in [CGJS00, CGS00] to simulate the 1D relaxation-time kinetic system for semiconductors. WENO approximations of the spatial and velocity derivatives were used coupled to the explicit Runge–Kutta method for time evolution due to the stability of the WENO approximation of the derivatives. This system is a first approximation of the Boltzmann–Poisson system for semiconductors; the right-hand side of the equation is approximated by $\frac{1}{\tau}(M_{T_L}(v)\rho(f) - f)$, where the relaxation time $\tau$ is assumed constant and related to standard values of the mobility $\mu$ by $\mu = \frac{e}{m^*}\tau$ and $M_{T_L}$ is the Maxwellian associated with this temperature

$$M_{T_L}(v) = \left(\frac{2\pi k_B T_L}{m^*}\right)^{-1/2} \exp\left(-\frac{m^* v^2}{2 k_B T_L}\right).$$

Therefore, we observe that the particularities of the material are reflected only in the parameters $\tau$ and $m^*$. In [CGS00] the authors analyzed and compared their results with several classical approximations to the Boltzmann–Poisson system (hydrodynamics, drift-diffusion systems). The results were also validated with Monte Carlo (DSMC) results for silicon diodes in [ACGS01, CGMS03-1] and a numerical validation for the Child–Langmuir limit for semiconductors for Si and GaAs devices was shown in [CCD02].

Later, in [CGMS03-2] all the scattering mechanisms considered in Si devices were taken into account. The complexity of the collision operator was overcome by considering the change of variables to suitable energy-type dimensionless variables. We write here the Boltzmann–Poisson system after the change of variables and the dimensionless process for the GaAs case [CCM06]. Since in this case two valleys have to be taken into account, we use the subindex $A$ to denote the different valleys: $\Gamma$ and $L$ (for Si we can remove this index) and denote with $^-$ the dimensionless variables and parameters. Equation (7.1.1) is written as

$$\frac{\partial \Phi_A}{\partial t} + \frac{\partial}{\partial z}\left(a_1^A \Phi_A\right) + \frac{\partial}{\partial w_A}\left(a_2^A \Phi_A\right) + \frac{\partial}{\partial \mu_A}\left(a_3^A \Phi_A\right) = \bar{Q}(\Phi_A), \ A = \Gamma, L,$$

$$(7.2.1)$$

where $\Phi_A$ is the new unknown function depending on dimensionless time $t \geq 0$, dimensionless spatial variable $z$, dimensionless energy $w_A$ and the cosine of the angle with respect to the field direction $\mu_A$. The flux coefficients are

$$a_1^A = a_1^A(w_A, \mu_A) = c_{1,A} \frac{\mu_A \sqrt{2\, w_A(1 + \bar{\alpha}_A w_A)}}{1 + 2\, \bar{\alpha}_A w_A},$$

$$a_2^A = a_2^A(w_A, \mu_A) = -c_{2,A}\, \bar{E}\, \mu_A \frac{\sqrt{2\, w_A(1 + \bar{\alpha}_A w_A)}}{1 + 2\bar{\alpha}_A w_A}, \qquad (7.2.2)$$

$$a_3^A = a_3^A(w_A, \mu_A) = -c_{2,A}\, \bar{E} \frac{1 - \mu_A^2}{\sqrt{2\, w_A(1 + \bar{\alpha}_A w_A)}},$$

with

$$c_{1,A} = \frac{t_*}{l_*} \sqrt{\frac{k_B T_L}{m_A^*}} \quad \text{and} \quad c_{2,A} = \sqrt{\frac{m_X^*}{m_A^*}}, \qquad (7.2.3)$$

where $t_*$, $l_*$ and $m_X^*$ are the characteristic time, length and mass respectively. The Poisson equation (7.1.3) is written as

$$\Delta_z \bar{V}(t, z) = c_p \left( c_\Gamma \int_0^\infty \int_{-1}^1 \Phi_\Gamma(t, z, w_\Gamma, \mu_\Gamma)\, d\mu_\Gamma\, dw_\Gamma \right.$$

$$(7.2.4)$$

$$\left. + c_L \int_0^\infty \int_{-1}^1 \Phi_L(t, z, w_L, \mu_L)\, d\mu_L\, dw_L - \bar{N}_D(z) \right)$$

with $c_p$, $c_\Gamma$ and $c_L$ defined as

$$c_p = \frac{\mathfrak{e}^2\, t_*\, l_*\, m_X^*\, k_B T_L}{\epsilon_v \epsilon_r\, \hbar^3}, \quad c_\Gamma = 2\,\pi\, Z_{L\Gamma}\left(\frac{m_\Gamma^*}{m_X^*}\right)^{3/2}, \quad c_L = 2\,\pi\, Z_{\Gamma L}\left(\frac{m_L^*}{m_X^*}\right)^{3/2},$$

(7.2.5)

where $Z_{\Gamma L}$ and $Z_{L\Gamma}$ are the degeneration numbers from the $\Gamma$-valley to the $L$-valley and vice versa. Returning to (7.2.1), we show the dimensionless collision term $\bar{Q}$:

$$\bar{Q}(\varPhi_A) = s_A(w_A)\int_{-1}^{1}\left\{2\,\pi\, c_{3,A}\varPhi_A(w_A, \mu_A') + c_{4,A}I_{4,A}(w_A, \mu_A, \mu_A')\varPhi_A(w_A, \mu_A')\right.$$

$$+ c_{5,A}\left[I_{5,A}^{+}(w_A, w_A - \alpha_p, \mu_A, \mu_A')\varPhi_A^{+}(w_A - \alpha_p, \mu_A')\right.$$

$$+ a_p I_{5,A}(w_A, w_A + \alpha_p, \mu_A, \mu_A')\varPhi_A(w_A + \alpha_p, \mu_A')\Big]$$

$$+ c_{6,A}\, 2\,\pi\left[\varPhi_A^{+}(w_A - \alpha_{np}, \mu_A') + a_{np}\,\varPhi_A(w_A + \alpha_{np}, \mu_A')\right]\bigg\}\, d\mu_A'$$

$$- \left\{4\,\pi\, c_{3,A}s_A(w_A) + 4\,\pi\, c_{4,A}\frac{s_A(w_A)}{\beta_A{}^2(4\, q_A(w_A) + \beta_A{}^2)}\right.$$

$$+ c_{5,A}\left[N_{5,A}(w_A, w_A + \alpha_p)s_A(w_A + \alpha_p)\right.$$

$$+ a_p\, N_{5,A}^{+}(w_A, w_A - \alpha_p)s_A^{+}(w_A - \alpha_p)\Big]$$

$$+ c_{6,A}\, 4\,\pi\left[s_A(w_A + \alpha_{np}) + a_{np}\, s_A^{+}(w_A - \alpha_{np})\right]\bigg\}\varPhi_A(w_A, \mu_A)$$

$$+ Z_{AB}s_A(w_A)\, c_{7,B}\, 2\pi\int_{-1}^{1}\left[\varPhi_B^{+}(w_A - \alpha_{np} - \bar{\varDelta}_{BA}, \mu_B)\right.$$

$$+ a_{np}\varPhi_B^{+}(w_A + \alpha_{np} - \bar{\varDelta}_{BA}, \mu_B)\Big]\, d\mu_B$$

$$- Z_{AB}\, c_{7,B}\, 4\pi\left[s_B^{+}(w_A + \alpha_{np} + \bar{\varDelta}_{AB})\right.$$

$$+ a_{np}\, s_B^{+}(w_A - \alpha_{np} + \bar{\varDelta}_{AB})\Big]\varPhi_A(w_A, \mu_A),$$

(7.2.6)

where $s_A(w_A)$ is the Jacobian of the coordinate transformation up to constants (the notation $^+$ means the positive part of the function), $I_{4,A}$, $I_{5,A}$, $N_{5,A}$ are functions which depend on the integral parts of the collision operator, $q_A(w) = 2w_A(1 + \bar{\alpha}_A w_A)$ and $c_{3,A}$, $c_{4,A}$, $c_{5,A}$, $c_{6,A}$ and $c_{7,A}$ are constants related to the different scatterings. The complexity of this term resides in the new, compared to Si, scattering mechanisms: impurity, optical polar phonon and optical non-polar phonon for inter-valleys. For a silicon device the collision operator is approximated by simple evaluations in mesh points. In GaAs the collision operator is more complicated since it involves the functions $I_4$ and $I_5$, from the impurities and optical phonon scattering, respectively, which present integrable singularities. This complexity of the collision operator produces a new numerical difficulty which is solved in [CCM06] by a suitable numerical approximation of the gain term.

The numerical scheme used is a conservative finite difference WENO scheme of fifth order in the transport variables and the evolution in time is done by a third-order total variance diminishing (TVD) Runge–Kutta method. For details we refer to [CGMS03-2] where the numerical scheme is explained. The boundary conditions are basically zero flux on the boundaries in energy and angle and inflow Maxwellian boundary conditions in space. See Section 7.3 for boundary conditions for the 2D case. The Poisson equation is discretized and numerically solved by a standard central finite difference scheme. Stability in time for the resulting scheme is ensured by computing time steps verifying a Courant–Friedrich–Levy (CFL) condition. The CFL condition may become severe whenever high field regions appear in the device. Investigations to avoid this problem are under way.

The main difference, from a numerical point of view, between Si and GaAs is the complexity of the latter, since two valleys have to be considered. This fact produces numerical difficulties for the preservation of total charge due to the inter-valley scattering mechanisms in the homogeneous case and the singular character of the kernels of the impurities and polar optical phonon scattering. The presence of two Boltzmann equations for the GaAs case produces a larger computational cost that can be overcome by a parallelized implementation [MCM05], in which the authors compute for the first time I-V curves for 2D devices by deterministic schemes.

In [CGMS03-2] the numerical scheme is tested for two $n^+ - n - n^+$ diodes of total length 1 $\mu$m and 0.25 $\mu$m, with 400 nm and 50 nm channels located in the middle of the device, respectively. Deterministic results are compared with DSMC and with other classical approximations of the Boltzmann–Poisson system: drift-diffusion, hydrodynamic and kinetic relaxation models, to determine how good the approximation of the real collision operator by a relaxation model [CGS00] is. Different phonon frequencies are taken into account in [MMM04] with comparisons to DSMC commercial solvers.

In [CCM06] the WENO solver proposed is validated by computing the electron valley occupancy, velocity and energy of the homogeneous case, the bulk material, and by comparing the DSMC results. The well-known experimental phenomena of the material is shown [To93]. This was the first step in analyzing the transport of electrons in $n^+ - n_i - n^+$ diodes and Gunn oscillators [CJSW98, MR05]. Here, $n_i$ is the intrinsic concentration of the material and the GaAs diode has a channel length of 0.25 $\mu$m for a total length of 0.55 $\mu$m.

As a summary of both works [CGMS03-2, CCM06], we can point out that the main contribution has been the derivation of a fully deterministic

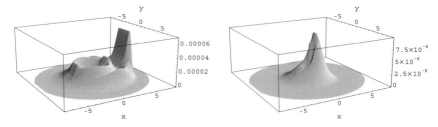

**Fig. 7.1.** GaAs device. Left: $\Gamma$-valley distribution. Right: $L$-valley distribution. Both in Cartesian coordinates at point $z = 0.3996674$ $\mu$m at time $t = 3$ ps with 0.75 V applied bias at room temperature 300 K ($x = \mathbf{k}_3$ and $y = \mp\sqrt{\mathbf{k}_1^2 + \mathbf{k}_2^2}$).

solver for the BTE for semiconductors including all the relevant scattering mechanisms of the considered material. This solver allows us to compute the noise-free evolution in time of the distribution function at every point of the device and consequently all the moments of the distribution function, i.e., hydrodynamical quantities and the stabilization in time towards a steady state.

As an example of these simulations, we emphasize here that the steady state in some points of the channel of the device is far from being a shifted Maxwellian distribution showing purely kinetic characteristics. Fig. 7.1 shows the distribution function in Cartesian coordinates for valleys $\Gamma$ and $L$ with 0.75 V applied bias at room temperature 300 K at point $z = 0.399667$ $\mu$m after $t = 3$ ps. Comparing Si results to GaAs results, the asymmetry in the distribution function is more relevant in GaAs for the $\Gamma$-valley distribution than in Si. Discrete points of the simulations were simply joined by lines.

This scheme has been adapted to bipolar p-n junctions [GCG1, GCG2]. Here, we deal again with a system of two kinetic equations for holes and electrons coupled through the Poisson equation, but we also have new terms due to recombination and generation of electron-hole pairs. .

Let us finally mention that improvements in the WENO methods in order to reduce its computational cost have been explored by improving the time discretization [AS05], by better adapting the energy discretization [ARS05] and by imposing different uniform discretizations in different parts of the domain [GCG2, AS05-2].

## 7.3 2D deterministic device simulations

The change of variables and the dimensionless process used in the 1D case can be extended to the 2D case. The Boltzmann equation (see [CGMS03-3]) is written, in terms of the new unknown $\Phi$, as

$$\frac{\partial \Phi}{\partial t} + \frac{\partial}{\partial x}(a_1 \Phi) + \frac{\partial}{\partial y}(a_2 \Phi) + \frac{\partial}{\partial \omega}(a_3 \Phi) + \frac{\partial}{\partial \mu}(a_4 \Phi) + \frac{\partial}{\partial \phi}(a_5 \Phi) = \bar{Q}(\Phi),$$

$$(7.3.1)$$

where the flux functions $a_i$ are given by

$$a_1(\omega) = \frac{1}{t_*} \frac{\mu s(\omega)}{(1 + 2\alpha_\kappa\omega)^2}$$

$$a_2(\omega, \mu, \phi) = \frac{1}{t_*} \frac{\sqrt{1 - \mu^2} s(\omega) \cos\phi}{(1 + 2\alpha_\kappa\omega)^2}$$

$$a_3(t, x, y, \omega, \mu, \phi) = -\frac{1}{t_*} \frac{2s(\omega)}{(1 + 2\alpha_\kappa\omega)^2}$$
$$\times \left[ E_x(t, x, y)\mu + E_y(t, x, y)\sqrt{1 - \mu^2} \cos\phi \right]$$

$$a_4(t, x, y, \omega, \mu, \phi) = -\frac{1}{t_*} \frac{1 + 2\alpha_\kappa\omega}{s(\omega)} \sqrt{1 - \mu^2}$$
$$\times \left[ E_x(t, x, y)\sqrt{1 - \mu^2} - E_y(t, x, y)\mu \cos\phi \right]$$

$$a_5(t, x, y, \omega, \mu, \phi) = \frac{E_y(t, x, y)}{t_*} \frac{\sin\phi}{\sqrt{1 - \mu^2}} \frac{1 + 2\alpha_\kappa\omega}{s(\omega)}.$$

Here, $t$ appears with dimensions in contrast to the previous section, from which the factor $t_*$ appears in the flux coefficients. We remove the subindex $A$ since we consider the Si case for simplicity. For GaAs, the differences, as in the 1D case, are the presence of other scattering mechanisms and two relevant valleys; consequently two Boltzmann equations have to be considered. Therefore, the 2D case for GaAs can be also simulated in analogous manner without further problems. In the 2D case the unknown $\Phi$ depends on $t$ (time), $x$, $y$ (space variables), $w$ (dimensionless energy), $\mu$ (cosine of the angle with respect to the $x$-axis) and $\phi$ (azimuthal angle).

Results of the 2D case were presented in [CGMS03-3] for metal semiconductor field effect transistor (MESFET) devices and in [MCM05] considering a parallelization of the numerical code. This deterministic solver was improved in [CGMS06] to clarify the implementation of the boundary conditions and the nature of boundary singularities in the electric field. The improved scheme was tested with a stochastic DSMC solver. Here we summarize both issues; see [CGMS06] for details.

### 7.3.1 Boundary conditions

We consider semiconductor devices whose boundaries can be separated in four kinds of regions: gates, contact areas, insulated areas and contact boundary between the semiconducting and the oxide regions (the last one for the metal oxide semiconductor field effect transistor (MOSFET) case only).

The boundary conditions associated to the electrostatic potential (i.e., solutions of the Poisson equation (7.1.3), (7.1.4)) are prescribed potential (Dirichlet type conditions) at the source, drain, and gate contacts, and insulating (homogeneous Neumann conditions) on the remainder of the boundary.

The boundary conditions imposed on the numerical probability density function (pdf) approximating the solution of the BTE are as follows:

*Source and drain contacts:* Numerical boundary conditions must approximate neutral charges. Here we employ a buffer layer of ghost points outside both the device contact source and drain areas and use conditions (2.1) from [CGMS06].

Under the assumption that highly doped regions $n^+(x, y)$ are slowly varying (i.e., of very small total variation norm), the corresponding small Debye length asymptotics yield neutral charges away from the endpoints of the contact boundary regions. Consequently, the total density $\rho$ (zero-order moment) and its variation take asymptotically, in Debye length, the values of $n^+(x, y)$ away from the contact endpoints. This is the well-known limit for drift-diffusion systems [MRS90], and we observe that this is also the case for the kinetic problem, provided the kinetic solution satisfies approximately neutral charge conditions at these contact boundaries.

*Gate contacts:* The numerical boundary condition yields an estimated incoming mean velocity which represents a transverse electric field effect due to the gate contacts. These conditions, given in (2.3) from [CGMS06], or (2.4) for its integrated form, are of Robin (or mixed) type, which are the natural conditions for a simulation of a gate contact in an oxide region. This form of boundary conditions follows from classical asymptotics corresponding to *thin shell* elliptic- and parabolic-type problems with Dirichlet data in the *thin outer shell* domain and a core larger-scale domain. Classical transmission conditions link the force fields corresponding to both domains, which combine into a Robin-type condition (mixed-type condition (2.8) from [CGMS06]). The 2D asymptotic boundary condition for the Poisson system linking the oxide and semiconductor region and yielding a Robin-type condition was rigorously studied in [Ga93-1, Ga93-2].

*Insulating walls or contact boundary between the semiconducting and the oxide regions:* In both of these cases we impose classical elastic specular boundary reflection conditions for the numerical pdf as in (2.10) from [CGMS06].

### 7.3.2 Appearance of singularities

It is well known that solutions of the Poisson equation develop singularities in a bounded domain, with a Lipschitz boundary and a boundary condition that changes type from Dirichlet to homogeneous Neumann or Robin-type data on both sides of a conical wedge, with an angle $0 \leq \vartheta \leq 2\pi$, whose vertex is at the point of data discontinuity. The solution will develop a singularity at the vertex point depending on the angle $\vartheta$. Indeed, for an elliptic problem in a domain with a data-type change, the solution remains bounded with a radial behavior of a power less than one if $\pi/2 \leq \vartheta \leq 2\pi$. Its gradient then becomes singular (see for instance Grisvard [Gr85] for a survey on boundary value problems of elliptic partial differential equations (PDEs) in a non-convex domain, and Gamba [Ga90], [Ga93-1] for a rigorous study of a drift-diffusion Poisson system in a MOSFET-type geometry). Therefore, in particular, the electric potential presents singularities and consequently singularities also appear in the electric field. Regardless, the deterministic solver presented is high-order accurate. Consequently, elliptic and parabolic descriptions of the density flow at the drift-diffusion level maintain the same singular boundary behavior, even though mobility functions are saturated quantities; that is, they are bounded functions of the magnitude of the electric field [Ga93-2]. In fact, these conditions have been used for 2D drift-diffusion simulations of MESFET devices and this singular boundary behavior was obtained [CC95].

We stress that our calculations resolve this singular boundary behavior as shown in Fig. 7.4 at the boundary points corresponding to the junction of the doping profile abruptly changing to the oxide regions and gate contacts.

In view of this boundary asymptotic behavior at the macroscopic level and the numerical results we obtain for the deterministic solutions of the BTE-Poisson system and its moments, we conjecture in [CGMS06] that the moments of the particle distribution function $f(t, \mathbf{x}, \mathbf{k})$ have *the same spatial asymptotic behavior as the density solution of the drift-diffusion problem* at the boundary points with a discontinuity in the data. In other words, the collision mechanism preserves the spatial regularity of the pdf as though its zeroth moment would satisfy the drift-diffusion equations, even though the numerical evidence indicates that the average of the pdf (i.e., density) will not evolve according to such a simple macroscopic model.

### 7.3.3 2D simulation example: double gate MOSFETs

The WENO-BTE solver is, again, a fifth-order finite difference WENO scheme coupled with a third-order TVD Runge–Kutta time discretization to solve (7.3.1), as used in [CGMS06].

The Poisson equations (7.1.3) for the potential $V$ and the electric field (7.1.4) are solved by the standard central difference scheme, with the given $V_{bias}$ boundary conditions at the source, drain and gate.

A realistic device used by electronic engineers for which the scheme presented above can be applied is the double-gate (DG) MOSFET. We show simulations for a DG-MOSFET whose structure is described in Fig. 7.2: two bands of oxide of thickness 2 nm sandwich a band of Si of 24 nm where electrons transit. Two gates (top and bottom) 50 nm long are considered at 50 nm of the source.

The voltage gate is 1.06 V for both, while the drain applied bias is 1.6 V. A "V"-shaped doping profile as in [BSF03] was taken into account: the doping concentration in the center of the channel is $10^{15}$ cm$^{-3}$, while at the edges of the channel it is $5 \times 10^{18}$ cm$^{-3}$, see Fig. 7.3. The acceptors concentration is $10^{10}$ cm$^{-3}$.

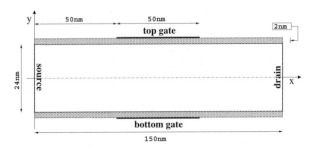

**Fig. 7.2.** Schematic representation of a DG-MOSFET.

The boundary conditions were described in Subsection 7.3.1. Due to the symmetry of the device in the $y$-component, the device can be simulated for $y > 0$ only and the results for the other part of the device can be obtained by symmetry. Electron transport only occurs in the semiconductor Si part of the device, therefore the BTE is only solved in this area, while the Poisson equation is solved in the whole device to take into account the presence of the gates.

As usual the BTE is simulated until the hydrodynamical quantities stabilize on time. In this sense, we consider that numerically the steady state has been achieved. In Fig. 7.4 we show the density, potential and energy of this stationary state. We observe two regions of the device with a high concentration of electrons close to the gates. Electrons are accelerated by the gate voltages and, thus, the energy achieves its maximum value around the end of the gates, at 100 nm of the source.

## 7.4 The multigroup-WENO model equations

This numerical technique consists of imposing an ansatz for the solution in the momentum variables: energy and angles, while keeping the dependence

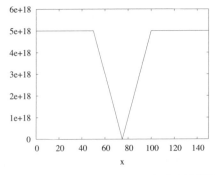

**Fig. 7.3.** Doping profile of the DG-MOSFET.

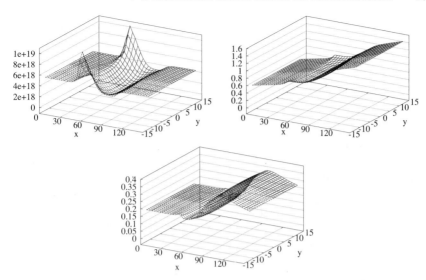

**Fig. 7.4.** DG-MOSFET device. Top left: the charge density $\rho$ (cm$^{-3}$); top right: the electric potential $V$; bottom: the energy $\mathcal{E}$ (eV). Units of $x$ and $y$ are nanometers.

on time and position through the expansion coefficients. Ideas of multigroup techniques are not far from the general series expansion methods analyzed in [R00]. In a first step, assuming that the distribution function depends only on **k** through the coordinates $w \in [0, w_{\max}]$, $\mu \in [-1, 1]$, we discretize them via

$$w_{i+1/2} = i\Delta w, \qquad i = 0, 1, \ldots, N, \ \Delta w = w_{\max}/N,$$
$$\mu_{j+1/2} = -1 + j\Delta\mu, \ j = 0, 1, \ldots, M, \ \Delta\mu = 2/M,$$

with two suitably chosen integers $N$ and $M$. Here, $w_{\max}$ is a maximum value for the dimensionless energy, related to the physically studied process, for which we have to check that $\Phi(t, \mathbf{x}, w_{\max}, \mu)$ is negligible for all $t$, $\mathbf{x}$ and $\mu$.

In the case of a unique valley, the unknown function $\Phi$ is approximated by the finite sum

$$\Phi(t, \mathbf{x}, w, \mu) \approx \sum_{i=1}^{N} \sum_{j=1}^{M} n_{ij}(t, \mathbf{x}) \lambda_{w_i}(w) \, \lambda_{\mu_j}(\mu) \tag{7.4.1}$$

containing $N \times M$ coefficients $n_{ij}(t, \mathbf{x})$ and where the functions $\lambda_{w_i}(w)$ and $\lambda_{\mu_j}(\mu)$ can be chosen in different ways. The first possibility is to assume that $\lambda_{w_i}(w) = \delta(w_i - w)$ and $\lambda_{\mu_j}(\mu) = \delta(\mu_j - \mu)$, where $\delta$ is the Dirac distribution. Alternatively, they can be defined by

$$\lambda_{w_i}(w) = \begin{cases} \dfrac{1}{\Delta w}, & \text{if } w \in [w_{i-1/2}, w_{i+1/2}], \\ 0, & \text{otherwise,} \end{cases} \tag{7.4.2}$$

and the other function analogously. Recently, a modified approximation for the distribution function has been proposed, introducing a weight function $p(w)$, such that

$$\Phi(t, \mathbf{x}, w, \mu) \approx p(w) \sum_{i=1}^{N} \sum_{j=1}^{M} n_{ij}(t, \mathbf{x}) \lambda_{w_i}(w) \lambda_{\mu_j}(\mu), \qquad (7.4.3)$$

where the function $\lambda$ can be defined as above. The main advantage of introducing the function $p$ is in taking into account the singularities or the shape of $\Phi$. In [GM06], $p(w) = s(w)$ was assumed; so, apart from a constant factor, $\Phi/s$ is the original distribution function $f$ and the singularity at $w = 0$ is removed. The evolution equations for the coefficients $n_{ij}$ are constructed as suggested by the method of weighted residuals. The ansatz (7.4.1) or (7.4.3) is inserted into the dimensionless Boltzmann equation and the result is integrated over the cells

$$\mathcal{Z}_{ij} = [w_{i-1/2}, w_{i+1/2}] \times [\mu_{j-1/2}, \mu_{j+1/2}].$$

This procedure yields a set of $N \times M$ partial differential equations for the $n_{ij}$. Therefore, WENO routines are used to obtain accurate numerical solutions. The flux through the cells $\mathcal{Z}_{ij}$, corresponding to the integration of the terms of the free streaming operator containing the partial derivatives with respect to the variables $w$ and $\mu$, is treated by using a simple formula based on the Min-Mod slope limiter. Details of the full general procedure are given in [Ga05, GM06].

This approach for solving the transport equation has some advantages and disadvantages with respect to the full WENO scheme. The multigroup-WENO (MW) technique is also simple to use in the case of a non-uniform grid in the $(w, \mu)$ space and the function $p(w)$ may be related to physical expectations (for instance, the shape or tail of $\Phi$). Moreover, it is requires less CPU time, due to the easy treatment of the force term in the Boltzmann equation. The main disadvantage is the loss of accuracy in the approximation of the force term. Although the order of error is comparable with those coming from the approximation of the collision operator, the presence of a strong electric field enlarges the numerical error. This can be meaningful, as the moments of the distribution functions are evaluated. In order to show this, we have solved the simple typical problem of the homogeneous charge transport in a bulk silicon device having a constant doping profile and a constant applied electric field. We use the full WENO scheme and the multigroup model based on Equation (7.4.3). The main thermodynamical quantities are the velocity and the energy. In Fig. 7.5 (left) we show the electron mean velocity for different grid sizes. The index $i$ of the $x$-axis corresponds to the values of the integer $N$ and $M$ ($w_{\max}$ is not changed) according to the following rule:

$$N = 20 + i \times 20 \quad M = 8 + i \times 4 \quad (i = 1, 2, \dots, 6).$$

Figure 7.5 (right) shows the mean energy results. The value of the electric field is 30 kV/cm.

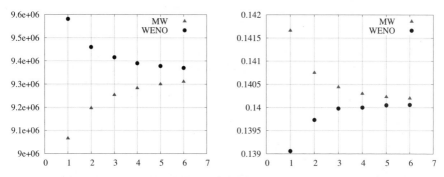

**Fig. 7.5.** Left: The velocity (cm/s). Right: The energy (eV).

It is evident that there is better accuracy for the WENO scheme than for the MW method. So, if we require highly accurate solutions, the WENO approach is preferable, since it needs fewer grid points than the MW method. We do not have the savings of CPU time of the MW technique in this case, because we must increase the number of grid points. When we use a coarse grid, both results are reasonable. In this test we have used a moderate electric field; if we increase its value, then the differences will be more emphasized.

## 7.5 Conclusions and future developments

The work developed during the last five years devoted to WENO methods shows that these methods lead to high-order accurate solutions of the BTE for different materials and geometries. They actually represent benchmarks for hydrodynamic or drift-diffusion solvers. However, they present a significant computational cost that can be reduced by using computer techniques, e.g., parallelization and dynamic memory, or by tuning different numerical aspects, e.g., time discretization, energy discretization, multilevel/multigrid methods, and multiresolution approaches. Some of them have already been tackled. We think multiresolution approaches could lead to a tremendous reduction in computational time and multigrid methods will ease the restriction to uniform grids. Summarizing, WENO-based methods present two main disadvantages: a large computational cost mainly due to the CFL condition and the need for working with uniform (or smooth) spatial grids.

The search for new deterministic numerical methods for the BTE should continue as we keep in mind the mentioned limitations of WENO-based methods. Methods allowing us to take larger advection steps, thus avoiding CFL conditions while maintaining good accuracy represent a good alternative to WENO methods. Semi-Lagrangian methods coupled to splitting techniques used for a long time in the simulation of Vlasov-type systems in plasma physics could be a good alternative [CV06].

On the other hand, to use non-structured grids is important in 2D real device simulation since the transport of electrons usually happens in narrow areas inside the device. Local discontinuous Galerkin methods are an alternate numerical technique in this direction. Recent developments in numerical and analytical studies for these type of solvers and simulations have been performed in [GP06].

### Acknowledgments

The research of the first two authors was supported by DGI-MEC (Spain) project MTM2005-08024. The research of the third author was supported by NSF grant DMS-0204568. The research of the fourth author was supported by the Italian COFIN 2004. The research of the last author was supported by ARO grant W911NF-04-1-0291 and NSF grant DMS-0510345. We thank the Institute of Computational Engineering and Sciences (ICES) at the University of Texas for partially supporting this research. We thank José A. Cañizo for editing Fig. 7.1.

# References

[ACGS01]    Anile, A.M.; Carrillo, J.A.; Gamba, I.M.; Shu, C.-W.: Approximation of the BTE by a relaxation-time operator: simulations for a 50nm-channel Si diode, VLSI Design, **13**, (2001), 349–354.

[AH02]      Anile, A.M.; Hern, S.D.: Two-population hydrodynamical models for carrier transport in gallium arsenide: simulation of gunn oscillations, VLSI Design, **15**, (2002), 681–693.

[AMR03]     Anile, A.M.; Mascali, G.; Romano, V.: Recent developments in hydrodynamical modeling of semiconductors, Mathematical problems in semiconductor physics, Lecture Notes in Math., **1823**, (2003), 1–56.

[ASK04]     Auer, C.; Schürrer, F.; Koller, W.: A semicontinuous formulation of the Bloch-Boltzmann-Peierls equations, SIAM J. Appl. Math., **64**, (2004), 1457–1475.

[AS05]      Auer, C.; Schürrer, F.: Efficient time integration of the Boltzmann–Poisson system applied to semiconductor device simulation, J. Comput. Electron., **5**, (2006), 5–14.

[AS05-2]    Auer, C.; Majorana, A.; Schürrer, F.: Numerical schemes for solving the non-stationary Boltzmann–Poisson system for two-dimensional semiconductor devices, ESAIM: Proceedings, **15**, (2005), 75–86.

[ARS05]     Auer, C.; Russo, G.; Schürrer, F.: Adaptive energy discretization of the semiconductor Boltzmann equation, preprint.

[BDMS01]    Ben Abdallah, N.; Degond, P.; Markowich, P.; Schmeiser, C.: High field approximations of the spherical harmonics expansion model for semiconductors, Z. Angew. Math. Phys., **52**, (2001), 201–230.

[BSF03]     Bufler, F.M.; Schenk, A.; Fichtner, W.: Monte Carlo hydrodynamic and drift-diffusions simulation of scaled double-gate MOSFETs, J. Comput. Electron., **2**, (2003), 81–84.

[CCD02]     Cáceres, M.J.; Carrillo, J.A.; Degond, P.: The Child–Langmuir limit
            for semiconductors: a numerical validation, ESAIM: Mathematical
            Modelling and Numerical Analysis, **36**, (2002), 1161–1176.
[CCM06]     Cáceres, M.J.; Carrillo, J.A.; Majorana, A.: Deterministic simula-
            tion of the Boltzmann–Poisson system in GaAs-based semiconductors,
            SIAM Journal of Scientific Computing, **27**, (2006), 1981–2009.
[CGS00]     Carrillo, J.A.; Gamba, I.M.; Shu, C.-W.: Computational macroscopic
            approximations to the 1D relaxation-time kinetic system for semicon-
            ductors, Physica D, **146**, (2000), 289–306.
[CGMS03-1]  Carrillo, J.A.; Gamba, I.M.; Muscato, O.; Shu, C.-W.: Comparison
            of Monte Carlo and deterministic simulations of a silicon diode, IMA
            Volume Series **135**, (2003), 75–84.
[CGMS02]    Carrillo, J.A.; Gamba, I.M.; Majorana, A.; Shu, C.-W.: A WENO-
            solver for the 1D non-stationary Boltzmann–Poisson system for semi-
            conductor devices, J. Comput. Electron., **1**, (2002), 365–370.
[CGMS03-2]  Carrillo, J.A.; Gamba, I.M.; Majorana, A.; Shu, C.-W.: A WENO-
            solver for the transients of Boltzmann–Poisson system for semiconduc-
            tor devices. Performance and comparisons with Monte Carlo methods,
            J. Comput. Phys., **184**, (2003), 498–525.
[CGMS03-3]  Carrillo, J.A.; Gamba, I.M.; Majorana, A.; Shu, C.-W.: A direct solver
            for 2D non-stationary Boltzmann–Poisson systems for semiconduc-
            tor devices: a MESFET simulation by WENO-Boltzmann schemes, J.
            Comput. Electron., **2**, (2003), 375–380.
[CGMS06]    Carrillo, J.A.; Gamba, I.M.; Majorana, A.; Shu, C.-W.: 2D semicon-
            ductor device simulations by WENO-Boltzmann schemes: efficiency,
            boundary conditions and comparison to Monte Carlo methods, J.
            Comput. Phys., **214**, (2006), 55–80.
[CV06]      Carrillo, J.A.; Vecil, F.: Non oscillatory interpolation methods applied
            to Vlasov-based models, preprint.
[CGJS00]    Cercignani, C.; Jerome, J.W.; Gamba, I.M.; Shu, C.-W.: Device
            benchmark comparisons via kinetic, hydrodynamic, and high-field
            models, Computer Methods in Applied Mechanics and Engineering,
            **181**, (2000), 381–392.
[CC95]      Chen, Z.; Cockburn, B.: Analysis of a finite element method for the
            drift-diffusion semiconductor device equations: the multidimensional
            case, Num. Math., **71**, (1995), 1–28.
[CJSW98]    Chen, G.-Q.; Jerome, J.W.; Shu, C.-W.; Wang, D.: Two carrier semi-
            conductor device models with geometric structure and symmetry
            properties, Modelling and Computation for Applications in Mathe-
            matics, Science, and Engineering (ed. J. Jerome), Oxford University
            Press, London, 103–140, 1998.
[DDM90]     Degond, P.; Delaurens, F.; Mustieles, F.J.: Semiconductor modelling
            via the Boltzmann equation, Computing Methods in Applied Sciences
            and Engineering, SIAM, (1990), 311–324.
[D04]       Degond, P.: Macroscopic limits of the Boltzmann equation: a review,
            Modeling and computational methods for kinetic equations, Model.
            Simul. Sci. Eng. Technol., Birkhäuser Boston, (2004), 3–57.
[DS04]      Domaingo, A.; Schürrer, F.: Simulation of Schottky barrier diodes
            with a direct solver for the Boltzmann–Poisson system, J. Comput.
            Electron., **3**, (2004), 221–225.

[DGS05]    Domaingo, A.; Galler, M.; Schürrer, F.: A combined multicell-WENO solver for the Boltzmann–Poisson system of 1D semiconductor devices, Compel, **24**, (2005), 1311–1327.

[ES05]    Ertler, C.; Schürrer, F.: A deterministic study of hot phonon effects in a 2D electron gas channel formed at an AlGaN/GaN heterointerface, J. Comput. Electron., **5**, (2006), 15–26.

[FO93]    Fatemi, E.; Odeh, F.: Upwind finite difference solution of Boltzmann equation applied to electron transport in semiconductor devices, J. Comput. Phys., **108**, (1993), 209–217.

[F91]    Ferry, D.K.: Semiconductors, Maxwell Macmillan, New York, 1991.

[GS04]    Galler, M.; Schürrer, F.: A deterministic solution method for the coupled system of transport equations for the electrons and phonons in polar semiconductors, J. Phys. A, **37**, (2004), 1479–1497.

[GS05]    Galler, M.; Schürrer, F.: A deterministic solver for the 1D non-stationary Boltzmann–Poisson system for GaAs devices: bulk GaAs and GaAs n+/ni/n+ diode, J. Comput. Electron., **4**, (2005), 261–273.

[GSb05]    Galler, M.; Schürrer, F.: A deterministic solver for the transport of the AlGaN/GaN 2D electron gas including hot-phonon and degeneracy effects, J. Comput. Phys., **210**, (2005), 519–534.

[Ga05]    Galler, M.: Multigroup equations for the description of the particle transport in semiconductors, Series on Advances in Mathematics for Applied Sciences 70, World Scientific Publishing, Singapore, 2005.

[GM06]    Galler, M.; Majorana, A.: Deterministic and stochastic simulations of electron transport in semiconductors, to appear in Transport Theory and Stat. Phys.

[GS06]    Galler, M.; Schürrer, F.: A direct multigroup-WENO solver for the 2D non-stationary Boltzmann–Poisson system for GaAs devices: GaAs-MESFET, J. Comput. Phys., **212**, (2006), 778–797.

[Ga90]    Gamba, I.M: Behavior of the potential at the pn-Junction for a model in semiconductor theory, Appl. Math. Lett., **3**, (1990), 59–63.

[Ga93-1]    Gamba, I.M: Asymptotic boundary conditions for an oxide region in a semiconductor device, Asymptotic Anal., **7**, (1993), 37–48.

[Ga93-2]    Gamba, I.M: Asymptotic behavior at the boundary of a semiconductor device in two space dimensions, Ann. Mat. Pura App. (IV), **CLXIII**, (1993), 43–91.

[GP06]    Gamba, I.M; Proft, J.: Local discontinuous Galerkin schemes to linear Boltzmann equations. Analysis and simulations, preprint.

[GCG1]    González, P.; Godoy, A.; Gámiz, F.; Carrillo, J.A.: Accurate deterministic numerical simulation of p-n junctions, Journal of Computational Electronics **3**, (2004), 235–238.

[GCG2]    González, P.; Carrillo, J.A.; Gámiz, F.: Deterministic Numerical Simulation of 1D kinetic descriptions of Bipolar Electron Devices, in Anile, A.M., Ali, G.; Mascali, G. (eds.) Scientific Computing in Electrical Engineering Series: Mathematics in Industry Subseries: The European Consortium for Mathematics in Industry, Vol. 9 Springer, Berlin, (2006), 339–344.

[Gr85]    Grisvard, P.: Elliptic problems in non-smooth domains, Monographs and Studies in Mathematics, **24**, Pitman, London 1985.

[JS94]      Jerome, J.W.; Shu, C.-W.: Energy models for one-carrier transport in semiconductor devices, IMA Volumes in Mathematics and Its Applications, **59**, (1994), 185–207.

[JS96]      Jiang, G.; Shu, C.-W.: Efficient implementation of weighted ENO schemes, J. Comput. Phys., **126**, (1996), 202–228.

[JPMRB06]   Jungemann, C.; Pham, A.; Meinerzhagen, B.; Ringhofer, C.; Boellhofer, M.: Stable discretization of the Boltzmann equation based on spherical harmonics, box integration and a maximum entropy dissipation principle, preprint 2006.

[L00]       Lundstrom, M.: Fundamentals of Carrier Transport, Cambridge University Press, Cambridge, 2000.

[MMM04]     Majorana, A.; Milazzo, C.; Muscato, O.: Charge transport in 1D silicon devices via Monte Carlo simulation and Boltzmann–Poisson solver, COMPEL, **23**, (2004), 410–425.

[MP01]      Majorana, A.; Pidatella, R.M.: A finite difference scheme solving the Boltzmann–Poisson system for semiconductor devices, J. Comput. Phys., **174**, (2001), 649–668.

[MCM05]     Mantas, J.M.; Carrillo J.A.; Majorana, A.: Parallelization of WENO-Boltzmann schemes for kinetic descriptions of 2D semiconductor devices, in Anile, A.M., Ali, G.; Mascali, G. (eds.) Scientific Computing in Electrical Engineering Series: Mathematics in Industry Subseries: The European Consortium for Mathematics in Industry, Vol. 9 Springer, Berlin, (2006), 357–362.

[MR05]      Mascali, G.; Romano, V.: Simulation of Gunn oscillations with a non-parabolic hydrodynamical model based on the maximum entropy principle, Compel, **24**, (2005), 35–54.

[MRS90]     Markowich, P.A.; Ringhofer, C.; Schmeiser, C.: Semiconductor Equations, Springer-Verlag, New York, 1990.

[R00]       Ringhofer, C.: Space-time discretization of series expansion methods for the Boltzmann transport equation, SIAM J. Numer. Anal., **38**, (2000), 442–465.

[R03]       Ringhofer, C.: A mixed spectral-difference method for the steady state Boltzmann–Poisson system, SIAM J. Numerical Analysis **41**, (2003), 64–89.

[Sel84]     Selberherr, S.: Analysis and Simulations of Semiconductor Devices, Springer, Vienna, 1984.

[Sh98]      Shu, C.-W.: Essentially non-oscillatory and weighted essentially non-oscillatory schemes for hyperbolic conservation laws, Lecture Notes in Mathematics **1697**, (1998), 325–432.

[To93]      Tomizawa, K.: Numerical Simulation of Submicron Semiconductor Devices, Artech House, Boston, 1993.

# Part III

# Miscellaneous Applications in Physics and Natural Sciences

# Methods and tools of mathematical kinetic theory towards modelling complex biological systems

Nicola Bellomo,[1] Abdelghani Bellouquid[2] and Marcello Delitala[1]

[1] Department of Mathematics, Politecnico, Torino, Italy
nicola.bellomo@polito.it   marcello.delitala@polito.it
[2] University Cadi Ayyad, Ecole Nationale des Sciences Appliquées, Safi, Maroc,
bellouq2002@yahoo.fr

## 8.1 Introduction

Methods of mathematical kinetic theory have been recently developed to describe the collective behavior of large populations of interacting individuals such that their microscopic state is identified not only by a mechanical variable (typically position and velocity), but also by a biological state (or socio-biological state) related to their organized, somehow intelligent, behavior. The interest in this type of mathematical approach is documented in the collection of surveys edited in [1], in the review papers [2], [3], and in the book [4].

A generalization of the classical Boltzmann equation has been proposed in [5] showing that it includes, as a particular case, the model of mathematical kinetic theory [6]. Following [7], the interacting entities are called *active particles*, while the variable which describes their state, called the *microscopic state*, includes a *mechanical state*, classically position and velocity, and an additional variable, called *activity*, which describes the specific functions of the particles related to their socio-biological functions. The overall behavior of the system is delivered by mathematical equations suitable to describe the evolution of the statistical distribution over the above microscopic state.

This survey is precisely focused on modelling of complex multicellular systems in biology by a suitable generalization of the mathematical kinetic theory. This approach was first introduced in [8] with specific reference to the competition between tumor and immune cells. The model has been subsequently developed by various authors, e.g., [9]–[17]. The content is organized into four more sections which follow this introduction. Section 8.2 describes a general mathematical framework which is proposed as a fundamental paradigm towards the modelling of specific biological systems with continuous and discrete representation of the activity. Section 8.3 deals with the analysis of a specific model of the competition between immune cells and abnormal cells or particle carriers of a pathology. Section 8.4 discusses the derivation of macroscopic equations from the underlying microscopic description offered by generalized kinetic theory. Section 8.5 develops a critical analysis addressed to show how the mathematical framework proposed in this chapter can be further developed towards relatively more accurate models of complex biological systems, possibly looking at a proper bio-mathematical theory.

## 8.2 Mathematical framework

The mathematical framework described in this section is proposed as a conceivable paradigm towards the derivation of specific models of complex biological systems. Specifically, we consider a large system of interacting cells organized into $n$ populations labelled by the index $i = 1, \ldots, n$; each population is characterized by a different way of organizing their peculiar activities as well as by the interactions with the other populations.

Modelling by methods of mathematical kinetic theory essentially means defining the microscopic state of the cells, the distribution function over that state, and deriving an evolution equation for the distribution. In detail, the physical variable charged to describe the state of each cell is called the *microscopic state*, denoted by: $\mathbf{w} = \{\mathbf{x}, \mathbf{v}, \mathbf{u}\}$, where $\mathbf{x} \in D_{\mathbf{x}}$ is the *position*, $\mathbf{v} \in D_{\mathbf{v}}$ is the *velocity*, and $\mathbf{u} \in D_{\mathbf{u}}$ is the *biological microscopic state* or *activity*. The space of the variable $\mathbf{w}$ is called the *space of the microscopic states*.

The description of the overall state of the system is given by the one-cell *generalized distribution function*:

$$f_i = f_i(t, \mathbf{w}) = f_i(t, \mathbf{x}, \mathbf{v}, \mathbf{u}), \qquad i = 1, \ldots, n, \qquad (8.2.1)$$

which is such that $f_i(t, \mathbf{w}) \, d\mathbf{w}$ denotes the number of cells whose state, at time $t$, is in the interval $[\mathbf{w}, \mathbf{w} + d\mathbf{w}]$.

In addition to the above continuous description, we also consider systems where the *activity* is a discrete variable: $\mathbf{u} = \{\mathbf{u}_1, \ldots, \mathbf{u}_h, \ldots, \mathbf{u}_H\}$ with components $\mathbf{u}_h$, where $h = 1, \ldots, H$. In this case, the statistical distribution suitable for describing the state of the system is the discrete one-particle distribution function

$$f_i^h = f_i^h(t, \mathbf{x}, \mathbf{v}), \qquad (8.2.2)$$

where $f_i^h$, corresponds to cells of the $i$th population with state $\mathbf{u}_h$, while the overall state of the system is described by the whole set $\mathbf{f}_i = \{f^h\}_{h=1}^{H}$ of all distribution functions $f_i^h$, and the space and velocity variables are still assumed to be continuous. A generalized kinetic theory for systems with discrete states has been proposed in [18] and further developed in [19]. Motivations for the assumptions of discrete states have been given in [18], while further critical analysis is proposed in the sections which follow.

The first step towards the derivation of an evolution equation for the above distribution function consists in the modelling of microscopic interactions between pairs of cells. Specifically, the following types of binary interactions are taken into account:

– *Conservative interactions*, between *candidate* or *test* cells and *field* cells, which modify the microscopic activity of the interacting cells, but not the size of the population;

– *Proliferating or destructive interactions*, between *test cells* and *field cells*, which generate death or birth of test particles;

– *Stochastic interactions* which modify the velocity of the particles according to a suitable velocity jump process.

The evolution equation is obtained by equating, in the elementary volume of the state space, the rate of increase of particles with microscopic state $\mathbf{w}$ to the net flux of particles which attain such a state due to microscopic interactions. The result of detailed calculations, in the continuous case, is as follows:

$$\left(\partial_t + \mathbf{v} \cdot \nabla_{\mathbf{x}}\right) f_i(t, \mathbf{x}, \mathbf{v}, \mathbf{u})$$

$$= \nu \sum_{j=1}^{n} \int_{D_{\mathbf{v}}} \left[ T(\mathbf{v}, \mathbf{v}^*) f_i(t, \mathbf{x}, \mathbf{v}^*, \mathbf{u}) - T(\mathbf{v}^*, \mathbf{v}) f_j(t, \mathbf{x}, \mathbf{v}, \mathbf{u}) \right] d\mathbf{v}^*$$

$$+ \sum_{j=1}^{n} \eta_{ij} \int_{D_{\mathbf{u}} \times D_{\mathbf{u}}} \mathcal{B}_{ij}(\mathbf{u}_*, \mathbf{u}^*; \mathbf{u}) f_i(t, \mathbf{x}, \mathbf{v}, \mathbf{u}_*) f_j(t, \mathbf{x}, \mathbf{v}, \mathbf{u}^*)\, d\mathbf{u}_*\, d\mathbf{u}^*$$

$$- f_i(t, \mathbf{x}, \mathbf{v}, \mathbf{u}) \sum_{j=1}^{n} \int_{D_{\mathbf{u}}} \eta_{ij} \left[1 - \mu_{ij}(\mathbf{u}, \mathbf{u}^*)\right] f_j(t, \mathbf{x}, \mathbf{v}, \mathbf{u}^*)\, d\mathbf{u}^*\,,$$

(8.2.3)

while for discrete activity analogous calculations yield

$$\left(\partial_t + \mathbf{v} \cdot \nabla_{\mathbf{x}}\right) f_i^h(t, \mathbf{x}, \mathbf{v})$$

$$= \nu \sum_{j=1}^{n} \int_{D_{\mathbf{v}}} \left[ T(\mathbf{v}, \mathbf{v}_*) f_i^h(t, \mathbf{x}, \mathbf{v}_*) - T(\mathbf{v}_*, \mathbf{v}) f_j^h(t, \mathbf{x}, \mathbf{v}) \right] d\mathbf{v}_*$$

$$+ \sum_{j=1}^{n} \sum_{p=1}^{H} \sum_{q=1}^{H} \eta_{ij} \mathcal{B}^{pq}(h) f_i^p(t, \mathbf{x}, \mathbf{v}) f_j^q(t, \mathbf{x}, \mathbf{v})$$

(8.2.4)

$$- f_i^h(t, \mathbf{x}, \mathbf{v}) \sum_{j=1}^{n} \sum_{q=1}^{H} \eta_{ij} f_j^q(t, \mathbf{x}, \mathbf{v}) [1 - \mu_{ij}^{hq}] f_j^q(t, \mathbf{x}, \mathbf{v})\,,$$

for $h = 1, \ldots, H$. The following definitions also apply.

• The linear transport term has been proposed by various authors to describe the dynamics of biological organisms modelled by a velocity-jump process, where $\nu$ is the turning rate or turning frequency (hence $\tau = 1/\nu$ is the mean run time), and $T(\mathbf{v}, \mathbf{v}^*)$ is the probability kernel for the new velocity $v \in D_{\mathbf{v}}$ assuming that the previous velocity was $\mathbf{v}^*$. This corresponds to the assumption that cells choose any direction with bounded velocity. Specifically, the set of possible velocities is denoted by $D_{\mathbf{v}}$, where $D_{\mathbf{v}} \subset \mathbb{R}^3$, and it is assumed that $D_{\mathbf{v}}$ is bounded and spherically symmetric (i.e., $\mathbf{v} \in D_{\mathbf{v}} \Rightarrow -\mathbf{v} \in D_{\mathbf{v}}$).

• The *interaction rate* between pairs of particles is denoted by $\eta_{ij}$ and is assumed to be a constant independent of the activities of the interacting particles.

• The *transition probability density*, in the continuous case $\mathcal{B}_{ij}(\mathbf{u}_*, \mathbf{u}^*; \mathbf{u})$, models the transition probability density of the candidate cell (of the $i$th population) with state $\mathbf{u}_*$ into the state $\mathbf{u}$ of the test particle after the interaction with the field cell (of the $j$th population) with state $u^*$. In the discrete case: $\mathcal{B}^{pq}(h) = \mathcal{B}(\mathbf{u}_p, \mathbf{u}_q; \mathbf{u}_h)$ models the transition probability density of a *candidate* particle with state $\mathbf{u}_p$ into the state $\mathbf{u}_h$ of the *test* particle after an interaction with a *field* particle with state $\mathbf{u}_q$. The above-defined transition density functions have the structure of a probability density with respect to the variables $\mathbf{u}$ and $\mathbf{u}_h$, respectively.

• The *proliferation rate* $\mu_{ij}(\mathbf{u}, \mathbf{u}^*)$ models, in the continuous case, the proliferation density due to encounters, with rate $\eta_{ij}$, between the test cell (of the $i$th population) with state $\mathbf{u}$ with the field cell (of the $j$th population) with state $u^*$. The analogous term in the discrete case is given by $\mu_{ij}^{hq}$.

If $f_i$ is known, then macroscopic gross variables can be computed, under suitable integrability properties, as moments weighted by the above distribution function. For instance, the *local size* of the $i$th population is given by

$$n_i(t, \mathbf{x}) = \int_{D_{\mathbf{v}} \times D_{\mathbf{u}}} f_i(t, \mathbf{x}, \mathbf{v}, \mathbf{u}) \, d\mathbf{v} \, d\mathbf{u}. \qquad (8.2.5)$$

The local initial size of the $i$th population, at $t = 0$, is denoted by $n_{i0}$, while the local size for all populations is denoted by $n_0$ and is given by

$$n_0(\mathbf{x}) = \sum_{i=1}^{n} n_{i0}(\mathbf{x}). \qquad (8.2.6)$$

Marginal densities may refer either to the generalized distribution over the mechanical state,

$$f_i^m(t, \mathbf{x}, \mathbf{v}) = \int_{D_{\mathbf{u}}} f_i(t, \mathbf{x}, \mathbf{v}, \mathbf{u}) \, d\mathbf{u}, \qquad (8.2.7)$$

or to the generalized distribution over the socio-biological state,

$$f_i^b(t, \mathbf{u}) = \int_{D_{\mathbf{x}} \times D_{\mathbf{v}}} f_i(t, \mathbf{x}, \mathbf{v}, \mathbf{u}) \, d\mathbf{x} \, d\mathbf{v} \,. \tag{8.2.8}$$

First-order moments provide either *linear mechanical macroscopic* quantities, or *linear socio-biological macroscopic* quantities. Focusing on biological functions, the linear moments related to each $j$th component of the state $\mathbf{u}$, related to the $i$th populations, can be called the *local activation* at the time $t$ in the position $\mathbf{x}$, and are computed as follows:

$$A_{ij} = A_j[f_i](t, \mathbf{x}) = \int_{D_{\mathbf{v}} \times D_{\mathbf{u}}} u_j f_i(t, \mathbf{x}, \mathbf{v}, \mathbf{u}) \, d\mathbf{v} \, d\mathbf{u} \,, \tag{8.2.9}$$

while the *local activation density* is given by

$$\mathcal{A}_{ij} = \mathcal{A}_j[f_i](t, \mathbf{x}) = \frac{A_j[f_i](t, \mathbf{x})}{n_i(t, \mathbf{x})} = \frac{1}{n_i(t, \mathbf{x})} \int_{D_{\mathbf{v}} \times D_{\mathbf{u}}} u_j f_i(t, \mathbf{x}, \mathbf{v}, \mathbf{u}) \, d\mathbf{v} \, d\mathbf{u} \,. \tag{8.2.10}$$

Global quantities are obtained by integrating over space. A different interpretation can be given for each of the above quantities. Large values of $A_{ij}$ may be due to a large number of cells with relatively small values of the $j$th biological function, but also to a small number of cells with relatively large values of the $j$th biological function, while $\mathcal{A}$ allows us to identify the size of the mean value of the activation.

Analogous calculations can be developed in the case of discrete activities. For instance, consider a system constituted by one population only. The density is given by

$$n[\mathbf{f}](t, \mathbf{x}) = \sum_{h=1}^{H} \int_{D_{\mathbf{v}}} f^h(t, \mathbf{x}, \mathbf{v}) \, d\mathbf{v} \,, \tag{8.2.11}$$

and, focusing on the activity terms, the linear moments related to each $h$th component of the state $\mathbf{u}$ are called the *activation* at the time $t$ in the position $\mathbf{x}$, and are computed as follows:

$$a^h = a^h[f^h](t, \mathbf{x}) = \int_{D_{\mathbf{v}}} \mathbf{u}_h f^h(t, \mathbf{x}, \mathbf{v}) \, d\mathbf{v} \,. \tag{8.2.12}$$

The *activation density* $\mathcal{A}^h[f^h]$ is obtained by dividing $a^h$ by the density $n$ given in (8.2.12).

Finally, we stress that the preceding mathematical structures offer a background for modelling which cannot, however, cover the whole variety of cellular phenomena. The sections which follow investigate how far modelling issues can take advantage of the structures offered in this section.

## 8.3 An example of the mathematical model

The first application of the mathematical framework proposed in Section 8.2 deals with the modelling of the competition between immune and progressing cells. Following Greller, Tobin, and Poste [20], the system is assumed to be constituted by two cellular populations. The microscopic state in the first population is the *progression*, which denotes how far cells are from the normal state. In more detail, $u \in \mathbb{R}$, where $u \leq 0$ identifies the state of normal cells, while $u > 0$ is the state of abnormal cells, where the degree of transformation from normal increases with increasing progression: growth autonomous, tissue invasive, metastatically competent. The microscopic state in the second population is the *activation*, which denotes how far immune cells are active to contrast the abnormal state. In this case, $u \in \mathbb{R}$, where $u \leq 0$ identifies the state of inhibited immune cells, while $u > 0$ is the state of active cells, where the degree of ability to contrast abnormal cells increases with increasing activation.

This system appears to be simpler than those considered in Section 8.2. Therefore the mathematical structure for modelling reduces to the following:

$$\partial_t f_i(t, u) = \sum_{j=1}^{2} \int \eta_{ij} \mathcal{B}_{ij}(u_*, u^*; u) f_i(t, u_*) f_j(t, u^*) \, du_* \, du^*$$

$$(8.3.1)$$

$$-f_i(t, u) \sum_{j=1}^{2} \int \eta_{ij} [1 - \mu_{ij}(u, u^*)] f_j(t, u^*) \, du^* ,$$

where, referring to the *transition probability density* $\mathcal{B}_{ij}$ related to *conservative interactions*, it is assumed that $\mathcal{B}_{ij}$ is a delta function over the most probable output $m_{ij}(u_*, u^*)$, which depends on the microscopic states $u_*$ and $u^*$ of the interacting pairs.

The following specific interactions are modelled by suitable phenomenological assumptions:

### • Interactions between cells of the first population

**H.1:** The most probable output of conservative interactions is given as follows:

$$u_*, u^* \in \mathbb{R} : m_{11} = u_* + \alpha_{11} ,$$

where $\alpha_{11}$ is a parameter related to the inner tendency of both a normal and progressing endothelial cell to degenerate and progress.

**H.2:** The proliferation rate of normal endothelial cells due to encounters with other endothelial cells is equal to zero. On the other hand, when $u_* \geq 0$, cells undergo uncontrolled mitosis stimulated by encounters with nonprogressing cells $u_* < 0$, which have a feeding ability

$$p_{11}(u_*, u^*) = \beta_{11} U_{[0,\infty)}(u_*) U_{(-\infty,0)}(u^*) ,$$

where $\beta_{11}$ is a parameter which characterizes the proliferating ability of tumor cells. Endothelial cells which are used for the proliferation are constantly replaced by the outer environment. Encounters between progressing cells do not lead to any proliferation or destruction.

**• Interactions between cells of the first with the second population**

**H.3:** The state of cells of the first population, if not progressing, does not change due to interactions with immune cells. If it is progressing, then its state does not change if the immune cell is not active.

**H.4:** The proliferation rate of nonprogressing cells due to encounters with immune cells is equal to zero. On the other hand, when $u_* \geq 0$, cells are partially destroyed due to encounters with active immune cells:

$$\mu_{12}(u_*, u^*) = -\beta_{12}U_{[0,\infty)}(u_*)U_{[0,\infty)}(u^*),$$

where $\beta_{12}$ is a parameter which characterizes the destructive ability of active immune cells.

**• Interactions between cells of the second with the first population**

**H.5:** Immune cells do not change state due to interactions with nonprogressing endothelial cells. Moreover, if the cell is inhibited, its state does not change on interacting with progressing endothelial cells. On the other hand, for positive values of $u^*$, the most probable output is given as follows:

$$u_* \geq 0, u^* \geq 0: \quad m_{21} = u_* - \alpha_{21},$$

where $\alpha_{21}$ is a parameter which indicates the ability of tumor cells to inhibit immune cells.

**H.6:** The proliferation rate of inhibited immune cells due to encounters with cells of the first population is equal to zero. On the other hand, when $u^* \geq 0$, cells proliferate due to encounters with progressing cells:

$$\mu_{21}(u_*, u^*) = \beta_{21}U_{[0,\infty)}(u_*)U_{[0,\infty)}(u^*), \qquad \mu_{22} = 0,$$

where $\beta_{21}$ is a parameter which characterizes the proliferating ability of tumor cells.

**• Interactions between cells of the second population.** It is assumed that this type of interactions does not modify the state of cells.

Substituting these microscopic models into Eq. (8.3.1) yields

$$
\begin{cases}
\partial_t f_1(t, u) = n_1(t)[f_1(t, u - \alpha_{11}) - f_1(t, u)] \\
\qquad + f_1(t, u)\left[\beta_{11}n_1^E(t) - \beta_{12}n_2^A(t)\right]U_{[0,\infty)}(u), \\
\\
\partial_t f_2(t, u) = n_1^T(t)\left[f_2(t, u + \alpha_{21})U_{[0,\infty)}(u + \alpha_{21})\right. \\
\qquad \left. + (\beta_{21} - 1)f_2(t, u)U_{[0,\infty)}(u)\right],
\end{cases}
\tag{8.3.2}
$$

where

$$n_1(t) = n_1^E(t) + n_1^T(t) = \int_{-\infty}^{0} f_1(t, u)\, du + \int_{0}^{\infty} f_1(t, u)\,, \qquad (8.3.3)$$

and

$$n_2^A(t) = \int_{0}^{\infty} f_2(t, u)\, du\,. \qquad (8.3.4)$$

The model is characterized by five phenomenological parameters, where the $\alpha$-type parameters are related to mass conservative encounters, while the $\beta$-type parameters are related to proliferating/destructive encounters. All parameters are positive quantities (eventually equal to zero) small with respect to unity. In detail:

$\alpha_{11}$ refers to the variation of the progression due to encounters between endothelial cells. It describes the tendency of a normal cell to degenerate and to increase its progression;

$\alpha_{21}$ corresponds to the ability of tumor cells to inhibit the active immune cells;

$\beta_{11}$ refers to the proliferation rate of tumor cells due to their encounters with normal endothelial cells;

$\beta_{12}$ refers to the ability of immune cells to destroy tumor cells;

$\beta_{21}$ corresponds to the proliferation rate of immune cells due to their interaction with progressed cells.

This model takes into account the ability of progressing cells to inhibit immune cells. When this ability cannot be expressed one has $\alpha_{21} = 0$, while the model express a pathology which is relatively less dangerous:

$$\begin{cases} \partial_t f_1(t, u) = n_1(t)[f_1(t, u - \alpha_{11}) - f_1(t, u)] \\[2mm] \qquad\qquad + f_1(t, u)[\beta_{11} n_1^E(t) - \beta_{12} n_2^A(t)] U_{[0,\infty)}(u)\,, \qquad (8.3.5) \\[2mm] \partial_t f_2(t, u) = \beta_{21} n_1^T(t) f_2(t, u) U_{[0,\infty)}(u)\,. \end{cases}$$

The most dangerous case is when the immune system is totally depressed. In this case $f_2 \cong 0$, and the model for the continuous growth of progressing cells is written as

$$\partial_t f_1(t, u) = n_1(t)\big(f_1(t, u - \alpha_{11}) - f_1(t, u)\big) + f_1(t, u)\beta_{11} n_1^E(t) U_{[0,\infty)}(u)\,. \qquad (8.3.6)$$

A qualitative analysis of the solutions to the initial value problem obtained by linking suitable initial conditions to the preceding models can be developed by application of the classical fixed point theorem. The following function spaces need to be defined:

- $L_1(\mathbb{R})$ is the Lebesgue space of measurable, real-valued functions which are integrable on $\mathbb{R}$. The norm is denoted by $\| \cdot \|_1$.
- $\mathcal{X} = L_1(\mathbb{R}) \times L_1(\mathbb{R}) = \{f = (f_1, f_2) : f_1 \in L_1(\mathbb{R}), f_2 \in L_1(\mathbb{R})\}$ is the Banach space equipped with the norm

$$\| f \| = \| f_1 \|_1 + \| f_2 \|_1 . \tag{8.3.7}$$

- $\mathcal{X}_+ = \{f = (f_1, f_2) \in \mathcal{X} : f_1 \geq 0, f_2 \geq 0\}$ is the positive cone of $\mathcal{X}$.
- $\mathcal{Y} = C([0,T], \mathcal{X})$ and $\mathcal{Y}_+ = C([0,T], \mathcal{X}_+)$ are the spaces of the functions continuous on $[0,T]$ with values, respectively, in a Banach space $\mathcal{X}$ and $\mathcal{X}_+$, equipped with the norm

$$\| f \|_{\mathcal{Y}} = \sup_{t \in [0,T]} \| f \| . \tag{8.3.8}$$

A detailed analysis is given in Chapter 3 of [4]. Local and global existence of the solutions to the value problem are stated by the following theorems.

**Theorem 1.** *Let $f_0 \in \mathcal{X}_+$. Then there exists a positive constants $T$ and $a_0$, such that the initial value problem (8.3.5) has a unique solution $f \in C([0,T], \mathcal{X}_+)$. The solution $f$ satisfies*

$$f(t) \in \mathcal{X}_+, \quad t \in [0,T], \tag{8.3.9}$$

*and*

$$\| f \| \leq a_0 \| f_0 \|, \quad \forall t \in [0,T]. \tag{8.3.10}$$

**Theorem 2.** *$\forall T > 0$, there exists a unique solution $f \in C([0,T], \mathcal{X})$ of (8.3.5) with the initial data, $f_0 \in \mathcal{X}_+$. The solution satisfies (8.3.9), and for some constant $C_T$ depending on $T$ and on the initial data,*

$$\sup_{t \in [0,T]} f(t) \leq C_T . \tag{8.3.11}$$

In general, it is interesting to analyze the influence of the parameters of the model and of the mathematical problem on the bifurcation separating two different behaviors:

i) "Blow up" of progressing cells, while the immune cells are inhibited;

ii) Destruction of progressing cells due to the action of the immune system which remains sufficiently active.

**Theorem 3.** *Consider the initial value problem for Eqs. (8.3.5) and let*

$$\lambda = (1 + \beta_{11}) n_1^E(0) - \beta_{12} n_2^A(0) .$$

*Then $n_2^A$ increases, $n_1^E$ decreases with time, and*

- If $\beta_{12} = 0$ then $n_1^T$ increases.
- If $\beta_{12} \neq 0$, then $n_1^T$ satisfies the following inequality which depends on initial data:

$$n_1^T \leq \exp(\lambda t)\left(n_{10}^T + \frac{(n_{10}^E)^2}{l}\right) - \frac{(n_{10}^E)^2}{l}. \tag{8.3.12}$$

In particular, if $\lambda < 0$, then the following estimate for $n_1^T(\infty)$ holds:

$$n_1^T(\infty) \leq -\frac{(n_{10}^E)^2}{l}. \tag{8.3.13}$$

*Proof.* The equations satisfied by $n_1^T$, $n_2^A$, and $n_1^E$ are the following:

$$\partial_t n_1^T = n_1 \int_{-\alpha_{11}}^0 f_1(t, u)du + n_1^T(\beta_{11}n_1^E - \beta_{12}n_2^A), \tag{8.3.14}$$

$$\partial_t n_1^E = -n_1(t) \int_{-\alpha_{11}}^0 f_1(t, u)du, \tag{8.3.15}$$

and

$$\partial_t n_2^A = \beta_{21} n_1^T n_2^A. \tag{8.3.16}$$

If $\beta_{12} = 0$, then, from (8.3.14)–(8.3.16) it follows that $n_1^T, n_2^A$ increases and $n_1^E$ decreases. Let $\beta_{12} \neq 0$, then using $n_1 = n_1^E + n_1^T$ and by

$$\int_{-\alpha_{11}}^0 f_1(t, u)du \leq n_1^E, \tag{8.3.17}$$

we prove the following estimate:

$$\partial_t n_1^T \leq (n_1^E)^2 + n_1^T((1 + \beta_{11})n_1^E(0) - \beta_{12}n_2^A(0)) = (n_1^E(0))^2 + l n_1^T. \tag{8.3.18}$$

Then, using Gronwall's lemma yields

$$n_1^T \leq \exp(lt)n_1^T(0) - \frac{(n_1^E(0))^2}{l}(1 - \exp(lt)) \tag{8.3.19}$$

which proves the theorem.     $\square$

From a biological point of view, the result of Theorem 3 predicts a trend to degenerate the normal cells which is not contrasted by immune cells, as $\alpha_{12} = 0$, and abnormal cells cannot inhibit immune cells, as $\alpha_{21} = 0$. Moreover, a reduction of abnormal cells is expected for specific sets of the $\beta$-type parameters and of the initial conditions.

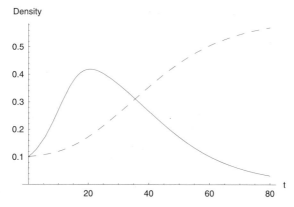

**Fig. 8.1.** Depletion of abnormal cells and activation of immune cells.

Simulations have confirmed and enhanced these results, giving a picture of the evolution not only of the densities but also of the distribution functions. Figure 8.1 shows the total depletion of abnormal cells with a growth in number of immune cells.

Conversely, if the nonconservative parameters and the initial conditions are chosen in such a way that $l > 0$, Theorem 1 gives no information and, from the computational analysis, we obtain an increase of the state of abnormal cells, while their density, after an initial growth, is reduced by the competition with immune cells which are stimulated to grow; see Figures 8.1, 8.2, and 8.3. Thus, the density of abnormal cells, eventually after a growth stage, is reduced by the immune cells.

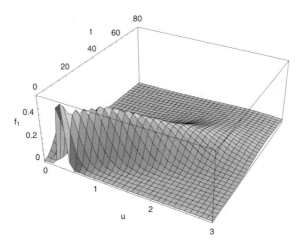

**Fig. 8.2.** $\alpha_{11} = 0.1$, $\alpha_{21} = 0$, and $l > 0$: Abnormal cells increase their progression, but finally are depleted.

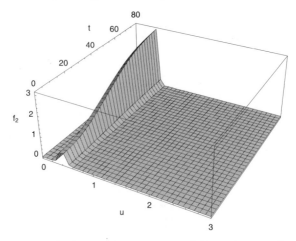

**Fig. 8.3.** $\alpha_{11} = 0.1$, $\alpha_{21} = 0$, and $l > 0$: Immune cells activate to destroy abnormal cells.

These results, from a biological viewpoint, show that when abnormal cells are not able to inhibit immune cells, they are asymptotically destroyed. We have to point out that once the density of abnormal cells reaches a certain threshold (which may be identified in comparison with suitable experimental and medical results), the host may not survive.

## 8.4 Asymptotic limit

One of the relevant issues of mathematical kinetic theory is the derivation of macroscopic equations from the underlying microscopic description. References to the existing literature in classical kinetic theory are reported in the surveys [21] and [22].

Analogous problems arise in generalized kinetic theory when applied to modelling multicellular systems. The above analysis plays an important role in understanding properties of biological tissues considering that the derivation of macroscopic equations is generally based on heuristic reasonings and that different models are often used to describe the same system. Surveys of different macroscopic models related to cancer modelling are proposed in [23], [24] as well as in the review paper [2].

The detailed analysis developed in [25] has shown that specific models of cancer growth [26] can be obtained from the underlying microscopic description under suitable assumptions on cellular interactions. Actually, different assumptions lead to different equations, This is not contradictory considering that the properties of biological tissues evolve in time as we have seen in Section 8.3.

The analysis developed in what follows aims to show how macroscopic equations can be derived from kinetic equations and that their structure depends on the ratio between the various interaction rates: mechanical, conservative biological, and proliferating/destructutive. Detailed calculations are proposed, with reference to [27] and [28], for the relatively simpler case of model (8.2.3), (8.2.4) for one population only. This issue is dealt with simply so that we can present a methodological approach to be properly developed with reference to specific models.

Bearing all this in mind, consider Eq. (8.2.4) in the case of one population only, where, to simplify its structure, we take $\mu_{pq} = \gamma\,\mu$, $\forall p, q$, and use the following scaling:

$$\eta = \varepsilon, \qquad \gamma\,\mu = \varepsilon^r, \qquad \nu = \frac{1}{\varepsilon^p},$$

where $\varepsilon$ is a small parameter that we let tend to zero. In addition, the diffusion scale time $\tau = \varepsilon t$ will be used so that the following scaled equation is obtained:

$$\varepsilon \partial_t f_\varepsilon^h(t, \mathbf{x}, \mathbf{v}) + \mathbf{v} \cdot \nabla_\mathbf{x} f_\varepsilon^h(t, \mathbf{x}, \mathbf{v}) = \frac{1}{\varepsilon^p}\,\mathcal{L} f_\varepsilon^h + \varepsilon\,\Gamma_h(\mathbf{f}_\varepsilon, \mathbf{f}_\varepsilon) + \varepsilon^r\,I_h(\mathbf{f}_\varepsilon, \mathbf{f}_\varepsilon), \tag{8.4.1}$$

for $h = 1, \ldots, H$, and where

$$\Gamma_h(\mathbf{f}, \mathbf{f}) = \left( \sum_{p=1}^{H} \sum_{q=1}^{H} \mathcal{B}^{pq}(h) f^p(t, \mathbf{x}, \mathbf{v}) f^q(t, \mathbf{x}, \mathbf{v}) - f^h(t, \mathbf{x}, \mathbf{v}) \sum_{q=1}^{H} f^q(t, \mathbf{x}, \mathbf{v}) \right)_h, \tag{8.4.2}$$

and

$$I_h(\mathbf{f}, \mathbf{f}) = \left( f^h(t, \mathbf{x}, \mathbf{v}) \sum_{q=1}^{H} f^q(t, \mathbf{x}, \mathbf{v}) \right)_h, \tag{8.4.3}$$

for $\mathbf{f} = (f^1, f^2, \ldots, f^H)$.

**Theorem 4.** *Suppose that there exists a bounded velocity distribution $M(\mathbf{v}) > 0$, independent of $x$ and $t$, such that the detailed balance*

$$T(\mathbf{v}_*, \mathbf{v}) M(\mathbf{v}) = T(\mathbf{v}, \mathbf{v}_*) M(\mathbf{v}_*) \tag{8.4.4}$$

*holds. The flow produced by this equilibrium distribution vanishes, and $M$ is normalized:*

$$\int_{D_\mathbf{v}} \mathbf{v} M(\mathbf{v})\, d\mathbf{v} = 0, \qquad \int_{D_\mathbf{v}} M(\mathbf{v})\, d\mathbf{v} = 1. \tag{8.4.5}$$

*The kernel $T(\mathbf{v}, \mathbf{v}_*)$ is bounded, and there exists a constant $\sigma > 0$ such that*

$$T(\mathbf{v}, \mathbf{v}_*) \geq \sigma M, \quad \forall (\mathbf{v}, \mathbf{v}_*) \in D_\mathbf{v} \times D_\mathbf{v}, \quad \mathbf{x} \in \mathbb{R}^3, t > 0. \tag{8.4.6}$$

*Let $f_\varepsilon^h(t, \mathbf{x}, \mathbf{v})$ be a sequence of solutions to the scaled kinetic equation (8.4.1) such that $f_\varepsilon^h$ converges, in the distributional sense, to a function $f^h$ as $\varepsilon$ goes to zero. Furthermore, assume that the moments*

$$\langle f_\varepsilon^h \rangle, \quad \left\langle \frac{k(\mathbf{v})}{M(\mathbf{v})} \otimes \mathbf{v} f_\varepsilon^h \right\rangle, \quad \langle \Gamma_h(\mathbf{f}_\varepsilon, \mathbf{f}_\varepsilon) \rangle, \quad \langle I_h(\mathbf{f}_\varepsilon, \mathbf{f}_\varepsilon) \rangle, \quad h = 1, \dots, H$$

$$(8.4.7)$$

*converge in $D'(t, \mathbf{x})$ to the corresponding moments*

$$\langle f^h \rangle, \quad \left\langle \frac{k(\mathbf{v})}{M(\mathbf{v})} \otimes \mathbf{v} f^h \right\rangle, \quad \langle \Gamma_h(\mathbf{f}, \mathbf{f}) \rangle, \quad \langle I_h(\mathbf{f}, \mathbf{f}) \rangle, \quad h = 1, \dots, H,$$

$$(8.4.8)$$

*and that all formally small terms vanish. Let now $n_\varepsilon(t, \mathbf{x})$ be the number density of active particles given by (8.2.11). Then in the limit $\varepsilon \longrightarrow 0$, $n_\varepsilon(t, \mathbf{x})$ converges to $n(t, \mathbf{x})$ in $D'(t, \mathbf{x})$, where $n(t, \mathbf{x})$ is the solution of the following:*

$$\partial_t n(t, \mathbf{x}) + \nabla_\mathbf{x} \cdot \langle k(\mathbf{v}) \otimes \mathbf{v} \cdot \nabla_\mathbf{x} n(t, \mathbf{x}) \rangle = \langle M^2 \rangle_\mathbf{v} n^2, \quad p = r = 1, \quad (8.4.9)$$

$$\partial_t n(t, \mathbf{x}) + \nabla_\mathbf{x} \cdot \langle k(\mathbf{v}) \otimes \mathbf{v} \cdot \nabla_\mathbf{x} n(t, \mathbf{x}) \rangle = 0, \quad p = 1, \quad r > 1, \quad (8.4.10)$$

$$\partial_t n(t, \mathbf{x}) = \langle M^2 \rangle_\mathbf{v} n^2, \quad p > 1, \quad r = 1, \quad (8.4.11)$$

*and*

$$\partial_t n(t, \mathbf{x}) = 0, \quad p > 1, \quad r > 1. \quad (8.4.12)$$

*Remark 1.* Equation (8.4.9) can be rewritten as a nonlinear diffusion equation

$$\partial_t n - \nabla_\mathbf{x} \cdot (D \cdot \nabla_\mathbf{x} n) = \langle M^2 \rangle_\mathbf{v} n^2, \quad p = r = 1, \quad (8.4.13)$$

where the diffusivity tensor is given by

$$D = -\int_{D_\mathbf{v}} \mathbf{v} \otimes k(\mathbf{v}) d\mathbf{v}. \quad (8.4.14)$$

Moreover, it can be proved that the tensor $D$ is symmetric and positive definite (see [27]).

*Proof of Theorem 4.* The proof of the theorem takes advantage of some preliminary results given in [27], and follows the method of proof of [28]. Specifically, under the assumptions of Theorem 4, the following properties and equalities related to the operator $\mathcal{L}$ hold true:

(i) $\mathcal{L}$ is a self-adjoint operator with respect to the scalar product in the space $L^2(\mathbf{v}, \frac{d\mathbf{v}}{M})$

(ii) $\langle \mathcal{L}f \rangle = 0$; and $N(\mathcal{L}) = \text{vect}(M(\mathbf{v}))$.

(iii) For $g \in L^2(D_{\mathbf{v}}, \frac{d\mathbf{v}}{M})$, the equation $\mathcal{L}(f) = g$ has a unique solution $f \in L^2(D_{\mathbf{v}}, \frac{d\mathbf{v}}{M})$ satisfying

$$\int_{D_{\mathbf{v}}} f \, d\mathbf{v} = 0 \quad \text{if and only if} \quad \int_{D_{\mathbf{v}}} g \, d\mathbf{v} = 0 \, .$$

In particular, the equation $\mathcal{L}(f) = M\mathbf{v}$ has a unique solution given by $f(\mathbf{v}) = k(\mathbf{v})$. Then, multiplying Eq. (8.4.1) by $\varepsilon^p$, letting $\varepsilon$ go to zero, and using the moment convergence assumptions yields $\mathcal{L}f^h = 0$. This implies that $f^h \in Ker(\mathcal{L})$ which consequently can be written

$$f^h = M(\mathbf{v})\rho^h, \quad h = 1, \dots, H. \tag{8.4.15}$$

Integrating Eq. (8.4.1) over $\mathbf{v}$, and using the fact that $\langle \mathcal{L}f_\varepsilon^h \rangle = 0$, yields

$$\partial_t \langle f_\varepsilon^h \rangle + \nabla_{\mathbf{x}} \cdot \left\langle \frac{\mathbf{v} f_\varepsilon^h}{\varepsilon} \right\rangle = \langle \Gamma_h(\mathbf{f}_\varepsilon, \mathbf{f}_\varepsilon) \rangle + \varepsilon^{r-1} \langle I_h(\mathbf{f}_\varepsilon, \mathbf{f}_\varepsilon) \rangle, \quad h = 1, \dots, H \, . \tag{8.4.16}$$

The limit in (8.4.16) is obtained for $r \geq 1$. The asymptotic limit of $\langle (1/\varepsilon)\mathbf{v} f_\varepsilon^h \rangle$ has to be estimated to recover the limit in (8.4.16). Then, using the above properties, and recalling that $\mathcal{L}$ is self-adjoint, yields

$$\left\langle \frac{\mathbf{v} f_\varepsilon^h}{\varepsilon} \right\rangle = \left\langle \frac{\mathbf{v} M f_\varepsilon^h}{\varepsilon M} \right\rangle = \left\langle \frac{\mathcal{L}f_\varepsilon^h}{\varepsilon}, \frac{k(\mathbf{v})}{M} \right\rangle, \quad h = 1, \dots, H \, . \tag{8.4.17}$$

Eliminating $\mathcal{L}f_\varepsilon^h$ yields

$$\frac{1}{\varepsilon} \mathcal{L}(f_\varepsilon^h) = \varepsilon^p \partial_t f_\varepsilon^h + \varepsilon^{p-1} \mathbf{v} \cdot \nabla_{\mathbf{x}} f_\varepsilon^h - \varepsilon^p \Gamma_h(\mathbf{f}_\varepsilon, \mathbf{f}_\varepsilon)$$
$$- \varepsilon^{p+r-1} I_h(\mathbf{f}_\varepsilon, \mathbf{f}_\varepsilon), \quad h = 1, \dots, H \, . \tag{8.4.18}$$

Finally, combining (8.4.17) and (8.4.18) yields

$$\nabla_{\mathbf{x}} \cdot \left\langle \mathbf{v} \frac{f_\varepsilon^h}{\varepsilon} \right\rangle \tag{8.4.19}$$

$$= \nabla_{\mathbf{x}} \cdot \left\langle \varepsilon^p \partial_t f_\varepsilon^h + \varepsilon^{p-1} \mathbf{v} \cdot \nabla_{\mathbf{x}} f_\varepsilon^h - \varepsilon^p \Gamma_h(\mathbf{f}_\varepsilon, \mathbf{f}_\varepsilon) - \varepsilon^{p+r-1} I_h(\mathbf{f}_\varepsilon, \mathbf{f}_\varepsilon), \frac{k(\mathbf{v})}{M} \right\rangle \, .$$

The limit for the term defined in (8.4.19) is obtained for $p \geq 1$; then, due to the hypothesis on the moments, it converges to

$$\nabla_{\mathbf{x}} \cdot \langle k(\mathbf{v}) \otimes \mathbf{v} \cdot \nabla_{\mathbf{x}} \rho^h \rangle, \quad p = 1, \quad \text{or} \quad 0 \quad \text{if} \quad p > 1 \, ,$$

while the asymptotic quadratic term of (8.4.16) converges to

$$\langle \Gamma_h(M\rho^h, M\rho^h) \rangle + \langle I_h(M\rho^h, M\rho^h) \rangle \quad \text{if} \quad r = 1 \, ,$$

$$\langle \Gamma_h(M\rho, M\rho)\rangle \quad \text{if} \quad r > 1.$$

Let $\varepsilon$ go to zero in (8.4.16), and using (8.4.15), one gets that $\rho^h(t, \mathbf{x})$ is the weak solution of the following equations:

$$\partial_t \rho^h + \nabla_{\mathbf{x}} \cdot \langle k(\mathbf{v}) \otimes \mathbf{v} \cdot \nabla_{\mathbf{x}} \rho^h\rangle = \langle M^2\rangle_{\mathbf{v}}(\Gamma_h + I_h)(\rho, \rho) \quad \text{if} \quad p = r = 1, \tag{8.4.20}$$

$$\partial_t \rho^h + \nabla_{\mathbf{x}} \cdot \langle k(\mathbf{v}) \otimes \mathbf{v} \cdot \nabla_{\mathbf{x}} \rho^h\rangle = \langle M^2\rangle_{\mathbf{v}}\Gamma_h(\rho, \rho) \quad \text{if} \quad p = 1, \quad r > 1, \tag{8.4.21}$$

$$\partial_t \rho^h = \langle M^2\rangle_{\mathbf{v}}(\Gamma_h + I_h)(\rho, \rho) \quad \text{if} \quad r = 1, \quad p > 1, \tag{8.4.22}$$

and

$$\partial_t \rho^h = \langle M^2\rangle_{\mathbf{v}}\Gamma_h(\rho, \rho) \quad \text{if} \quad p > 1, \quad r > 1, \tag{8.4.23}$$

where $\rho = (\rho^1, \dots, \rho^H)$.

Under the assumptions of the theorem, one deduces that

$$n_\varepsilon(t, \mathbf{x}) \to n(t, \mathbf{x}) = \sum_{h=1}^{H} \rho^h \quad \text{in} \quad D'(t, \mathbf{x})$$

as $\varepsilon \to 0$. Then summing over $h$ in (8.4.20)–(8.4.23) and using $\sum_{h=1}^{H} \Gamma_h(\rho, \rho) = 0$, one completes the proof. $\qquad\square$

*Remark 2.* If $T(\mathbf{v}, \mathbf{v}^*) = T_1(\mathbf{v})$, one can compute the solution of the equation $\mathcal{L}(f) = M\mathbf{v}$. Indeed, since $\mathcal{L}(M\mathbf{v}) = -M\mathbf{v}\langle T_1\rangle_{\mathbf{v}}$, the solution is given by $h(\mathbf{v}) = -(1/\langle T_1\rangle_{\mathbf{v}})M\mathbf{v}$. In particular, if $T_1(\mathbf{v}) = \sigma M(\mathbf{v})$, then $f(\mathbf{v}) = -(1/\sigma)M\mathbf{v}$.

## 8.5 Looking forward

A detailed analysis of a large class of equations of mathematical kinetic theory, modelling large systems of interacting cell populations, has been examined in the preceding sections. The microscopic state of cells is identified by both mechanical variables and biological ones. Specifically, the following topics have been discussed:

(i) Modelling phenomena of immune competition;
(ii) Deriving macroscopic equations from the underlying microscopic description.

One certainly cannot claim that the mathematical structures which have been analyzed here have the ability to describe the whole variety of multicellular phenomena. However, some specific characteristics can be derived with reference to suggestions from biologists. Specifically Hartwell et al. [31], deeply analyze the conceptual differences that arise in dealing with inert and living matter:

*Biological systems are very different from the physical or chemical systems analyzed by statistical mechanics or hydrodynamics. Statistical mechanics typically deals with systems containing many copies of a few interacting components, whereas cells contain from millions to a few copies of each of thousands of different components, each with very specific interactions.*

*In addition, the components of physical systems are often simple entities, whereas in biology each of the components is often a microscopic device in itself, able to transduce energy and work far from equilibrium.*

*A system in biology cannot be simply observed and interpreted at a macroscopic level. A system constituted by millions of cells shows at the macroscopic level only the output of the cooperative and organized behaviors which may not be, or are not, singularly observed.*

Actually, the mathematical structure proposed in Section 8.2 is consistent with these statements considering that the number of cell populations may be large, that the microscopic state includes biological functions, and that interactions modify, far from equilibrium, biological states including proliferating and/or destructive events.

The applications proposed in Sections 8.3 and 8.4 have shown that interesting biological phenomena can be described by models derived according to the framework of Section 8.2. On the other hand, we remark that the structure can be made more general. For instance, we can include in the term which models the velocity jumps also the biological state according to the ability of cells to modify their dynamics [17].

So far the analysis can be regarded as a first step towards the fascinating, however difficult, effort of designing a mathematical approach to living matter. Some of the difficulties of this project are well documented in [32], while various projects are developed by applied mathematicians to analyze the role that mathematics can play in building a bridge towards biological sciences [33]. Biologists appear to be interested in this effort and speculate, from their viewpoint, about how a new biology can be designed which looks for rigorous theories to replace the traditional pragmatic approach.

# References

[1]   N. Bellomo and M. Pulvirenti, Eds., *Modeling in Applied Sciences: A Kinetic Theory Approach*, Birkhäuser, Boston, 2000.

[2]   N. Bellomo, E. De Angelis, and L. Preziosi, Multiscale modelling and mathematical problems related to tumor evolution and medical therapy, *J. Theor. Medicine*, **5**, 111–136, (2003).

[3]   N. Bellomo, A. Bellouquid, and M. Delitala, Mathematical topics in the modelling complex multicellular systems and tumor immune cells competition, *Math. Mod. Meth. Appl. Sci.*, **14**, 1683–1733, (2004).

[4]   A. Bellouquid and M. Delitala, *Modelling Complex Systems in Biology—A Kinetic Theory Approach*, Birkhäuser, Boston, 2006)

[5]   L. Arlotti, N. Bellomo, and E. De Angelis, Generalized kinetic (Boltzmann) models: Mathematical structures and applications, *Math. Mod. Meth. Appl. Sci.*, **12**, 579–604, (2002).

[6]   C. Cercignani, R. Illner, and M. Pulvirenti, *Theory and Application of the Boltzmann Equation*, Springer, Heidelberg, 1993.

[7]   F. Schweitzer, *Brownian Agents and Active Particles*, Springer, Berlin, 2003.

[8]   N. Bellomo and G. Forni, Dynamics of tumor interaction with the host immune system, *Math. Comp. Mod.*, **20**, 107–122, (1994).

[9]   L. Arlotti, M. Lachowicz, and A. Gamba, A kinetic model of tumor/immune system cellular interactions, *J. Theor. Medicine*, **4**, 39–50, (2002).

[10]  E. De Angelis and P.E. Jabin, Qualitative analysis of a mean field model of tumor-immune system competition, *Math. Mod. Meth. Appl. Sci.*, **13**, 187–206, (2003).

[11]  M. Kolev, Mathematical modeling of the competition between acquired immunity and cancer, *Appl. Math. Comp. Science*, **13**, 289–297, (2003).

[12]  A. Bellouquid and M. Delitala, Kinetic (cellular) models of cell progression and competition with the immune system, *Z. Agnew. Math. Phys.*, **55**, 295–317, (2004).

[13]  L. Derbel, Analysis of a new model for tumor-immune system competition including long time scale effects, *Math. Mod. Meth. Appl. Sci.*, **14**, 1657–1682, (2004).

[14]  M. Willander, E. Mamontov, and Z. Chiragwandi, Modelling living fluids with the subdivision into the components in terms of probability distributions, *Math. Mod. Meth. Appl. Sci.*, **14**, 1495–1521, (2004).

[15]  M. Kolev, A mathematical model of cellular immune response to leukemia, *Math. Comp. Mod.*, **41**, 1071–1082, (2005).

[16]  M. Kolev, E. Kozlowska, and M. Lachowicz, Mathematical model of tumor invasion along linear or tubular structures, *Math. Comp. Mod.*, **41**, 1083–1096, (2005).

[17]  A. Bellouquid and M. Delitala, Mathematical methods and tools of kinetic theory towards modelling of complex biological systems, *Math. Mod. Meth. Appl. Sci.*, **15**, (2005), 1619–1638.

[18]  M.L. Bertotti and M. Delitala, From discrete kinetic and stochastic game theory to modeling complex systems in applied sciences, *Math. Mod. Meth. Appl. Sci.*, **14**, 1061–1084, (2004).

[19]  A. Chauviere and I. Brazzoli, On the discrete kinetic theory for active particles—Mathematical tools, *Math. Comp. Mod.*, **43**, 933–944, (2006).

[20]  L. Greller, F. Tobin, and G. Poste, Tumor heterogeneity and progression: conceptual foundation for modeling, *Invasion and Metastasis*, **16**, 177–208, (1996).

[21]  B. Perthame, Mathematical tools for kinetic equations, *Bull. Am. Math. Society*, **41**, 205–244, (2004).

[22] C. Villani, Recent advances in the theory and applications of mass transport, *Contemp. Math. Amer. Math. Soc.*, **353**, 95–109, (2004).

[23] J. Adam and N. Bellomo, Eds., *A Survey of Models on Tumor Immune Systems Dynamics*, Birkhäuser, Boston, 1997.

[24] L. Preziosi, *Modeling Cancer Growth*, CRC Press, Boca Raton, 2003.

[25] M. Lachowicz, Micro and meso scales of description corresponding to a model of tissue invasion by solid tumours, *Math. Mod. Meth. Appl. Sci.*, **15**, 1667–1684, (2005).

[26] M.A.J. Chaplain and G. Lolas, Spatio-temporal heterogeneity arising in a mathematical model of cancer invasion of tissue, *Math. Mod. Meth. Appl. Sci.*, **15**, 1685–1778, (2005).

[27] N. Bellomo and A. Bellouquid, From a class of kinetic models to macroscopic equations for multicellular systems in biology, *Discrete Contin. Dyn. Syst. B*, **4**, 59–80, (2004).

[28] N. Bellomo and A. Bellouquid, On the mathematical kinetic theory of active particles with discrete states—the derivation of macroscopic equations, *Math. Comp. Mod.*, (2006).

[29] T. Hillen and H. Othmer, The diffusion limit of transport equations derived from velocity jump processes, *SIAM J. Appl. Math.*, **61**, (2000), 751–775.

[30] A. Stevens, The derivation of chemotaxis equations as limit dynamics of moderately interacting stochastic many-particles systems, *SIAM J. Appl. Math.*, **61**, 183–212, (2000).

[31] H.L. Hartwell, J.J. Hopfield, S. Leibner, and A.W. Murray, From molecular to modular cell biology, *Nature*, **402**, c47–c52, (1999).

[32] R. Reed, Why is mathematical biology so hard?, *Notices of the American Mathematical Society*, **51**, 338–342, (2004).

[33] R.M. May, Uses and abuses of mathematics in biology, *Science*, **303**, 790–793, (2004).

[34] C.R. Woese, A new biology for a new century, *Microbiology and Molecular Biology Reviews*, **68**, 173–186, (2004).

# 9

# Kinetic modelling of late stages of phase separation

Guido Manzi[1] and Rossana Marra[2]

[1] Max-Planck-Institut für Mathematik in den Naturwissenschaften, 04103 Leipzig, Germany. manzi@mis.mpg.de
[2] Dipartimento di Fisica, Università di Roma Tor Vergata e INFN, 00133 Roma, Italy. marra@roma2.infn.it

## 9.1 Introduction

We want to provide some tools for studying the behaviour of a fluid where two phases are present separated by a sharp layer which moves according to the dynamics of the system. We specialize to the situation of a mixture of two fluids 1 and 2, that, if quenched below the coexistence curve, start to segregate in domains, some of which are rich in fluid 1 and the others in fluid 2.Rℂ

A classical approach to phase segregation distinguishes below the coexistence curve a region of unstable states from one of metastable states. If the system is cooled inside the metastable region, it begins to develop finite amplitude fluctuations like droplets. This process is known as nucleation. On the other hand, if we quench the system into the unstable region, infinitesimal fluctuations appear that yield macroscopic pattern. This is called spinodal decomposition. In late stages of phase segregation, the two processes tend to coincide.

There are several reasons for choosing a kinetic model. First of all there is a clear physical intuition underlying the mathematical structure of the equations, because we can still think in terms of interacting particles. Second those particles move in the continuous space and not on a lattice. Moreover there are convincing arguments relating Newtonian dynamics to the kinetic ones, and there are powerful tools to derive hydrodynamics from kinetic equations.

Phase segregation can occur in a single-component fluid when the long-range interaction between the particles is attractive. Then when the temperature is sufficiently low, different phases appear. Of course, one of them will be characterized by high density. Often kinetic models have problems when densities are too high. Hence, it is advantageous to use a two-component mixture. In this case particles of different types repel each other. Thus the densities in all the phases can be now kept low. We will present two models corresponding to two different physical situations.

### 9.1.1 Vlasov–Fokker–Planck model

The physical situation we have in mind is the following: we have two different kinds of particles, 1 and 2, in a torus $\Omega_\epsilon = \epsilon^{-1}\Omega$, $\Omega \subset \mathbb{R}^3$, a torus of size 1. Particles of the same type do not interact at all. Particles 1 are repelled by particles 2 and vice versa. The system is in contact with a thermal bath, which keeps the temperature constant. The thermalization mechanism is not characterized by collisions between particles of the fluid, but by the interaction with a somewhat idealized entity (the reservoir), whose internal structure is unknown and that is unchanged by the dynamics of the mixture. In this case we can also consider high densities and thus the model is suitable to describe systems such as polymer blends. Indeed there is only one type of conserved quantity, namely the mass of each component of the mixture. Since energy and momentum are dissipated on a much shorter time scale, we expect that the late stages of the phase segregation are not influenced by hydrodynamical effects; on the contrary the motion will be quasi-static.

Each component of the mixture is described by the distribution function $f_i$, $i = 1, 2$, in the one-particle phase space. The interaction is modelled through an auto-consistent Vlasov term. It means that the force $F_i$ acting on each particle is the gradient of the average potential generated by all the others. In the underlying microscopic model, particles of different species interact through a two-body Kac potential $U_\gamma$, which is repulsive, weak and long ranged. The action of the heat reservoir on the system is translated in terms of a Fokker–Planck operator $L_\beta$.

The equations are the following:

$$\partial_\tau f_i + v \cdot \nabla_x f_i + F_i \cdot \nabla_v f_i = L_\beta f_i, \tag{9.1.1}$$

where

$$L_\beta f_i = \nabla_v \cdot \left( M_\beta \nabla_v \left( \frac{f_i}{M_\beta} \right) \right),$$

$M_\beta$ is a Maxwellian with mean zero and variance $\beta^{-1}$,

$$M_\beta = \left( \frac{\beta}{2\pi} \right)^{\frac{3}{2}} e^{-\frac{\beta}{2} v^2},$$

and $\beta^{-1}$ is the temperature of the reservoir. The auto-consistent Vlasov force is

$$F_i = -\nabla_x \int_{\Omega_\epsilon} dx' U_\gamma(x - x') \int_{\mathbb{R}^3} dv f_j(x', v, \tau)$$

and the Kac potential $U_\gamma(x) = \gamma^3 U(|\gamma x|)$ has range and intensity modulated by the parameter $\gamma > 0$. The function $U$ has compact support, it is smooth and its integral over the whole space is equal to one. We assume that the functions

$f_i$ are normalized in such a way that $\int_{\Omega_\epsilon} dx \int_{\mathbb{R}^3} dv f_i(x, v)$ gives the total mass in $\Omega_\epsilon$ of the component $i$. Thus $\rho_i(x) := \int_{\mathbb{R}^3} dv f_i(x, v)$ can be interpreted as mass density.

Equations (9.1.1) have stationary solutions of the form $f_i = \rho_i M_\beta$, where $\rho_i$ are functions only of the position solving

$$\ln \rho_i(x) + \beta \int_{\Omega_\epsilon} dx' U_\gamma(x - x') \rho_j(x') = C_i, \quad x \in \Omega_\epsilon, \ i, j = 1, 2, \ i \neq j \quad (9.1.2)$$

and $C_i$ are arbitrary constants, related to the masses of the components of the mixture and to their concentrations. It is proved in [CCELM], under the assumption of a monotone $U_\gamma(x)$, that at low temperature there are nonhomogeneous solutions to (9.1.2), thermodynamically stable in the sense that they minimize the free energy functional

$$\mathcal{F}(\rho_1, \rho_2) = \int_{\Omega_\epsilon} dx (\rho_1 \ln \rho_1 + \rho_2 \ln \rho_2) + \beta \int_{\Omega_\epsilon \times \Omega_\epsilon} dx dy U_\gamma(x - y) \rho_1(x) \rho_2(y),$$
$$(9.1.3)$$

which is obtained (apart from unimportant constants) by computing

$$\mathcal{G}(f_1, f_2) = \int_{\Omega_\epsilon \times \mathbb{R}^3} dx dv (f_1 \ln f_1 + f_2 \ln f_2) + \frac{\beta}{2} \int_{\Omega_\epsilon \times \mathbb{R}^3} dx dv (f_1 + f_2) v^2$$
$$+ \beta \int_{\Omega_\epsilon \times \Omega_\epsilon} dx dy U_\gamma(x - y) \int_{\mathbb{R}^3} dv f_1(x, v) \int_{\mathbb{R}^3} dv' f_2(y, v')$$

on functions $f_i = \rho_i M_\beta$. $\mathcal{G}$ is a Lyapunov functional for the evolution equations (9.1.1).

The structure of the minimizers is such that, when $\gamma$ is much smaller than the typical size of the container, the volume $\Omega_\epsilon$ is divided into two regions where the densities are those of the pure phases (the equilibrium values given by intersecting with a horizontal line the coexistence curve at the temperature of the mixture) with an interpolating region called the *interface* whose size is again of order $\gamma$. On the infinite line it is possible to characterize the equilibrium states in terms of the excess free energy functional: the non-homogeneous minimizers $w_i(z)$, $z \in \mathbb{R}$ and $i = 1, 2$, of the excess free energy arise by prescribing asymptotic values for the densities corresponding to the two different phases coexisting at equilibrium. They are known as *fronts*; they have monotonicity properties and interpolate smoothly over a region of size $\gamma$ between the asymptotic constant values.

In late stages of phase segregation, when domains of different phases are well established and separated by sharp interfaces, the change in density across them is described by those stationary profiles. The dynamics of the system can be fully understood in terms of the following quasi-static problem:

$$\begin{cases} \triangle_r \psi(r, t) = 0 & r \in \Omega \setminus \Gamma_t \\ \psi(r, t) = c_1 SK(r, t) & r \in \Gamma_t \end{cases} \qquad \begin{cases} \triangle_r \zeta(r, t) = 0 & r \in \Omega \setminus \Gamma_t \\ c_2 SK(r, t) = [\zeta]_-^+ & r \in \Gamma_t \\ 0 = [\nu_t \cdot \nabla_r \zeta]_-^+ \end{cases}$$

$$V = c_3[\nu_t \cdot \nabla_r \psi]_-^+ + c_4 \nu_t \cdot \nabla_r \zeta,$$

where $\Gamma_t$ is the interface at time $t$ and $\nu_t$ its outward normal. The quantities $c_i$, $i = 1, \ldots, 4$, depend on the initial data and on the temperature; $S$ is the surface tension and $K$ is the sum of the principal curvatures of $\Gamma_t$. $V$ is the normal velocity of the interface. Finally the symbol $[\cdot]_-^+$ means the difference between the values of the argument on the two sides of the interface. We will comment later on the physical interpretation of $\psi$ and $\zeta$, which are related to the chemical potentials.

These dynamics conserve the volume of the inner domain, while meanwhile decreasing the area of its surface. We will show how to derive that dynamics from our kinetic model, via a limiting procedure in which $\epsilon$ plays the role of the vanishing parameter and $\gamma$ has to be chosen properly.

### 9.1.2 Vlasov–Boltzmann model

We replace in the previous model the interaction of each species with the reservoir by elastic collisions between particles, independent of the species, that we model by a Boltzmann collision kernel. The evolution is then ruled by two coupled Vlasov–Boltzmann (VB) equations for the one-particle distributions $f_i(x, v, \tau), i = 1, 2$. These equations, which conserve the mass of each species, and the total momentum and energy, have the form

$$\partial_\tau f_i + v \cdot \nabla_x f_i + F_i \cdot \nabla_v f_i = J(f_i, f_1 + f_2), \quad i = 1, 2, \qquad (9.1.4)$$

where $J(f, g)$ is the non-symmetric Boltzmann collision operator for hard spheres [Ce],

$$J(f, g) = \int_{\mathbb{R}^3} dv_* \int_{S_2} d\omega B(|v - v_*|, \omega)[f(v')g(v_*') - f(v)g(v_*)],$$

where $S_2$ is the 2D sphere in $\mathbb{R}^3$, $d\omega$ is the surface measure on it, the vectors $v$, $v_*$ are the outgoing velocities of a binary elastic collision between two particles with incoming velocities $v'$ and $v_*'$ and $B(|v - v_*|, \omega) = \frac{1}{2}|(v - v_*) \cdot \omega|$.

The equilibrium states for this system can be characterized by looking to the (negative of the) entropy functional

$$\cdot \; \mathcal{H}(f_1, f_2) = \sum_{i=1}^{2} \int_{\Omega_\epsilon \times \mathbb{R}^3} dx dv f_i \log f_i, \qquad (9.1.5)$$

which is a Lyapunov functional for (9.1.4) in the sense that

$$\frac{d}{dt} \mathcal{H}(f_1, f_2) \leq 0. \qquad (9.1.6)$$

The equilibrium states are then determined by imposing the equality in (9.1.6). They are local Maxwellians with mean value $u = 0$, variance $\beta^{-1}$ and densities $\rho_i = \int dv f_i(x, v, \tau)$ satisfying (9.1.2).

An alternative way to obtain equilibrium states is to minimize the entropy functional under the constraints on the total energy and total masses. The densities will be determined [CCELM] as the minimizers of the free energy functional $\mathcal{F}(\rho_1, \rho_2)$ defined in (9.1.3).

Equations (9.1.2) are indeed the Euler–Lagrange equations for this minimization problem. $\beta$ is determined by the constraint on the energy and $C_i$ by the constraints on the masses.

Since this dynamics conserves masses, momentum and energy the hydrodynamics effects in the late stages of the coarsening process become relevant. When the fluid is well segregated with sharp interfaces between different phases the interface moves in its normal direction following the incompressible velocity field solution of the Navier–Stokes equation, while the pressure satisfies Laplace's law relating it to the surface tension and curvature.

This limiting evolution is ruled by the following free boundary problem. Let $\Gamma^0$ (the interface at time zero) be a regular surface in a 3D torus $\Omega$ dividing the torus in two regions $\Omega^+$ and $\Omega^-$. For each $t$ one has to find a surface $\Gamma_t$, moving with velocity $V$, a continuous velocity field $u(\cdot, t)$ and a pressure function $p(\cdot, t)$ such that

$$\begin{cases} \partial_t u + (u \cdot \nabla)u + \nabla p = \eta \triangle u \\ V = -u \cdot \nu \\ [p]_-^+ = K_{\Gamma_t} S \qquad \qquad \text{on } \Gamma_t \\ \nabla \cdot u = 0 \\ \Gamma_0 = \Gamma^0, \quad u(\cdot, 0) = u_0(\cdot), \end{cases} \qquad (9.1.7)$$

where $\eta$ is the kinematic viscosity, $S$ denotes surface tension, $\nu(\cdot, t)$ is the normal to the surface pointing towards $\Omega^+$, $K_{\Gamma_t}$ stands for the curvature of $\Gamma_t$ and $[h]_-^+ = h^+ - h^-$ stands for the jump of the observable $h$ across $\Gamma_t$.

This free boundary problem was first formulated to describe the oscillations of an impermeable interface separating two viscous fluids in [H]. Chandrasekar [Cha] then studied the linear stability of this system. We mention that recently Coutand and Shkoller [CS] have obtained existence, uniqueness and regularity results for the one-sided case, namely the flow of an incompressible Navier–Stokes fluid confined in a region with a free boundary where suitable surface tension boundary conditions are specified. Such results can be extended to the present two-sided case [Sh].

This flow diminishes the length of the boundary while conserving the volume and at the same time forces the velocity field to decay to zero. Hence, the stationary solution should be characterized by $u = 0$ and a surface $\Gamma$ determined by the isoperimetric problem on the torus, separating $\Omega$ in two phase regions with different values of the pressure such that $[p]_-^+ = K_\Gamma S$.

## 9.2 Some words on the method

When we try to describe a macroscopic system starting from a kinetic model, it is clear that at some point we will exploit the scale separation. The latter usually provides a law of large numbers annihilating most of the strange fluctuations of the microscopic landscape. In other words we expect, at least, that if the system is initially close to a local equilibrium, it will evolve smoothly, with variations appreciable only on the macroscopic scale. Over a long span of space and time we will observe microscopically a lot of fluctuations, but none of these will be able to modify the macroscopic picture. If we want to get rid of all this meaningless information, we can look at the solution of (9.1.1) only after macroscopic times and distances. This is why we define functions $f_i^\epsilon(r, v, t) := f_i(\epsilon^{-1}r, v, \epsilon^{-a}t)$, where $\epsilon$ is the ratio between the kinetic and macroscopic scales and $(r, t)$ are the macroscopic values of space and time. Choosing different $a > 0$ we obtain different behaviours. For example if we chose $a = 1$ we would not see any dissipation phenomena, because diffusion starts to be effective only after a time of the order of the square of the distance. Another important parameter in the theory is the range of the interaction $\gamma^{-1}$. Since the size of the interface is linearly related to the range of the interaction, by choosing $\gamma = 1$, the width of the interface on the macroscopic scale is of order $\epsilon$, so that in the limit $\epsilon \to 0$ the interface becomes sharp. In the context of the Vlasov–Fokker–Planck (VFP) model, we will describe the dynamics in the case of a sharp interface at time $\epsilon^{-3}t$, when the system has almost completely relaxed to equilibrium, namely in each domain the values of the densities are those of the thermodynamic equilibrium; only the shape of the domains can still change. Another possible choice is $a = 2$: the limiting equation is a nonlinear diffusion equation with Dirichlet boundary conditions on the interface (Stefan problem).

From (9.1.1) we can derive an equation for $f_i^\epsilon$ where the parameter $\epsilon$ appears explicitly. The simple and powerful idea consists in seeking a solution as an asymptotic series of powers of $\epsilon$. Then we obtain a hierarchy of equations that can be solved step by step.

## 9.3 Sharp interface limit for the VFP model

Here we begin the analysis of the late stages dynamics provided by the VFP model.

### 9.3.1 The initial data

We assume that the system at time zero is already characterized by macroscopic domains filled by different phases and that the layers dividing them are sharp. In other words their width is of order $\epsilon$ in macroscopic units. The interface can so be approximated by a geometrical surface $\Gamma_0$ in the sense that the transition between different phases takes place in a layer of width $\epsilon$ around $\Gamma_0$. We will show that this picture is preserved by the evolution of the system; namely, there is a well-defined surface $\Gamma_t^\epsilon$ that can define the interface within errors of order $\epsilon$. We are definitely interested in such a $\Gamma_t^\epsilon$.

As explained above, we introduce functions $f_i^\epsilon(r, v, t) = f_i(\epsilon^{-1}r, v, \epsilon^{-3}t)$, $r \in \Omega$, a 3D torus of size 1, $t > 0$. $f_i^\epsilon$ describes the large-scale variations of the system, when observed after very long time intervals. We recall that the fronts are the nonhomogeneous stationary solutions of the one-dimensional version of the problem (9.1.1). They are unique up to translations, once the initial physical conditions are given. We denote by $w_i$ the fronts centered at the origin, that is $w_1(0) = w_2(0)$. The asymptotic values are $w_i(\pm\infty) = \rho_i^\pm$ with the additional symmetry $\rho_1^\pm = \rho_2^\mp$.

The initial data are given by $\bar{f}_i^\epsilon = \rho_i^\epsilon M_\beta$, where

$$\rho_i^\epsilon(r) = w_i(\epsilon^{-1}d(r, \Gamma_0)) + O(\epsilon) \tag{9.3.1}$$

and $d$ is the signed distance from the interface.

## 9.3.2 The rescaled equation

The functions $f_i^\epsilon$ solve

$$\partial_t f_i^\epsilon + \epsilon^{-2} v \cdot \nabla_r f_i^\epsilon + \epsilon^{-2} F_i^\epsilon \cdot \nabla_v f_i^\epsilon = \epsilon^{-3} L_\beta f_i^\epsilon. \tag{9.3.2}$$

$$F_i^\epsilon(r, t) = -\nabla_r \int_\Omega dr' \epsilon^{-3} U(\epsilon^{-1}|r - r'|) \int dv' f_j^\epsilon(r', v', t) =: -\nabla_r g_i^\epsilon.$$

The function $g_i^\epsilon$ is the Vlasov mean potential experienced by particles of type $i$.

Equations (9.3.2) complemented with the initial condition $f_i^\epsilon(r, v, 0) = \bar{f}_i^\epsilon(r, v)$ are the mathematical object of our study.

## 9.3.3 Notation and definitions

We assume that during its evolution the interface does not develop singularities within a time interval $[0, T]$, where $T > 0$ can be eventually very small. More precisely we define the interface as the set

$$\Gamma_t^\epsilon = \{r \in \Omega : \rho_1^\epsilon(r, t) = \rho_2^\epsilon(r, t)\}.$$

Then we suppose that in any point $x \in \Gamma_t^\epsilon$ it is possible to compute the principal curvatures. Then let $k(x)$ be the maximum between those principal curvatures in $x$. We can now introduce $k(\Gamma_t^\epsilon)$ as the supremum of $\{k(x) : x \in \Gamma_t^\epsilon\}$.

Our main assumption is the existence of $\epsilon_0$ and of a time $T > 0$ such that the quantity $\delta$, defined by $\delta^{-1} = \sup_{0 \leq t \leq T, 0 \leq \epsilon \leq \epsilon_0} k(\Gamma_t^\epsilon)$, is strictly greater than zero. Under that hypothesis let $d^\epsilon(r,t)$ be the signed distance of $r$ from $\Gamma_t^\epsilon$; then, if

$$\mathcal{N}(\delta, t) = \{r \in \Omega : |d^\epsilon(r,t)| < \delta\}$$

and $r \in \mathcal{N}$, there exists a unique point $s(r)$ on $\Gamma_t^\epsilon$ such that

$$\nu^\epsilon(s(r))d^\epsilon(r,t) + s(r) = r, \qquad (9.3.3)$$

where $\nu^\epsilon(s(r)) = \nabla_r d^\epsilon(r,t)$ is the unit normal vector to $\Gamma_t^\epsilon$. We will use $s(r)$ to build a local system of reference moving with the interface.

In terms of $d^\epsilon$ we can also define the normal velocity $V^\epsilon$ of the interface and its curvature $K^\epsilon$ (actually the sum of the principal curvatures):

$$V^\epsilon(s(r)) = \partial_t d^\epsilon(r,t), \quad K^\epsilon = \triangle_r d^\epsilon(r,t), \quad r \in \Gamma_t^\epsilon.$$

The interface $\Gamma_t^\epsilon$ divides the domain $\Omega_\epsilon$ into two subdomains $\Omega_{\epsilon,t}^\pm$ such that $d^\epsilon > (<) 0$ in $\Omega_{\epsilon,t}^{+(-)}$. Clearly $\Omega_\epsilon = \Gamma_t^\epsilon \cup \Omega_{\epsilon,t}^+ \cup \Omega_{\epsilon,t}^-$. From now on we will drop the apex $\epsilon$.

Since the transition layer between different phases has a width of order $\epsilon$ and we expect that the density profile will approach the equilibrium values exponentially fast in the bulk, in $\mathcal{N}$ there is plenty of space to perform the transition if $\epsilon$ is sufficiently small (i.e., $\epsilon \ll \delta$). Because of the boundedness assumption on the curvature of the interface, locally it will appear flat; this means, on a first approximation, that only in the normal direction can something interesting happen. The macroscopic units are not able to reveal these fine structures of the interface; then it is quite natural to introduce for any point $r \in \mathcal{N}$ a fast-varying variable $z = \epsilon^{-1}d(r,t)$ and for any function $h(r,t)$ a new function $\tilde{h}(z,r,t)$ such that $\tilde{h}(z,r,t) = h(\epsilon z \nu(s(r)) + s(r), t)$ and $\tilde{h}(z, r + \ell\nu(s(r)), t) = \tilde{h}(z,r,t), \forall \ell$ such that $r + \ell\nu(s(r)) \in \mathcal{N}$. In other words, if $h$ has a fast-varying behaviour, then it is more convenient to replace it with $\tilde{h}$ which is able to zoom around the interface; so, with respect to $z$ the transition will appear smooth. In $\tilde{h}(z,r,t)$ the second argument contains all the information on the dependence on $r$ except that on the distance from the interface, which is encoded in the first argument. Of course, we can add velocity as an additional argument. From $h(r,t) = \tilde{h}(\epsilon^{-1}d(r,t), r, t)$ we can derive the following relations:

$$\nabla_r h = \frac{1}{\epsilon}\nu\partial_z\tilde{h} + \overline{\nabla}_r\tilde{h}; \quad \partial_t h = \frac{1}{\epsilon}V\partial_z\tilde{h} + \partial_t\tilde{h}; \qquad (9.3.4)$$

$$\triangle_r h = \frac{1}{\epsilon^2}\partial_z^2\tilde{h} + \frac{1}{\epsilon}(\nabla_r \cdot \nu)\partial_z\tilde{h} + \overline{\triangle}_r\tilde{h},$$

where a bar on a derivative operator means a derivative with respect to the second argument $r$, keeping the other variables fixed. Note that $\nu \cdot \overline{\nabla}_r h(z,r,t) = 0$.

### 9.3.4 Hilbert expansion

We follow the approach based on the truncated Hilbert expansion introduced by Caflish [C]: we try to find a solution of (9.3.2) as a power series in $\epsilon$,

$$f_i^\epsilon = \sum_{n=0}^{\infty} \epsilon^n f_i^{(n)}. \tag{9.3.5}$$

Of course we also need to expand any other quantity depending on $\epsilon$. We start writing $U^\epsilon \star \sum_{n=0}^{\infty} \epsilon^n \rho_j^{(n)} = \sum_{n=0}^{\infty} \epsilon^n g_i^{(n)}$ and $F_i^{(n)} = -\nabla_r g_i^{(n)}$. Then

$$d(r,t) = \sum_{n=0}^{\infty} \epsilon^n d^{(n)}(r,t) \tag{9.3.6}$$

and $\nu^{(n)}$ is the gradient $\nabla_r d^{(n)}$, $\bar\nu := \nu^{(0)}$. The condition $|\nabla_r d| = 1$ implies that $|\nabla_r d^{(0)}| = 1$ as well; so $d^{(0)}$ can be interpreted as a signed distance from an interface that we denote by $\bar\Gamma_t = \{r \in \Omega : |d^{(0)}(r,t) = 0\}$. Similarly,

$$\mathcal{N}^{(0)}(\delta') = \{r \in \Omega : |d^{(0)}(r,t)| < \delta'\}, \quad \Omega^{+(-)} = \{r \in \Omega : d^{(0)}(r,t) > (<)0\}.$$

For any $\epsilon$ sufficiently small the assumptions made on the smoothness of the full interface still hold for $\bar\Gamma_t$.

As pointed out at the end of the previous section, near the interface it is convenient to use as coefficients in the series (9.3.5) functions $\tilde f_i^{(n)} = \tilde f_i^{(n)}(z,r,v,t)$. In the bulk it is sufficient to use only macroscopic coordinates. However, in order to avoid misunderstandings, we call $\hat f_i^{(n)} = f_i^{(n)}(r,v,t)$ the coefficients of the series in the bulk. In other words we write in $\Omega \setminus \mathcal{N}^{(0)}(\delta')$

$$f_i^\epsilon = \sum_{n=0}^{\infty} \epsilon^n \hat f_i^{(n)} \tag{9.3.7}$$

and in $\mathcal{N}^0(\delta')$

$$f_i^\epsilon = \sum_{n=0}^{\infty} \epsilon^n \tilde f_i^{(n)}. \tag{9.3.8}$$

On the border of $\mathcal{N}^{(0)}$, the two expansions, from now on denoted by the inner and the outer expansion, have to be matched. Then we choose $\delta' = \epsilon^c$, $c \in (0,1)$ and we require that as $\epsilon \to 0$ ([CF])

$$\tilde f_i^{(0)} = (\hat f_i^{(0)})^\pm + O(e^{-\alpha|z|})$$
$$\tilde f_i^{(1)} = (\hat f_i^{(1)})^\pm + \nu^{(0)} \cdot (\nabla_r \hat f_i^{(0)})^\pm (z - d^{(1)}) + O(e^{-\alpha|z|})$$

$$\tilde{f}_i^{(2)} = (\hat{f}_i^{(2)})^\pm + \nu^{(0)} \cdot (\nabla_r \hat{f}_i^{(1)})^\pm (z - d^{(1)})$$
$$+ (\nabla_r \hat{f}_i^{(0)})^\pm \cdot (-\nu^{(0)} d^{(2)} + \nu^{(1)}(z - d^{(1)}))$$
$$+ \frac{1}{2}(\partial_{r_h} \partial_{r_k} \hat{f}_i^{(0)})^\pm \nu_h^{(0)}(z - d^{(1)}) \nu_k^{(0)}(z - d^{(1)}) + O(e^{-\alpha|z|})$$

$\ldots$

where the symbol $(\hat{h})^\pm$ stands for $\lim_{\ell \to 0^\pm} \hat{h}(r + \bar{\nu}\ell)$, $r \in \bar{\Gamma}_t$. We will refer to the above relations as *matching conditions*.

We replace (9.3.7) and (9.3.8) in the equations and equate terms of the same order in $\epsilon$ separately in $\Omega^\pm \setminus \mathcal{N}^0(\delta')$ and $\mathcal{N}^0(\delta')$. We will use the notation $\rho_i^{(n)} = \int dv f_i^{(n)}$, and we denote by $\hat{h}, \tilde{h}$ a function $h(f_i^{(n)})$ whenever it is evaluated on $\hat{f}_i^{(n)}, \tilde{f}_i^{(n)}$.

### 9.3.5 Outer expansion

At the lowest order ($\epsilon^{-3}$),

$$L_\beta \hat{f}_i^{(0)} = 0,$$

which implies that $\hat{f}_i^{(0)}$ has to be Maxwellian in velocity with variance $\beta^{-1}$ times a function $\hat{\rho}_i^{(0)}(r, t)$ which will be determined through the $\epsilon^{-1}$ order equation. At order $\epsilon^{-2}$,

$$v \cdot \nabla_r \hat{f}_i^{(0)} + \hat{F}_i^{(0)} \cdot \nabla_v \hat{f}_i^{(0)} = L_\beta \hat{f}_i^{(1)}. \tag{9.3.9}$$

The solution can be found explicitly by simply trying with a Maxwellian multiplied by a polynomial of degree one in $v$. Of course the solution is not unique because the kernel of $L_\beta$ is not void. So $\hat{f}_i^{(1)}$ can be written as

$$\hat{f}_i^{(1)} = \hat{\rho}_i^{(1)} M_\beta - M_\beta \hat{\rho}_i^{(0)} v \cdot \nabla_r \hat{\mu}_i^{(0)}, \tag{9.3.10}$$

where $\mu_i^\epsilon(\rho^\epsilon) = \frac{1}{\beta} \ln \rho_i^\epsilon + U^\epsilon \star \rho_j^\epsilon$ and $\mu_i^\epsilon = \sum_{n=0}^\infty \epsilon^n \mu_i^{(n)}$ and the function $\hat{\rho}_i^{(1)}$ will be determined by imposing the solvability of the $\epsilon^0$ order equation.

The order $\epsilon^{-1}$ equation is

$$v \cdot \nabla_r \hat{f}_i^{(1)} + \hat{F}_i^{(0)} \cdot \nabla_v \hat{f}_i^{(1)} + \hat{F}_i^{(1)} \cdot \nabla_v \hat{f}_i^{(0)} = L_\beta \hat{f}_i^{(2)}. \tag{9.3.11}$$

The solvability condition for this equation says that the integral of the velocity on the left-hand side has to be zero. By integrating over the velocity and using the explicit expression for $\hat{f}_i^{(1)}$ we get

$$-\frac{1}{\beta} \nabla_r \cdot (\hat{\rho}_i^{(0)} \nabla_r \hat{\mu}_i^{(0)}) = 0.$$

The choice of the initial data implies that the only solution of that equation is the piecewise constant function equal to $\rho_i^\pm$ in $\Omega^\pm \setminus \mathcal{N}(\delta')$.

$\hat{f}_i^{(2)}$ is determined by substituting (9.3.10) in equation (9.3.11). The result is

$$\hat{f}_i^{(2)} = -M_\beta \hat{\rho}_i^{(0)} v \cdot \nabla_r \hat{\mu}_i^{(1)} + \hat{\rho}_i^{(2)} M_\beta, \qquad (9.3.12)$$

where $\hat{\mu}_i^{(1)} = \hat{\rho}_i^{(1)}/(\beta \hat{\rho}_i^{(0)}) + \hat{g}_i^{(1)}$.

As above, by integrating the $\epsilon^0$ order equation over $v$ and taking into account that $\hat{f}_i^{(0)}$ is Maxwellian in velocity, we get the following condition on $\hat{u}^{(2)}$, where $u_i^{(n)} = \int dv v f_i^{(n)}$:

$$\nabla_r \cdot \hat{u}_i^{(2)} = 0. \qquad (9.3.13)$$

Now we use $\hat{f}_i^{(2)}$ as given by (9.3.12) to get $\hat{u}_i^{(2)} = -\frac{1}{\beta}\hat{\rho}_i^{(0)}\nabla_r \hat{\mu}_i^{(1)}$ which we substitute into (9.3.13) to get the equation for $\hat{\mu}_i^{(1)}$:

$$\Delta_r \hat{\mu}_i^{(1)} = 0.$$

## 9.3.6 Inner expansion

At the lowest order $(\epsilon^{-3})$

$$v \cdot \bar{\nu} \partial_z \tilde{f}_i^{(0)} - \bar{\nu} \cdot \nabla_v \tilde{f}_i^{(0)} \partial_z \tilde{g}_i^{(0)} = L_\beta \tilde{f}_i^{(0)}.$$

It can be proved that any solution of this equation has the form $M_\beta(v)\tilde{\rho}_i^{(0)}$, with $\tilde{\rho}_i^{(0)}$ a function of $z$. Substituting back in the equation we have

$$\partial_z \tilde{\rho}_i^{(0)} + \beta \tilde{\rho}_i^{(0)}\partial_z(\tilde{U} \star \tilde{\rho}_j^{(0)}) = 0 \iff \partial_z \tilde{\mu}_i^{(0)} = 0, \qquad (9.3.14)$$

where $\tilde{U}$ is the potential $U$ integrated over all the coordinates but one. We solve this equation and the following ones as $z$ was defined on the whole line; the asymptotic values $\rho_i^\pm$ are provided by the matching conditions. The solution is given exactly by the fronts $w_i$. It is clear that for finite $\epsilon$ this solution cannot be matched with the outer expansion, but it should be possible to prove the exponential convergence of $w_i$ through standard methods (see for example [DOPT]). In that case we would only commit an exponentially small error.

We now find $\tilde{f}_i^{(1)}$ by examining the $\epsilon^{-2}$ order:

$$v \cdot \bar{\nu} \partial_z \tilde{f}_i^{(1)} - \bar{\nu} \cdot \nabla_v \tilde{f}_i^{(0)} \partial_z \tilde{g}_i^{(1)} - \bar{\nu} \cdot \nabla_v \tilde{f}_i^{(1)} \partial_z \tilde{g}_i^{(0)} = L_\beta \tilde{f}_i^{(1)}. \qquad (9.3.15)$$

Note that $\nu^{(1)} \cdot (v\partial_z \tilde{f}_i^{(0)} - \nabla_v \tilde{f}_i^{(0)}\partial_z \tilde{g}_i^{(0)}) = \beta v \cdot \nu^{(1)} M_\beta \tilde{\rho}_i^{(0)}\partial_z \tilde{\mu}_i^{(0)} = 0$, because $\tilde{f}_i^{(0)}$ is a solution of the lowest order equation and the bar operators vanish because $\tilde{\rho}_i^{(0)}$ is a function of $z$ only. Again the solution has to be necessarily Maxwellian in velocity, so that we can write $\tilde{f}_i^{(1)} = \tilde{\rho}_i^{(1)}M_\beta$ with $\tilde{\rho}_i^{(1)}$ to be determined by the following equation:

$$\partial_z \tilde{\rho}_i^{(1)} + \beta \tilde{\rho}_i^{(0)} \partial_z \tilde{g}_i^{(1)} + \beta \tilde{\rho}_i^{(1)} \tilde{U} \star \partial_z \tilde{\rho}_j^{(0)} = 0. \qquad (9.3.16)$$

Taking into account that $-\beta \tilde{U} \star \partial_z \tilde{\rho}_j^{(0)} = \partial_z \ln w_i$, from the equation for the front, we get

$$\partial_z \left( \frac{1}{\beta} \tilde{\rho}_i^{(1)} (w_i)^{-1} + \tilde{g}_i^{(1)} \right) = 0 \iff \partial_z \tilde{\mu}_i^{(1)} = 0. \qquad (9.3.17)$$

Hence, the value of $\tilde{\mu}_1^{(1)} - \tilde{\mu}_2^{(1)}$ in $z = 0$ is enough to find $\tilde{\mu}_1^{(1)} - \tilde{\mu}_2^{(1)}$ for any $z$. It can be shown (see [MM]) that

$$(\tilde{\mu}_1^{(1)} - \tilde{\mu}_2^{(1)})(0, r, t)[w_1]_{-\infty}^{+\infty} = \bar{K}(r, t)S, \qquad (9.3.18)$$

where $\bar{K}(r, t)$ is the sum of the principal curvature of $\bar{\Gamma}_t$ at point $r$ and time $t$ and $S$ is the surface tension for this model:

$$S = \sum_{(i,j) \in \{(1,2),(2,1)\}} \int dz dz' w_i'(z)(z - z') \tilde{U}(z - z') w_j(z'). \qquad (9.3.19)$$

Now we integrate the order $\epsilon^{-2}$ equation in $v$ to obtain, after several cancellations due to the fact that $\tilde{f}_i^{(0)}$ and $\tilde{f}_i^{(1)}$ are Maxwellian in velocity,

$$w_i' \bar{V} + \partial_z (\bar{\nu} \cdot \tilde{u}_i^{(2)}) = 0. \qquad (9.3.20)$$

As $\epsilon \to 0$ we can define the hat functions until the interface. Then to determine their values exactly on $\bar{\Gamma}$ we use the matching conditions imposing that $\tilde{\mu}_1^{(1)} - \tilde{\mu}_2^{(1)} \to (\hat{\mu}_1^{(1)})^\pm - (\hat{\mu}_2^{(1)})^\pm$ for $z \to \pm\infty$, so that for $r \in \bar{\Gamma}_t$

$$[(\hat{\mu}_1^{(1)})^\pm - (\hat{\mu}_2^{(1)})^\pm][w_1]_{-\infty}^{+\infty} = \bar{K}(r, t)S. \qquad (9.3.21)$$

Moreover $\tilde{u}_i^{(2)} \to (\hat{u}_i^{(2)})^\pm$ when $z \to \pm\infty$, so that

$$-\bar{V}[w_i]_{-\infty}^{+\infty} = [\bar{\nu} \cdot \hat{u}_i^{(2)}]_-^+, \qquad r \in \bar{\Gamma}_t. \qquad (9.3.22)$$

### 9.3.7 Limiting equations

We can now collect the results of the previous two subsections, considering $\epsilon = 0$ because all quantities involved are independent of $\epsilon$. Then the system is fully described by the hat functions, because the layer $\mathcal{N}$ has shrunk to the surface $\bar{\Gamma}_t$. We have

$$\begin{cases} \triangle \hat{\mu}_i^{(1)} = 0 & r \in \Omega \setminus \bar{\Gamma}_t \\ (\hat{\mu}_1^{(1)} - \hat{\mu}_2^{(1)})[\bar{\rho}_1^+ - \bar{\rho}_1^-] = \bar{K}(r, t)S & r \in \Omega \setminus \bar{\Gamma}_t \\ \bar{V}(\bar{\rho}_i^+ - \bar{\rho}_i^-) = [\frac{1}{\beta} \bar{\rho}_i \bar{\nu} \cdot \nabla_r \hat{\mu}_i^{(1)}]_-^+ & r \in \bar{\Gamma}_t \end{cases}, \qquad (9.3.23)$$

where $\bar{\rho}_i = \hat{\rho}_i^{(0)} = \hat{\rho}^+ \chi_{\Omega^+} + \hat{\rho}_i^- \chi_{\Omega^-}$, $\chi_A$ being the characteristic function of the set $A$.

Those equations can be put in a more physically meaningful format, but we need some additional notation. First of all we introduce

$$\bar{\rho}(r) = \frac{\bar{\rho}_1(r) + \bar{\rho}_2(r)}{2} \quad \text{and} \quad \bar{\varphi}(r) = \frac{\bar{\rho}_1(r) - \bar{\rho}_2(r)}{2}.$$

Then let

$$\psi = \hat{\mu}_1^{(1)} - \hat{\mu}_2^{(1)} \quad \text{and} \quad \zeta = \bar{\rho}_1 \hat{\mu}_1^{(1)} + \bar{\rho}_2 \hat{\mu}_2^{(1)}.$$

After some algebra we can recast equations (9.3.23) in the following way:

$$\begin{cases} \triangle_r \psi(r,t) = 0 & r \in \Omega \setminus \bar{\Gamma}_t \\ \psi(r,t) = \dfrac{S\bar{K}(r,t)}{\bar{\rho}_1^+ - \bar{\rho}_1^-} & r \in \bar{\Gamma}_t \\ \bar{V} = \dfrac{1}{2\beta[\bar{\rho}_1^+ - \bar{\rho}_1^-]} \left[ \dfrac{1}{\bar{\rho}}(\bar{\rho}^2 - |\bar{\varphi}|^2)[\bar{\nu} \cdot \nabla_r \psi]_-^+ + \dfrac{1}{\bar{\rho}}[\bar{\varphi}\bar{\nu} \cdot \nabla_r \zeta]_-^+ \right] & r \in \bar{\Gamma}_t \end{cases} \tag{9.3.24}$$

and

$$\begin{cases} \triangle_r \zeta(r,t) = 0 & r \in \Omega \setminus \bar{\Gamma}_t \\ [\zeta]_-^+ = 2|\bar{\varphi}|S\bar{K}(r,t)/(\bar{\rho}_1^+ - \bar{\rho}_1^-) & r \in \bar{\Gamma}_t \\ 0 = [\bar{\nu} \cdot \nabla_r \zeta]_-^+ & r \in \bar{\Gamma}_t. \end{cases} \tag{9.3.25}$$

The limiting motion is a superposition of a Mullins–Sekerka type flow, described by the first set of equations, and of a Hele–Shaw problem, ruled by the second set of equations. We note that the velocity is composed of two terms: one depending on $\psi$, the other on $\zeta$. Let us call $V_{MS}$ the former and $V_{HS}$ the latter. $V_{MS}$ is the velocity of an interface in the Mullins–Sekerka motion and $V_{HS}$ is the velocity of an interface in the Hele–Shaw problem.

We remark that the Hele–Shaw motion has more conserved quantities than the Mullins–Sekerka motion. In fact, the former conserves the volume of each connected component of both phases, while the latter conserves only the total volume.

The relative importance of the two contributions $V_{HS}$ and $V_{MS}$ is ruled by the coefficients: if $(\bar{\rho}^-)^{-1} - (\bar{\rho}^+)^{-1} \ll 1$ (near the critical point of the coexistence curve) the $V_{MS}$ term dominates, while for deep quenches the $V_{HS}$ term prevails.

Equations (9.3.24) and (9.3.25) are identical to the equations in [OE], describing the sharp interface arising in a polymer blend. In that paper the hydrodynamical equation is a modification of the Cahn–Hilliard equation for a mixture of two fluids, where a Lagrangian multiplier $p$ ("pressure") appears to take into account the constraint of constant total density:

$$\partial_t \rho_i = \nabla \cdot (\rho_i \nabla(\mu_i + p)) \quad i = 1, 2$$

$$\rho_1 + \rho_2 = 1.$$

It is exactly $p$ which gives rise to the Hele–Shaw contribution to the velocity in the sharp interface limit. If we perform a parabolic scaling on the VFP equations in order to derive the hydrodynamics, we can argue that $\zeta$ has the same role as $p$. Then we conclude that $\zeta$ describes the pressure in the VPF model.

## 9.4 Sharp interface limit for the VB model

We introduce again the macroscopic coordinate $r = \epsilon x$, $x \in \epsilon^{-1}\Omega$, $r \in \Omega$, the torus of size 1. We wish to study the small $\epsilon$ behaviour of a solution of the VB equations (9.1.4). In order to observe diffusive effects one has to consider very long times of order $\epsilon^{-2}t$, with $t$ the macroscopic time. We also choose $\gamma = \epsilon$ for reasons that will be explained later on. Setting $f_i^\epsilon(r, v, t) = f_i(\epsilon^{-1}r, v, \epsilon^{-2}t)$, $\rho_i^\epsilon(r, t) = \int dv f_i^\epsilon(r, v, t)$, the VB equation, in this space-time scaling, becomes

$$\partial_t f_i^\epsilon + \epsilon^{-1} v \cdot \nabla_r f_i^\epsilon + \epsilon^{-1} F_i^\epsilon \cdot \nabla_v f_i^\epsilon = \epsilon^{-2} J(f_i^\epsilon, f_1^\epsilon + f_2^\epsilon) \qquad (9.4.1)$$

$$F_i^\epsilon(r, t) = -\nabla_x \int_{T_1} dr' \epsilon^{-6} U(\epsilon^{-2}|r - r'|) \int_{T_1} dv' f_j^\epsilon(r', v', t) \qquad (9.4.2)$$

$$=: -\nabla_r U^\epsilon * \rho_j^\epsilon.$$

Consider now a situation in which there is, at the initial time, an interface separating the system in two regions with densities corresponding to the equilibrium values at temperature $T$ (i.e., with coexistence of two phases, one richer in species 1 and the other richer in species 2). For $\epsilon$ finite, we approximate the density profiles by one-dimensional fronts in the direction orthogonal to the interface at each point. The fronts interpolate between the two phases on a scale $\epsilon^2$. If the interface were flat, this would be a stationary solution of the VB equations. Since the interface is not flat, the fluid starts to move because of the imbalance of the pressure on the two sides of the interface (surface tension). This pushes the interface to move with the component of the fluid velocity in the direction of the normal at each point of the surface. Since the initial density in the bulk is the equilibrium and the space-time scaling is diffusive, we expect that the fluid in the bulk will evolve as an incompressible Navier–Stokes (INS) fluid. We recall that the INS equations can be obtained from an equation of the type (9.4.1) when the average velocity is small (low Mach numbers). Therefore, the effect of the surface tension has to be suitably small in order not to get velocities that are too big. The surface tension effect is proportional to the size of the interface and this is the reason why we choose in this case $\gamma = \epsilon$.

We construct an expansion in the bulk (outer) and a different expansion close to the interface (inner) and impose matching conditions on an intermediate region. At the first order we find the following free boundary problem for the velocity field $u$.

$$\partial_t u + (u \cdot \nabla)u + \nabla p = \eta \Delta u.$$

The kinematic viscosity $\eta$ is obtained from the Boltzmann equation as in [Ce], $u$ is continuous across the interface $\Gamma_t$ whose normal velocity is given by

$$v_{\Gamma_t}(r) = -u(r,t) \cdot \nu(r,t)$$

while the pressure is discontinuous at the surface and satisfies Laplace's law

$$(p_+ - p_-) = SK.$$

Here $(p_+)$ $p_-$ is the pressure on the side of $\Gamma_t$ (not) containing the normal $\nu$, $K$ is the mean curvature of $\Gamma_t$ and $S$ is the surface tension given in terms of the fronts $w_i$ by (9.3.19).

Moreover, we get equations for the first correction to the temperature $T^{(1)}$ (which at order zero is the constant $\bar{T}$) and concentration $\phi^{(1)}$

$$\begin{aligned} D_t \phi^{(1)} &= D A_1 \Delta \phi^{(1)} + D A_2 \Delta T^{(1)} \\ D_t T^{(1)} &= k \Delta T^{(1)} + A_3 D_t \phi^{(1)}, \end{aligned} \qquad (9.4.3)$$

where the diffusion coefficient $D$, the heat conductivity $k$ and the constants $A_i$ are explicit functions of $\bar{\rho}, \bar{T}$. These equations are similar to the ones in the phase field models [CF], except that in (9.4.3) the nonlinear term in the concentration is missing and the term proportional to the temperature is replaced by the Laplacian of the temperature times a possibly negative coefficient.

## 9.5 Open problems

### 9.5.1 Stability

Consider the excess free energy functional in one dimension on the infinite line defined as

$$\mathcal{G}(\rho_1, \rho_2) - \mathcal{G}(M\rho^+, M\rho^-). \qquad (9.5.1)$$

Here $\rho^\pm$ are the values of the densities at infinity, $\lim_{x \to \pm\infty} \rho_1 = \rho^\pm = \lim_{x \to \mp\infty} \rho_2$. These values are functions of the parameters $\beta$ and $C$ (the latter related to the chemical potential) through

$$\ln \rho^+ + \beta\rho^- = C, \quad \ln \rho^- + \beta\rho^+ = C.$$

We choose $\beta$ and $C$ so that $\rho^+ - \rho^- > 0$, namely in the phase transition region of the phase diagram. It is not difficult to extend the arguments in

[CCELM] to prove that there exist nonhomogeneous minimizers of the excess free energy such that $\lim_{x \to \pm\infty} f_1(x) = M\rho^{\pm}$, $\lim_{x \to \pm\infty} f_2(x) = M\rho^{\mp}$. They are monotone and regular and are solutions of the Euler–Lagrange equations for the functional, namely equations (9.1.1). We have called them fronts and their stability properties play an important role in the motion of the interfaces. Since this functional is invariant by translations we construct a one parameter family of fronts by translating a given solution. The same is true for the stationary solutions of (9.1.1).

Here we show that all the stationary solutions which are monotone and regular (with first derivative) are thermodynamically stable in the following sense. Consider a small perturbation $(\alpha_1(r), \alpha_2(r))$, $r \in \mathbb{R}^d$ of a given stationary solution $Mw_i$:

$$f_1 = Mw_1 + \epsilon\alpha_1 \qquad \text{and} \qquad f_2 = Mw_2 + \epsilon\alpha_2 ,$$

such that $\int dx dv \alpha_i = 0$. We have that

$$\begin{aligned}
\mathcal{G}(Mw_1 + \epsilon\alpha_1, Mw_2 + \epsilon\alpha_2) &- \mathcal{G}(Mw_1, Mw_2) \\
&= S(f/M) + \mathcal{F}(w_1 + \epsilon\eta_1, w_2 + \epsilon\eta_2) - \mathcal{F}(w_1, w_2),
\end{aligned} \qquad (9.5.2)$$

where $\eta_i = \int dv \alpha_i$, $\mathcal{F}$ is defined in (9.1.3) and $S(f/M)$ is the relative entropy between the state $(f_1, f_2)$ and the Maxwellian state $(M \int dv f_1, M \int dv f_2)$ defined as

$$S(f/M) = \sum_{i=1,2} \int dx dv f_i \left( \log f_i - \log(M \int dv f_i) \right).$$

We easily compute that

$$\begin{aligned}
\mathcal{F}(w_1 + \epsilon\eta_1, w_2 &+ \epsilon\eta_2) - \mathcal{F}(w_1, w_2) \\
&= \epsilon^2 \left[ \frac{1}{2\beta} \left[ \int_{\mathbb{R}} \frac{\eta_1^2(r)}{\bar{\rho}_1(z)} dr + \int_{\mathbb{R}} \frac{\eta_2^2(r)}{\bar{\rho}_2(z)} dr \right] \right. \\
&\quad \left. + \int_{\mathbb{R}} \int_{\mathbb{R}} U(|r - r'|) \eta_1(r) \eta_2(r') dr dr' \right] + \mathcal{O}(\epsilon^3).
\end{aligned}$$

We now define the operator $\mathcal{A}$ through the quadratic form

$$\begin{aligned}
\langle (\eta_1, \eta_2), \mathcal{A}(\eta_1, \eta_2) \rangle &= \frac{1}{2\beta} \left[ \int_{\mathbb{R}} \frac{\eta_1^2(r)}{w(z)} dr + \int_{\mathbb{R}} \frac{\eta_2^2(r)}{w_2(z')} dr' \right] \\
&\quad + \int_{\mathbb{R}} \int_{\mathbb{R}} U(|r - r'|) \eta_1(r) \eta_2(r') dr dr'.
\end{aligned}$$

We have the following [wip].

**Proposition.** *The operator $\mathcal{A}$ is positive with*

$$\mathcal{A}(\eta_1, \eta_2) = (0, 0)$$

*if and only if $(\eta_1, \eta_2)$ is a scalar multiple of $(w_1', w_2')$.*

The main ingredients in the proof are the Euler–Lagrange equations and the monotonicity properties. Moreover, it is well known that the relative entropy is nonnegative and is zero if and only if $f_i$ is Maxwellian. The conclusion is that the difference

$$\mathcal{G}(Mw_1 + \epsilon\alpha_1, Mw_2 + \epsilon\alpha_2) - \mathcal{G}(Mw_1, Mw_2)$$

is nonnegative for $\epsilon$ small and is zero if and only if $f_i = Mw_i'$.

A natural and relevant question is if the stationary solutions are also dynamically stable versus the evolution (9.1.1). The fact that the excess free energy decreases should force a small perturbation to tend to the family of fronts and the conservation law should select one of the family. The invariance by translation of the theory is one of the main difficulties in establishing the result. A similar difficulty has been faced in [CCO2] for a conservative dynamics describing the evolution on a mesoscopic scale of the magnetization for a Kac–Ising model. Relaxation to the stationary state and rate of convergence have been established in this case. On the other hand, a new approach based on a suitable micro-macro decomposition of the distribution function has been successful in providing many stability results for kinetic models [LY], [G], [SG]. We think that a suitable merging of these methods could provide stability results for the VFP and also for the VB dynamics.

### 9.5.2 Surface tension

In more than one dimension the geometric structure of the nonuniform minimizers is not clear. We can consider an approximation for a large volume of the free energy, computed on the nonuniform minimizers. The first term is given by the bulk free energy, obtained in the infinite volume limit, the next term is a correction proportional

to the surface area separating the two phases (divided by the volume) which corresponds to the surface tension:

$$\inf_{\rho_i : \int dx \rho_i = n_i} \mathcal{F}_\Omega(\{\rho_1, \rho_2\}) = f(n_1, n_2)|\Omega| + b(n_1, n_2)|\Omega|^{1-1/d} + \text{ lower order terms}$$

where $b(n_1, n_2) = S|\Gamma|$, with $|\Gamma| = da(d)r_0^{d-1}$ the surface of the sphere with equimolar radius $r_0$ (depending on $n_1, n_2$), $a(d)$ being the volume of the unit sphere in $\mathbb{R}^d$, and $S$ is the surface tension (9.3.19). Moreover,

$$f(n_1, n_2) = CE\left[\sum_i n_i \log n_i + \alpha n_1 n_2\right],$$

where $\alpha = \int_{\mathbb{R}^d} U(x)dx$ , and $CEf(n_1, n_2)$ denotes the maximal convex function lying below $f$.

Then, the phase boundaries will be arranged to have the minimum surface area, and their shape will be determined by the solution of the isoperimetric problem on the torus. The surface tension correction has been computed for the case of the Cahn–Hilliard functional in the literature. In [CCELM2] the analysis has been extended to also include the possibility that in some cases the minimizing profile of the density $\rho$ might correspond to the "dissolution" of a droplet of the minority phase, as proved for the 2D Ising model in [BCK]. In [CCELM1], [CCELM2] the critical density for droplet formation for the Cahn–Hilliard functional has been exactly determined. The Cahn–Hilliard functional is defined as

$$\mathcal{F}_\Omega(\{m\}) = \int_\Omega \left( \frac{\vartheta^2}{2} |\nabla m|^2 + F(m(x)) \right) \mathrm{d}x , \qquad (9.5.3)$$

where $F(t) = (t^2 - 1)^2/4$ is a symmetric double-well potential which has minima at $t = \pm 1$.

One of the ingredients is a Hilbert expansion for the minimizers in the parameter $\frac{1}{L}$ ($L$ is the linear dimension of the box). This approach could also be useful for studying the case of functionals involving nonlocal interactions, like $\mathcal{F}$, because there is a close formal connection between the Cahn–Hilliard free energy functional and the nonlocal free energy functionals. To see that, let us write the functional (9.1.3) as

$$\mathcal{F}(\rho_1, \rho_2) = \int_\Omega \mathrm{d}x \left[ (\rho_1 \ln \rho_1)(x) + (\rho_2 \ln \rho_2)(x) + \beta \alpha \rho_1(x)\rho_2(y) \right]$$
$$- \frac{1}{2} \int_{\Omega \times \Omega} \mathrm{d}x \mathrm{d}y \gamma^3 U(\gamma(x-y))[\rho_1(x) - \rho_1(y)][\rho_2(x) - \rho_2(y)].$$

Making the approximation that the densities vary sufficiently slowly on the range of the potential we replace in $\mathcal{F}(\rho_1, \rho_2)$ the first-order Taylor approximation

$$\rho_i(y) \approx \rho_i(x) + \nabla \rho_i(x) \cdot (y - x),$$

obtaining

$$\mathcal{F}(\rho_1, \rho_2) \approx \int_\Omega \left[ f(\rho_1, \rho_2) - \frac{\vartheta^2}{2} \nabla \rho_1(x) \nabla \rho_2(x) \right] \mathrm{d}x ,$$

with $\vartheta^2 = \int_{\mathbb{R}^d} x^2 U(x) \mathrm{d}x$. In terms of the variables $\phi$ and $\rho$

$$\mathcal{F}(\rho, \phi) \approx \int_\Omega \left[ f(\rho, \phi) + \frac{\vartheta^2}{2} [(\nabla \phi(x))^2 - (\nabla \rho(x))^2] \right] \mathrm{d}x ,$$

$$f(\rho, \phi) = \frac{\rho + \phi}{2} \log \frac{\rho + \phi}{2} + \frac{\rho - \phi}{2} \log \frac{\rho - \phi}{2} + \frac{1}{4}\alpha(\rho^2 - \phi^2).$$

This functional reduces to a Cahn–Hilliard functional type (9.5.3) for $\rho$ constant with $f(\rho, \phi)$ for $\alpha$ low enough—a double-well potential whose essential features are represented by $F(m) = (m^2 - 1)^2/4$.

# References

[BCK]     M. Biskup, L. Chayes, and R. Kotecky, On the formation/dissolution of equilibrium droplets, *Europhys. Lett.* **60**, 21–27 (2002); Critical region for droplet formation in the two-dimensional Ising model, *Comm. Math. Phys.* **242**, 137–183 (2003).

[C]       R. Caflish, The fluid dynamical limit of the nonlinear Boltzmann equation, *Commun. Pure and Appl. Math.* **33**, 651–666 (1980).

[CF]      G. Caginalp and P. C. Fife, Phase-field methods for interfacial boundaries, *Phys Rev. B* **33**, 7792–7794 (1986); Dynamics of layered interfaces arising from phase boundaries, *SIAM J. Appl. Math.* **48**, 506–518 (1988).

[CCELM]   E. A. Carlen, M. C. Carvahlo, R. Esposito, J. L. Lebowitz, and R. Marra, Free energy minimizers for a two-species model with segregation and liquid-vapor transition, *Nonlinearity* **16**, 1075–1105 (2003).

[CCELM1]  E. Carlen, M. C. Carvahlo, R. Esposito, J. L. Lebowitz, and R. Marra, Phase transitions in equilibrium systems: Microscopic models and mesoscopic free energies, *Molecular Physics* **103**, 3141–3151 (2005).

[CCELM2]  E. Carlen, M. C. Carvahlo, R. Esposito, J. L. Lebowitz, and R. Marra, Droplet minimizers for the Cahn–Hilliard free energy functional, *Journal of Geometric Analysis* **16** n. 2 (2006).

[CCO1]    E. A. Carlen, M. C. Carvalho, and E. Orlandi, Approximate solution of the Cahn–Hilliard equation via corrections to the Mullins–Sekerka motion, *Arch. Rat. Mechanics* **178**, 1–55, (2005).

[CCO2]    E. A. Carlen, M. C. Carvalho, and E. Orlandi, Algebraic rate of decay for the excess free energy and stability of fronts for a non local phase kinetics equation, I and II , *Jour. Stat. Phys.* **95**, 1069–1117 (1999) and *Comm. PDE* **25**, 847–886 (2000).

[Ce]      C. Cercignani, R. Illner, and M. Pulvirenti, *The Mathematical Theory of Dilute Gases*, Springer-Verlag, New York (1994).

[Cha]     S. Chandrasekar, *Hydrodynamic and Hydromagnetic Stability*, Chapter X, Clarendon Press, Oxford (1961).

[CS]      D. Coutand and S. Shkoller, Unique solvability of the free-boundary Navier–Stokes equations with surface tension, arXiv:math.AP/0212116 v2 (2003).

[DOPT]    A. De Masi, E. Orlandi, E. Presutti, and L. Triolo, Stability of the interface in a model of phase separation, *Proc. Royal Soc. Edinburgh* **124**, (1994).

[Ge]      P. G. de Gennes, Dynamics of fluctuations and spinodal decomposition in polymer blends, *J. Chem. Phys.* **72**, 4756–4763 (1980).

[GL]      G. Giacomin and J. L. Lebowitz, Exact macroscopic description of phase segregation in model alloys with long range interactions, *Phys. Rev. Lett.* **76**, 1094–1098 (1996); Phase segregation dynamics in particle systems with long range interaction. I: Macroscopic limits, *J. Stat. Phys.* **87**, 37–61 (1997); Phase segregation dynamics in particle systems with long range interaction. II: Interface motion, *SIAM J. Appl. Math.* **58**, 1707–1729 (1998).

[GLP]     G. Giacomin, J. L. Lebowitz, and E. Presutti, Deterministic and stochastic hydrodynamic equations arising from simple microscopic model systems, in *Stochastic Partial Differential Equations, Six Perspectives*, pp. 107–152, Math. Survey Monograph **64**, Amer. Math. Soc., Providence, RI (1999).

[G]     Y. Guo, The Boltzmann equation in the whole space, *Indiana Univ. Math. J.* **53**, 1081–1094 (2004).

[H]     W. J. Harrison, The influence of viscosity on the oscillations of superposed fluids, *Proc. London Math. Soc.* **6**, 396–405 (1908).

[LY]    T.-P. Liu and S.-H. Yu, Boltzmann equation: Micro-Macro decompositions and positivity of shock profiles, *Commun. Math. Phys.* **246**, 133–179 (2004).

[MM]    G. Manzi and R. Marra, Phase segregation and interface dynamics in kinetic systems, *Nonlinearity* **19**, 115–147 (2006).

[OE]    F. Otto and E. Weinan, Thermodynamically driven incompressible fluid mixtures, *J. Chem. Phys.* **107(23)**, 10177–10184 (1997).

[Sh]    S. Shkoller, Private communication.

[SG]    R. M. Strain and Y. Guo, Almost exponential decay near Maxwellian, *Comm. Partial Differential Equations* **31**, 417–429 (2006).

[wip]   Work in progress (E. A. Carlen, M. C. Carvahlo, R. Esposito, J. L. Lebowitz, and R. Marra).

# 10

# Ground states and dynamics of rotating Bose–Einstein condensates

Weizhu Bao[1]

Department of Mathematics and Center for Computational Science and Engineering, National University of Singapore, Singapore 117543.
bao@cz3.nus.edu.sg

## 10.1 Introduction

Since its realization in dilute bosonic atomic gases [7, 23], Bose–Einstein condensation of alkali atoms and hydrogen has been produced and studied extensively in the laboratory [1], and has permitted an intriguing glimpse into the macroscopic quantum world. In view of potential applications [38, 61, 63], the study of quantized vortices, which are well-known signatures of superfluidity, is one of the key issues. In fact, bulk superfluids are distinguished from normal fluids by their ability to support dissipationless flow. Such persistent currents are intimately related to the existence of quantized vortices, which are localized phase singularities with integer topological charge [39]. The superfluid vortex is an example of a topological defect that is well known in superconductors [52] and in liquid helium [33]. The occurrence of quantized vortices in superfluids has been the focus of fundamental theoretical and experimental work [33]. Different research groups have obtained quantized vortices in Bose–Einstein condensates (BECs) experimentally, e.g., the JILA group [35, 57], the ENS group [56] and the MIT group [1, 32]. Currently, there are at least two typical ways to generate quantized vortices from a BEC ground state: (i) impose a laser beam rotating with an angular velocity on the magnetic trap holding the atoms to create a harmonic anisotropic potential [51, 4, 75]; or (ii) add to the stationary magnetic trap a narrow, moving Gaussian potential, representing a far-blue detuned laser [49, 50, 24, 25, 10, 12]. The recent experimental and theoretical advances in the exploration of quantized vortices in BEC have spurred great excitement in the atomic physics community and renewed interest in studying superfluidity.

The properties of a BEC in a rotational frame at temperatures $T$ much smaller than the critical condensation temperature $T_c$ are usually well modeled by a nonlinear Schrödinger equation (NLSE) for the macroscopic wave function known as the Gross-Pitaevskii equation (GPE) [62, 63, 52], which incorporates the trap potential, rotational frame and the interactions among the atoms. The effect of the interactions is described by a mean field which leads to a nonlinear term in the GPE. The cases of repulsive and attractive interactions—which can both be realized in the experiment—correspond to defocusing and focusing nonlinearities in the GPE, respectively.

There has been a series of recent analytical and numerical studies of ground states in a rotating BEC. For example, Aftalion and Du [4] and Aftalion and Riviere [6] studied numerically and asymptotically the ground state, critical angular velocity and energy diagram in the Thomas–Fermi (TF) or semiclassical regime. Aftalion and Danaila [3] and Modugno et al. [59] reported bent vortices, e.g., the S-shaped vortex and U-shaped vortex, numerically in cigar-shaped condensation and compared then with the experimental results [65]. Garcia-Ripoll and Perez-Garcia [41, 40, 43] and Bao and Zhang [21] studied stability of the central vortex, and Tsubota et. al [74] reported vortex lattice formation. Bao et al. [9, 20] presented a continuous normalized gradient flow with backward Euler finite difference discretization to compute the ground state, provided asymptotics of the energy and chemical potential of the ground state in the semiclassical regime and showed that the ground state is a global minimizer of the energy functional over the unit sphere and that all excited states are saddle points in the linear case. Moreover, Svidzinsky and Fetter [73] have studied the dynamics of a vortex line depending on its curvature. For an analysis of the GP-functional in a rotational frame, we refer to [66]. For a numerical and theoretical review of quantized vortices, we refer to [39] and the recent book [63].

In order to study effectively the dynamics of BEC, especially in the strong repulsive interaction regime, an efficient and accurate numerical method is a key issue. For a nonrotating BEC, many numerical methods were proposed in the literature. For example, Bao et al. [12, 17, 21] proposed a fourth-order time-splitting sine or Fourier pseudo-spectral (TSSP) method, and Bao and Shen [17] presented a fourth-order time-splitting Laguerre–Hermite (TSLH) pseudo-spectral method for the GPE when the external trapping potential is radially or cylindrically symmetric in two (2D) or three dimensions (3D). The key ideas for the numerical methods in [12, 11, 21, 17, 14, 15] are based on the following: (i) a time-splitting technique is applied to decouple the nonlinearity in the GPE [12, 11, 14, 15]; (ii) proper spectral basis functions are chosen for a linear Schrödinger equation with a potential such that the ordinary differential equation (ODE) system in phase space is diagonalized and thus can be integrated exactly [12, 17]. These methods are explicit, unconditionally stable, and of spectral accuracy in space and fourth-order accuracy in time. Thus they are very efficient and accurate for computing the dynamics of a nonrotating BEC in 3D [13] and for the multicomponent case [18], which are very challenging problems in the numerical simulation of BEC. Some other numerical methods for nonrotating BEC include the finite difference method [27, 58], the particle-inspired scheme [28, 58] and the Runge–Kutta pseudo-spectral method [58]. Due to the appearance of the angular momentum rotation term in the GPE, new numerical difficulties must be overcome in designing efficient and accurate numerical methods for a rotating BEC. Currently, the numerical methods used in the physics literature for studying dynamics of a rotating BEC remain limited [4, 51], and they usually are low-order finite difference methods. Recently, some efficient and accurate numerical methods have been designed for computing the dynamics of a rotating BEC. For example, Bao, Du and Zhang [10] proposed a numerical method that applies a time-splitting technique for decoupling the nonlinearity in the GPE and adopting the polar coordinates or cylindrical coordinates so as to make the coefficient of the angular momentum rotation term constant. The method is time reversible, time transverse invariant, unconditionally stable and implicit in 1D but can be solved very efficiently, and it conserves the total density. It is of spectral accuracy in the transverse direction, but usually of second- or fourth-order accuracy in the radial direction. Zhang and Bao [76] used the leap-frog spectral method for studying vortex lattice dynamics in a rotating BEC in which Cartesian coordinates are adopted. This method is explicit, time reversible, of spectral accuracy in space and of second-order accuracy in time. It is stable under a stability constraint for time steps [76]. Bao and Wang [19] presented a time-splitting spectral (TSSP) method by applying a time-splitting technique for decoupling the nonlinearity and using the alternating direction implicit (ADI) technique for the coupling in the angular momentum rotation term in the GPE. Thus at every time step, the GPE in the rotational frame is decoupled into a nonlinear ODE and two partial differential equations with constant coefficients. This allowed them to develop new TSSP methods for computing the dynamics of BECs in a rotational frame. The new numerical method is explicit, unconditionally stable and of spectral accuracy in space and second-order accuracy in time. Moreover, it is time reversible and time transverse invariant, and conserves the position density in the discretized level.

The main aim of this chapter is to review the preceding results and methods for rotating BECs. The chapter is organized as follows. In Section 10.2, we take the 3D GPE with an angular momentum term, scale it to get a four-parameter model, reduce it to a 2D problem in a limiting regime, and present its semiclassical scaling and geometrical optics. In Section 10.3, we discuss existence/nonexistence of the ground state in a rotating BEC and provide an approximate ground state in limiting parameter regimes. In Section 10.4, we review the continuous normalized gradient flow and its backward Euler finite difference discretization for computing the ground and vortex states of a rotating BEC and report some numerical results. Some analytical results for the dynamics of a rotating BEC are reviewed in Section 10.5, and several efficient and accurate numerical methods for computing the dynamics of a rotating BEC are discussed in Section 10.6. Finally, in Section 10.7, some conclusions are drawn.

## 10.2 GPE in a rotational frame

At temperatures $T$ much smaller than the critical temperature $T_c$ [52], a BEC in a rotational frame is well described by the macroscopic wave function $\psi(\mathbf{x}, t)$, whose evolution is governed by a self-consistent, mean field nonlinear Schrödinger equation known as the Gross–Pitaevskii equation (GPE) with an angular momentum rotational term [26, 36, 41], without loss of generality (w.l.o.g.) assuming the rotation is around the $z$-axis:

$$i\hbar\partial\psi(\mathbf{x}, t) \over \partial t} = \frac{\delta E(\psi)}{\delta\psi^*} := H \ \psi \tag{10.2.1}$$

$$= \left(-\frac{\hbar^2}{2m}\nabla^2 + V(\mathbf{x}) + NU_0|\psi(\mathbf{x}, t)|^2 - \Omega L_z\right) \psi(\mathbf{x}, t),$$

where $\mathbf{x} = (x, y, z)^T \in \mathbb{R}^3$ is the spatial coordinate vector, $m$ is the atomic mass, $\hbar$ is the Planck constant, $N$ is the number of atoms in the condensate, $\Omega$ is an angular velocity, and $V(\mathbf{x})$ is an external trapping potential. When a harmonic trap potential is considered, $V(\mathbf{x}) = \frac{m}{2}\left(\omega_x^2 x^2 + \omega_y^2 y^2 + \omega_z^2 z^2\right)$ with $\omega_x$, $\omega_y$ and $\omega_z$ being the trap frequencies in the $x$-, $y$- and $z$-directions respectively. $U_0 = \frac{4\pi\hbar^2 a_s}{m}$ describes the interaction between atoms in the condensate with the $s$-wave scattering length $a_s$ (positive for repulsive interactions and negative for attractive interactions) and

$$L_z = xp_y - yp_x = -i\hbar\left(x\partial_y - y\partial_x\right) \tag{10.2.2}$$

is the $z$-component of the angular momentum $\mathbf{L} = \mathbf{x} \times \mathbf{P}$ with the momentum operator $\mathbf{P} = -i\hbar\nabla = (p_x, p_y, p_z)^T$. The energy functional per particle $E(\psi)$ is defined as

$$E(\psi) = \int_{\mathbb{R}^3} \left[\frac{\hbar^2}{2m}|\nabla\psi|^2 + V(\mathbf{x})|\psi|^2 + \frac{NU_0}{2}|\psi|^4 - \Omega\psi^* L_z\psi\right] d\mathbf{x}. \tag{10.2.3}$$

Here we use $f^*$ to denote the conjugate of a function $f$. It is convenient to normalize the wave function by requiring

$$\int_{\mathbb{R}^3} |\psi(\mathbf{x}, t)|^2 \, d\mathbf{x} = 1. \tag{10.2.4}$$

### 10.2.1 Dimensionless GPE in a rotational frame

Under the normalization condition (10.2.4), by introducing the dimensionless variables: $t \to t/\omega_m$ with $\omega_m = \min\{\omega_x, \omega_y, \omega_z\}$, $\mathbf{x} \to \mathbf{x}a_0$ with $a_0 = \sqrt{\hbar/m\omega_m}$, $\psi \to \psi/a_0^{3/2}$, $\Omega \to \Omega\omega_m$ and $E(\cdot) \to \hbar\omega_m E_{\beta,\Omega}(\cdot)$, we get the dimensionless GPE

$$\frac{i\,\partial\psi(\mathbf{x},t)}{\partial t} = \frac{\delta E_{\beta,\Omega}(\psi)}{\delta\psi^*} := H\,\psi$$

$$= \left(-\frac{1}{2}\nabla^2 + V(\mathbf{x}) + \beta\,|\psi(\mathbf{x},t)|^2 - \Omega L_z\right)\psi(\mathbf{x},t), \quad (10.2.5)$$

where

$$\beta = \frac{U_0 N}{a_0^3 \hbar \omega_m} = \frac{4\pi a_s N}{a_0},$$

$$L_z = -i(x\partial_y - y\partial_x),$$

$$V(\mathbf{x}) = \frac{1}{2}\left(\gamma_x^2 x^2 + \gamma_y^2 y^2 + \gamma_z^2 z^2\right)$$

with $\gamma_x = \omega_x/\omega_m$, $\gamma_y = \omega_y/\omega_m$ and $\gamma_z = \omega_z/\omega_m$, and the dimensionless energy functional per particle $E_{\beta,\Omega}(\psi)$ is defined as

$$E_{\beta,\Omega}(\psi) = \int_{\mathbb{R}^3} \left[\frac{1}{2}|\nabla\psi(\mathbf{x},t)|^2 + V(\mathbf{x})|\psi|^2 + \frac{\beta}{2}|\psi|^4 - \Omega\psi^* L_z\psi\right] d\mathbf{x}. \quad (10.2.6)$$

In a disk-shaped condensation with parameters $\omega_x \approx \omega_y$ and $\omega_z \gg \omega_x$ ($\Longleftrightarrow$ $\gamma_x = 1$, $\gamma_y \approx 1$ and $\gamma_z \gg 1$ choosing $\omega_m = \omega_x$), the 3D GPE (10.2.5) can be reduced to a 2D GPE with $\mathbf{x} = (x,y)^T$ [12, 8, 18]:

$$\frac{i\,\partial\psi(\mathbf{x},t)}{\partial t = -\frac{1}{2}\nabla^2\psi + V_2(x,y)\psi + \beta_2|\psi|^2\psi - \Omega L_z\psi,} \quad (10.2.7)$$

where $\beta_2 \approx \beta_2^a = \beta\sqrt{\gamma_z/2\pi}$ and $V_2(x,y) = \frac{1}{2}\left(\gamma_x^2 x^2 + \gamma_y^2 y^2\right)$ [12, 18, 4]. Thus here we consider the dimensionless GPE in a rotational frame in $d$ dimensions ($d = 2, 3$):

$$\frac{i\,\partial\psi(\mathbf{x},t)}{\partial t = -\frac{1}{2}\nabla^2\psi + V_d(\mathbf{x})\psi + \beta_d|\psi|^2\psi - \Omega L_z\psi,} \quad \mathbf{x} \in \mathbb{R}^d, \ t \geq 0, \quad (10.2.8)$$

$$\psi(\mathbf{x},0) = \psi_0(\mathbf{x}), \qquad \mathbf{x} \in \mathbb{R}^d; \quad (10.2.9)$$

where $\beta_3 = \beta$ and $V_3(x,y,z) = V(x,y,z)$.

Two important invariants of (10.2.8) are the *normalization of the wave function*

$$N(\psi) = \int_{\mathbb{R}^d} |\psi(\mathbf{x},t)|^2 \, d\mathbf{x} \equiv \int_{\mathbb{R}^d} |\psi(\mathbf{x},0)|^2 \, d\mathbf{x} = 1, \qquad t \geq 0 \quad (10.2.10)$$

and the *energy*

$$E_{\beta,\Omega}(\psi) = \int_{\mathbb{R}^d} \left[\frac{1}{2}|\nabla\psi(\mathbf{x},t)|^2 + V_d(\mathbf{x})|\psi|^2 + \frac{\beta_d}{2}|\psi|^4 - \Omega\psi^* L_z\psi\right] d\mathbf{x}.$$
$$(10.2.11)$$

## 10.2.2 Stationary states

To find a stationary solution of (10.2.8), we write

$$\psi(\mathbf{x}, t) = e^{-i\mu t} \phi(\mathbf{x}), \tag{10.2.12}$$

where $\mu$ is the chemical potential of the condensate and $\phi$ is independent of time. Inserting (10.2.12) into (10.2.8) gives the following equation for $\phi(\mathbf{x})$:

$$\mu \, \phi(\mathbf{x}) = -\frac{1}{2}\Delta\phi(\mathbf{x}) + V_d(\mathbf{x}) \, \phi(\mathbf{x}) + \beta_d|\phi(\mathbf{x})|^2\phi(\mathbf{x}) - \Omega L_z\phi(\mathbf{x}), \ \mathbf{x} \in \mathbb{R}^d, \tag{10.2.13}$$

under the normalization condition

$$\|\phi\|^2 = \int_{\mathbb{R}^d} |\phi(\mathbf{x})|^2 \, d\mathbf{x} = 1. \tag{10.2.14}$$

This is a nonlinear eigenvalue problem with a constraint and any eigenvalue $\mu$ can be computed from its corresponding eigenfunction $\phi$ by

$$\begin{aligned}
\mu &= \mu_{\beta,\Omega}(\phi) \\
&= \int_{\mathbb{R}^d} \left[ \frac{1}{2}|\nabla\phi(\mathbf{x})|^2 + V_d(\mathbf{x})\,|\phi(\mathbf{x})|^2 + \beta_d\,|\phi(\mathbf{x})|^4 - \Omega\phi^*(\mathbf{x})L_z\phi(\mathbf{x}) \right] d\mathbf{x} \\
&= E_{\beta,\Omega}(\phi) + \int_{\mathbb{R}^d} \frac{\beta_d}{2}\,|\phi(\mathbf{x})|^4 \, d\mathbf{x}.
\end{aligned} \tag{10.2.15}$$

In fact, the eigenfunctions of (10.2.13) under the constraint (10.2.14) are the critical points of the energy functional $E_{\beta,\Omega}(\phi)$ over the unit sphere $S = \{\phi \in \mathbb{C} \mid \|\phi\| = 1, E_{\beta,\Omega}(\phi) < \infty\}$. Furthermore, (10.2.13) is the Euler–Lagrange equation of the energy functional (10.2.11) with $\psi = \phi$ under the constraint (10.2.14).

## 10.2.3 Semiclassical scaling and geometrical optics

When $\beta_d \gg 1$, i.e., in a strongly repulsive interacting condensation or in a semiclassical regime, another scaling (under the normalization (10.2.10) with $\psi = \psi^\varepsilon$) for the GPE (10.2.8) is also very useful in practice. We choose $\mathbf{x} \to \varepsilon^{-1/2}\mathbf{x}$ and $\psi = \psi^\varepsilon \, \varepsilon^{d/4}$ with $\varepsilon = \beta_d^{-2/(d+2)}$:

$$\begin{aligned}
\frac{i\varepsilon \, \partial\psi^\varepsilon(\mathbf{x}, t)}{\partial t} &= \frac{\delta E_{\varepsilon,\Omega}(\psi^\varepsilon)}{\delta(\psi^\varepsilon)^*} := H^\varepsilon \, \psi^\varepsilon \\
&= -\frac{\varepsilon^2}{2}\nabla^2\psi^\varepsilon + V_d(\mathbf{x})\psi^\varepsilon + |\psi^\varepsilon|^2\psi^\varepsilon - \varepsilon\Omega L_z\psi^\varepsilon, \quad \mathbf{x} \in \mathbb{R}^d,
\end{aligned} \tag{10.2.16}$$

where the energy functional $E_{\varepsilon,\Omega}(\psi^\varepsilon)$ is defined as

$$E_{\varepsilon,\Omega}(\psi^\varepsilon) = \int_{\mathbb{R}^3} \left[ \frac{\varepsilon^2}{2} |\nabla \psi^\varepsilon|^2 + V_d(\mathbf{x})|\psi^\varepsilon|^2 + \frac{1}{2}|\psi^\varepsilon|^4 - \varepsilon\Omega(\psi^\varepsilon)^* L_z \psi^\varepsilon \right] d\mathbf{x}$$
$$= O(1),$$

assuming that $\psi^\varepsilon$ is $\varepsilon$-oscillatory and "sufficiently" integrable such that the integral of each term in the above energy functional is $O(1)$. Similarly, the non-linear eigenvalue problem (10.2.13) (under the normalization (10.2.14) with $\phi = \phi^\varepsilon$) reads

$$\mu^\varepsilon \phi^\varepsilon(\mathbf{x}) = -\frac{\varepsilon^2}{2}\Delta\phi^\varepsilon + V_d(\mathbf{x})\phi^\varepsilon + |\phi^\varepsilon|^2\phi^\varepsilon - \varepsilon\Omega L_z\phi^\varepsilon, \quad \mathbf{x} \in \mathbb{R}^d, \quad (10.2.17)$$

where any eigenvalue $\mu^\varepsilon$ can be computed from its corresponding eigenfunction $\phi^\varepsilon$ by

$$\mu^\varepsilon = \mu_{\varepsilon,\Omega}(\phi^\varepsilon) = \int_{\mathbb{R}^d} \left[ \frac{\varepsilon^2}{2}|\nabla\phi^\varepsilon|^2 + V_0(\mathbf{x})|\phi^\varepsilon|^2 + |\phi^\varepsilon|^4 - \varepsilon\Omega(\psi^\varepsilon)^* L_z\psi^\varepsilon \right] d\mathbf{x}$$
$$= O(1).$$

Furthermore, it is easy to get the leading asymptotics of the energy functional $E_{\beta,\Omega}(\psi)$ in (10.2.11) and the chemical potential $\mu_{\beta,\Omega}(\phi)$ in (10.2.15) when $\beta_d \gg 1$ from this scaling:

$$E_{\beta,\Omega}(\psi) = \varepsilon^{-1} E_{\varepsilon,\Omega}(\psi^\varepsilon) = O\left(\varepsilon^{-1}\right) = O\left(\beta_d^{2/(d+2)}\right), \qquad (10.2.18)$$

$$\mu_{\beta,\Omega}(\phi) = \varepsilon^{-1}\mu_{\varepsilon,\Omega}(\phi^\varepsilon) = O\left(\varepsilon^{-1}\right) = O\left(\beta_d^{2/(d+2)}\right), \beta_d \gg 1. \qquad (10.2.19)$$

These asymptotic results were confirmed by the numerical results in [16, 20]. When $0 < \varepsilon \ll 1$, i.e., $\beta_d \gg 1$, we set

$$\psi^\varepsilon(\mathbf{x},t) = \sqrt{\rho^\varepsilon(\mathbf{x},t)} \exp\left(\frac{i}{\varepsilon}S^\varepsilon(\mathbf{x},t)\right), \qquad (10.2.20)$$

where $\rho^\varepsilon = |\psi^\varepsilon|^2$ and $S^\varepsilon$ is the phase of the wave function. Inserting (10.2.20) into (10.2.16) and collecting real and imaginary parts, we get the transport equation for $\rho^\varepsilon$ and the Hamilton–Jacobi equation for the phase $S^\varepsilon$:

$$\partial_t \rho^\varepsilon + \operatorname{div}\left(\rho^\varepsilon \nabla S^\varepsilon\right) + \Omega\widehat{L}_z\rho^\varepsilon = 0, \qquad (10.2.21)$$

$$\partial_t S^\varepsilon + \frac{1}{2}|\nabla S^\varepsilon|^2 + V_d(\mathbf{x}) + \rho^\varepsilon + \Omega\widehat{L}_z S^\varepsilon = \frac{\varepsilon^2}{2}\frac{1}{\sqrt{\rho^\varepsilon}}\nabla^2\sqrt{\rho^\varepsilon}, \qquad (10.2.22)$$

where the operator $\widehat{L}_z = (x\partial_y - y\partial_x)$. Eq. (10.2.21) is the transport equation for the atom density and (10.2.22) the Hamilton–Jacobi equation for the phase. Furthermore, by defining the current density [20, 15]

$$\mathbf{J}^\varepsilon = \rho^\varepsilon \nabla S^\varepsilon = \varepsilon \operatorname{Im}\left((\psi^\varepsilon(\mathbf{x},t))^* \nabla\psi^\varepsilon(\mathbf{x},t)\right), \qquad (10.2.23)$$

we can get the quantum-hydrodynamic Euler system with a third-order dispersion term:

$$\partial_t \rho^\varepsilon + \mathrm{div}\mathbf{J}^\varepsilon + \Omega \widehat{L}_z \rho^\varepsilon = 0, \tag{10.2.24}$$

$$\partial_t \mathbf{J}^\varepsilon + \mathrm{div}\left(\frac{\mathbf{J}^\varepsilon \otimes \mathbf{J}^\varepsilon}{\rho^\varepsilon}\right) + \nabla P(\rho^\varepsilon) + \rho^\varepsilon \nabla V_d(\mathbf{x})$$

$$+\Omega\left(\widehat{L}_z + \mathbf{G}\right)\mathbf{J}^\varepsilon = \frac{\varepsilon^2}{4}\nabla\left(\rho^\varepsilon \nabla^2 \ln \rho^\varepsilon\right), \tag{10.2.25}$$

where $P(\rho) = \rho^2/2$ is the hydrodynamic pressure and the symplectic matrix $\mathbf{G}$ is defined as

$$\mathbf{G} = \begin{pmatrix} 0 & 1 \\ -1 & 0 \end{pmatrix}, \ \text{for } d = 2, \qquad \mathbf{G} = \begin{pmatrix} 0 & 1 & 0 \\ -1 & 0 & 0 \\ 0 & 0 & 0 \end{pmatrix}, \ \text{for } d = 3. \tag{10.2.26}$$

By formally passing to the limit $\varepsilon \to 0+$ in (10.2.21),(10.2.22), we obtain the system

$$\partial_t \rho^0 + \mathrm{div}\left(\rho^0 \nabla S^0\right) + \Omega \widehat{L}_z \rho^0 = 0, \tag{10.2.27}$$

$$\partial_t S^0 + \frac{1}{2}\left|\nabla S^0\right|^2 + V_d(\mathbf{x}) + \rho^0 + \Omega \widehat{L}_z S^0 = 0. \tag{10.2.28}$$

Similarly, letting $\varepsilon \to 0^+$ in (10.2.24), (10.2.25), we can formally obtain the following Euler system:

$$\partial_t \rho^0 + \mathrm{div}\mathbf{J}^0 + \Omega \widehat{L}_z \rho^0 = 0, \tag{10.2.29}$$

$$\partial_t \mathbf{J}^0 + \mathrm{div}\left(\frac{\mathbf{J}^0 \otimes \mathbf{J}^0}{\rho^0}\right) + \nabla P(\rho^0) + \rho^0 \nabla V_d(\mathbf{x}) + \Omega\left(\widehat{L}_z + \mathbf{G}\right)\mathbf{J}^0 = 0, \tag{10.2.30}$$

which is the isotropic Euler system (velocity given by $v^0 = \nabla s^0$) with quadratic pressure-density constitutive relation in the rotational frame. The formal asymptotics is supposed to hold up to the caustic onset time!

## 10.3 Ground state

The ground state wave function $\phi^g(\mathbf{x}) := \phi^g_{\beta,\Omega}(\mathbf{x})$ of a rotating BEC is found by minimizing the energy functional $E_{\beta,\Omega}(\phi)$ over the unit sphere $S$:
  Find $(\mu^g_{\beta,\Omega}, \phi^g_{\beta,\Omega} \in S)$ such that

$$E^g := E^g_{\beta,\Omega} = E_{\beta,\Omega}(\phi^g_{\beta,\Omega}) = \min_{\phi \in S} E_{\beta,\Omega}(\phi), \quad \mu^g := \mu^g_{\beta,\Omega} = \mu_{\beta,\Omega}(\phi^g_{\beta,\Omega}). \tag{10.3.1}$$

Any eigenfunction $\phi(\mathbf{x})$ of (10.2.13) under the constraint (10.2.14) whose energy $E_{\beta,\Omega}(\phi) > E_{\beta,\Omega}(\phi^g_{\beta,\Omega})$ is usually called an excited state in the literature [63].

Existence/nonexistence results for the ground state, depending on the magnitude $|\Omega|$ of the angular velocity relative to the trapping frequencies, are known and availeble [66].

### 10.3.1 Existence of the ground state when $|\Omega| < \gamma_{xy} := \min\{\gamma_x, \gamma_y\}$

To study the existence of the ground state in a rotating BEC, we first present some properties of the energy functional [20].

**Lemma 1.** (i) *In 2D, we have*

$$E_{\beta,-\Omega}(\phi(x,-y)) = E_{\beta,\Omega}(\phi(x,y)),$$
$$E_{\beta,-\Omega}(\phi(-x,y)) = E_{\beta,\Omega}(\phi(x,y)), \qquad \phi \in S. \tag{10.3.2}$$

(ii) *In 3D, we have*

$$E_{\beta,-\Omega}(\phi(x,-y,z)) = E_{\beta,\Omega}(\phi(x,y,z)),$$
$$E_{\beta,-\Omega}(\phi(-x,y,z)) = E_{\beta,\Omega}(\phi(x,y,z)), \qquad \phi \in S. \tag{10.3.3}$$

(iii) *In 2D and 3D, we have*

$$\int_{\mathbb{R}^d} \left[ \frac{1-|\Omega|}{2} |\nabla\phi(\mathbf{x})|^2 + \left( V_d(\mathbf{x}) - \frac{|\Omega|}{2}(x^2+y^2) \right) |\phi|^2 + \frac{\beta_d}{2}|\phi|^4 \right] d\mathbf{x} \le E_{\beta,\Omega}(\phi)$$

$$\le \int_{\mathbb{R}^d} \left[ \frac{1+|\Omega|}{2} |\nabla\phi(\mathbf{x})|^2 + \left( V_d(\mathbf{x}) + \frac{|\Omega|}{2}(x^2+y^2) \right) |\phi|^2 + \frac{\beta_d}{2}|\phi|^4 \right] d\mathbf{x}. \tag{10.3.4}$$

From this lemma, since $\gamma_y \ge \gamma_x = \gamma_{xy}$ and $\gamma_z > 0$, when $\beta_d \ge 0$ and $|\Omega| < \gamma_{xy}$, we know that the energy functional $E_{\beta,\Omega}(\phi)$ is positive, coercive and weakly lower semicontinuous on $S$. Thus the existence of a minimum follows from the standard theory [68] and we have the following.

**Theorem 1.** (i) *In 2D, if* $\phi_{\beta,\Omega}(x,y) \in S$ *is a ground state of the energy functional* $E_{\beta,\Omega}(\phi)$, *then* $\phi_{\beta,\Omega}(x,-y) \in S$ *and* $\phi_{\beta,\Omega}(-x,y) \in S$ *are ground states of the energy functional* $E_{\beta,-\Omega}(\phi)$. *Furthermore,*

$$E^g_{\beta,\Omega} = E^g_{\beta,-\Omega}, \qquad \mu^g_{\beta,\Omega} = \mu^g_{\beta,-\Omega}. \tag{10.3.5}$$

(ii) *In 3D, if* $\phi_{\beta,\Omega}(x,y,z) \in S$ *is a ground state of the energy functional* $E_{\beta,\Omega}(\phi)$, *then* $\phi_{\beta,\Omega}(x,-y,z) \in S$ *and* $\phi_{\beta,\Omega}(-x,y,z) \in S$ *are ground states of the energy functional* $E_{\beta,-\Omega}(\phi)$, *and (10.3.5) is also valid.*

(iii). *When* $\beta_d \ge 0$ *and* $|\Omega| < \gamma_{xy}$, *there exists a minimizer for the minimization problem (10.3.1), i.e., there exist ground states.*

To understand the uniqueness question, note that $E_{\beta,\Omega}(\alpha\phi^g_{\beta,\Omega}) = E_{\beta,\Omega}(\phi^g_{\beta,\Omega})$ for all $\alpha \in \mathbb{C}$ with $|\alpha| = 1$. Thus an additional constraint has to be introduced to show uniqueness. For a nonrotating BEC, i.e., $\Omega = 0$, the unique positive minimizer is usually taken as the ground state. In fact, the ground state is unique up to a constant $\alpha$ with $|\alpha| = 1$, i.e., the density of the ground state is unique, when $\Omega = 0$. For a rotating BEC under $|\Omega| < \gamma_{xy}$, several numerical methods were proposed in the literature [4, 20] for computing a minimizer of the minimization problem (10.3.1). From the numerical results [4, 20], the density of the ground state may no longer be unique when $|\Omega| > \Omega^c$ with $\Omega^c$ a critical angular rotation speed.

## 10.3.2 Nonexistence of ground states when $|\Omega| > \gamma^{xy} := \max\{\gamma_x, \gamma_y\}$

Denote $\gamma_r := \gamma^{xy}$ and note that $\frac{1}{2}(\gamma_x^2 x^2 + \gamma_y^2 y^2) \le \frac{1}{2}\gamma_r^2 r^2$ with $r = \sqrt{x^2 + y^2}$. We have

$$E_{\beta,\Omega}(\phi) \le \frac{1}{2}\int_0^{2\pi}\int_0^\infty \left[|\partial_r\phi|^2 + \frac{1}{r^2}|\partial_\vartheta\phi|^2 + \gamma_r^2 r^2|\phi|^2 \right.$$
$$\left. + \beta_2|\phi|^4 + 2i\Omega\phi^*\partial_\vartheta\phi\right] r\, dr d\vartheta, \quad d = 2, \qquad (10.3.6)$$

$$E_{\beta,\Omega}(\phi) \le \frac{1}{2}\int_{-\infty}^\infty\int_0^{2\pi}\int_0^\infty \left[|\partial_r\phi|^2 + \frac{1}{r^2}|\partial_\vartheta\phi|^2 + |\partial_z\phi|^2 + (\gamma_r^2 r^2 + \gamma_z^2 z^2)|\phi|^2 \right.$$
$$\left. + \beta_2|\phi|^4 + 2i\Omega\phi^*\partial_\vartheta\phi\right] r\, dr d\vartheta dz, \quad d = 3 \qquad (10.3.7)$$

where $(r, \vartheta)$ and $(r, \vartheta, z)$ are polar (in 2D) and, resp., cylindrical coordinates (in 3D). In 2D, let

$$\phi_m(\mathbf{x}) = \phi_m(r, \vartheta) = \phi_m(r)\, e^{im\vartheta}, \quad \text{with } \phi_m(r) = \frac{\gamma_r^{(|m|+1)/2}}{\sqrt{\pi|m|!}} r^{|m|} e^{-\frac{\gamma_r r^2}{2}},$$
$$(10.3.8)$$

where $m$ is an integer. In fact, $\phi_m(\mathbf{x})$ is the central vortex state with winding number $m$ of the GPE (10.2.8) with $d = 2$, $\beta_d = 0$ and $\Omega = 0$. It is very easy to check that $\phi_m$ satisfies

$$\|\phi_m\| = 2\pi\int_0^\infty |\phi_m(r)|^2\, r\, dr = 1, \qquad m \in \mathbb{Z}, \qquad (10.3.9)$$

$$\frac{1}{2}\left[-\frac{1}{r}\frac{d}{dr}\left(r\frac{d}{dr}\right) + r^2 + \frac{m^2}{r^2}\right]\phi_m(r) = (|m| + 1)\gamma_r\phi_m(r), \quad 0 < r < \infty.$$
$$(10.3.10)$$

Thus $\phi_m \in S$ and we compute

$$E_{\beta,\Omega}(\phi_m(\mathbf{x})) \le (|m|+1)\gamma_r - \Omega m + \beta_2 2\pi \int_0^\infty |\phi_m(r)|^4 r \, dr$$

$$= (|m|+1)\gamma_r - \Omega m + \frac{\beta_2 \gamma_r (2|m|)!}{4\pi(2^{|m|}(|m|!))^2}. \qquad (10.3.11)$$

Thus when $|\Omega| > \gamma_r$, we have

$$\inf_{\phi \in S} E_{\beta,\Omega}(\phi) \le \begin{cases} \lim_{m\to\infty} E_{\beta,\Omega}(\phi_m) & \Omega > 0, \\ \lim_{m\to\infty} E_{\beta,\Omega}(\phi_{-m}) & \Omega < 0 \end{cases}$$

$$= \lim_{m\to\infty} (\gamma_r - |\Omega|)|m| + \gamma_r + \frac{\beta_2 \gamma_r (2|m|)!}{4\pi(2^{|m|}(|m|!))^2} \qquad (10.3.12)$$

$$= -\infty.$$

This implies that there is no minimizer of the minimization problem (10.3.1) when $|\Omega| > \gamma^{xy}$ in 2D.

Similarly, in 3D, the argument proceeds with the central vortex line state with winding number $m$ of the GPE (10.2.8) with $d = 3$, $\beta_d = 0$ and $\Omega = 0$,

$$\phi_m(\mathbf{x}) = \phi_m(r, \vartheta, z) = \phi_m(r, z) \, e^{im\vartheta},$$

$$\phi_m(r, z) = \frac{\gamma_r^{(|m|+1)/2} \gamma_z^{1/4}}{\pi^{3/4}\sqrt{|m|!}} r^{|m|} e^{-\frac{\gamma_r r^2 + \gamma_z z^2}{2}}, \qquad (10.3.13)$$

and we conclude that there is no minimizer of the minimization problem (10.3.1) when $|\Omega| > \gamma^{xy}$ in 3D.

Remark 1. When $\gamma_{xy} < |\Omega| \le \gamma^{xy}$ in an anisotropic trap, although no rigorous mathematical justification, the numerical results in [20] show that there is no ground state of the energy functional $E_{\beta,\Omega}(\phi)$.

### 10.3.3 Stationary states as minimizer/saddle points in the linear case

For the stationary states of (10.2.13), we have the following lemma, valid in the linear case $\beta_d = 0$.

**Lemma 2.** *Suppose $\beta_d = 0$, $|\Omega| < \gamma_{xy}$ and $V_d(\mathbf{x}) \ge 0$ for $\mathbf{x} \in \mathbb{R}^d$; then we have:*
   (i) *The ground state $\phi^g$ is a global minimizer of $E_{0,\Omega}(\phi)$ over $S$.*
   (ii) *Any excited state $\phi^e$ is a saddle point of $E_{0,\Omega}(\phi)$ over $S$.*

*Proof.* Let $\phi_e$ be an eigenfunction of the eigenvalue problem (10.2.13) and (10.2.14). The corresponding eigenvalue is $\mu_e$. For any function $\phi$ such that $E_{0,\Omega}(\phi) < \infty$ and $\|\phi_e + \phi\| = 1$, notice by (10.2.14) that we have

$$\|\phi\|^2 = \|\phi + \phi_e\|^2 - \|\phi_e\|^2 - \int_{\mathbb{R}^d} (\phi^* \phi_e + \phi \phi_e^*) \, d\mathbf{x}$$

$$= - \int_{\mathbb{R}^d} (\phi^* \phi_e + \phi \phi_e^*) \, d\mathbf{x}. \tag{10.3.14}$$

From (10.2.11) with $\psi = \phi_e + \phi$, using (10.2.14) and (10.3.14) and integration by parts, we get

$$E_{0,\Omega}(\phi_e + \phi)$$

$$= \int_{\mathbb{R}^d} \left[ \frac{1}{2} |\nabla \phi_e + \nabla \phi|^2 + V_d(\mathbf{x}) |\phi_e + \phi|^2 - \Omega (\phi_e + \phi)^* L_z (\phi_e + \phi) \right] d\mathbf{x}$$

$$= \int_{\mathbb{R}^d} \left( \frac{1}{2} |\nabla \phi_e|^2 + V_d(\mathbf{x}) |\phi_e|^2 - \Omega \phi_e^* L_z \phi_e \right) d\mathbf{x}$$

$$+ \int_{\mathbb{R}^d} \left( \frac{1}{2} |\nabla \phi|^2 + V_d(\mathbf{x}) |\phi|^2 - \Omega \phi^* L_z \phi \right) d\mathbf{x}$$

$$+ \int_{\mathbb{R}^d} \left( -\frac{1}{2} \Delta \phi_e + V_d(\mathbf{x}) \phi_e - \Omega L_z \phi_e \right)^* \phi \, d\mathbf{x}$$

$$+ \int_{\mathbb{R}^d} \left( -\frac{1}{2} \Delta \phi_e + V_d(\mathbf{x}) \phi_e - \Omega L_z \phi_e \right) \phi^* \, d\mathbf{x}$$

$$= E_{0,\Omega}(\phi_e) + E_{0,\Omega}(\phi) - \mu_e \|\phi\|^2$$

$$= E_{0,\Omega}(\phi_e) + [E_{0,\Omega}(\phi/\|\phi\|) - \mu_e] \|\phi\|^2. \tag{10.3.15}$$

(i) Taking $\phi_e = \phi_g$ and $\mu_e = \mu_g$ in (10.3.15) and noting that $E_{0,\Omega}(\phi/\|\phi\|) \geq E_{0,\Omega}(\phi_g) = \mu_g$ for any $\phi \neq 0$, we get immediately that $\phi_g$ is a global minimizer of $E_{0,\Omega}$ over $S$.

(ii) Taking $\phi_e = \phi_j$ and $\mu_e = \mu_j$ in (10.3.15), since $E_{0,\Omega}(\phi_g) < E_{0,\Omega}(\phi_j)$ and it is easy to find an eigenfunction $\phi$ of (10.2.13) such that $E_{0,\Omega}(\phi) > E_{0,\Omega}(\phi_j)$, we get immediately that $\phi_j$ is a saddle point of the functional $E_{0,\Omega}(\phi)$ over $S$. $\qquad\square$

### 10.3.4 Approximate ground state

When $\beta_d = 0$ and $\Omega = 0$, the ground state solution is given explicitly [18]:

$$\mu_{0,0}^g = \frac{1}{2} \begin{cases} \gamma_x + \gamma_y, & d = 2, \\ \gamma_x + \gamma_y + \gamma_z, & d = 3, \end{cases}$$

$$\phi_{0,0}^g(\mathbf{x}) = \frac{1}{\pi^{d/4}} \begin{cases} (\gamma_x \gamma_y)^{1/4} e^{-\frac{\gamma_x x^2 + \gamma_y y^2}{2}}, & d = 2, \\ (\gamma_x \gamma_y \gamma_z)^{1/4} e^{-\frac{\gamma_x x^2 + \gamma_y y^2 + \gamma_z z^2}{2}}, & d = 3. \end{cases} \tag{10.3.16}$$

In fact, this solution can be viewed as an approximation of the ground state for a weakly interacting slowly rotating condensate, i.e., $|\beta_d| \ll 1$ and $|\Omega| \approx 0$.

For a condensate with strong repulsive interactions, i.e., $\beta_d \gg 1$, $|\Omega| \approx 0$, $\gamma_x = O(1)$, $\gamma_y = O(1)$ and $\gamma_z = O(1)$, the ground state can be approximated by the TF approximation in this regime [12, 18, 4, 21]:

$$\phi_\beta^{TF}(\mathbf{x}) = \begin{cases} \sqrt{(\mu_\beta^{TF} - V_d(\mathbf{x}))/\beta_d}, & V_d(\mathbf{x}) < \mu_\beta^{TF}, \\ 0, & \text{otherwise,} \end{cases} \quad (10.3.17)$$

$$\mu_\beta^{TF} = \frac{1}{2} \begin{cases} (4\beta_2\gamma_x\gamma_y/\pi)^{1/2} & d = 2, \\ (15\beta_3\gamma_x\gamma_y\gamma_z/4\pi)^{2/5} & d = 3. \end{cases} \quad (10.3.18)$$

Clearly $\phi_\beta^{TF}$ is not differentiable at $V_d(\mathbf{x}) = \mu_\beta^{TF}$, thus $E_{\beta,\Omega}(\phi_\beta^{TF}) = \infty$ and $\mu_{\beta,\Omega}(\phi_\beta^{TF}) = \infty$ [12, 21]. This shows that one can't use (10.2.11) to define the energy of the TF approximation (10.3.17). How to define the energy of the TF approximation is not clear in the literature. Using (10.2.15), (10.3.18) and (10.3.17), and following [21] for a nonrotating BEC, here we use one method to define the energy of the TF approximation (10.3.17):

$$E_{\beta,\Omega}^{TF} = \mu_{\beta,\Omega}^{TF} - \int_{\mathbb{R}^d} \frac{\beta_d}{2} |\phi_\beta^{TF}(\mathbf{x})|^4 \, d\mathbf{x} = \frac{d+2}{d+4} \mu_\beta^{TF}, d = 2, 3. \quad (10.3.19)$$

The numerical results in [20] show that the TF approximation (10.3.17) is very accurate for the density of the ground state, except at the vortex core, when $\beta_d \gg 1$ and $|\Omega| < \gamma_{xy}$, and (10.3.18) and (10.3.19) converge to the chemical potential and energy respectively only when $|\Omega| \approx 0$, but diverge when $|\Omega|$ is near $\gamma_{xy}$.

**10.3.5 Critical angular velocity in symmetric trap**

In 2D with radial symmetry and in 3D with cylindrical symmetry, for any $\beta_d \geq 0$, when $\Omega = 0$, the ground state satisfies $\phi_{\beta,0}^g(\mathbf{x}) = \phi_{\beta,0}^0(r)$ in 2D and $\phi_{\beta,0}^g(\mathbf{x}) = \phi_{\beta,0}^0(r, z)$ in 3D with $\phi_{\beta,0}^0(r)$ and $\phi_{\beta,0}^0(r, z)$ the symmetric state of the problem (10.2.13), (10.2.14) in 2D and 3D respectively, i.e., the ground state is radially symmetric. When $\Omega$ increases to a critical angular velocity, $\Omega_\beta^c$, defined as

$$\Omega^c := \Omega_\beta^c = \max\left\{\Omega \mid E_{\beta,\Omega}(\phi_{\beta,\Omega}^g) = E_{\beta,\Omega}(\phi_{\beta,\Omega}^0) = E_{\beta,0}(\phi_{\beta,0}^0)\right\},$$

the energy of the ground state will be less than that of the symmetric state, i.e., symmetry breaking occurs in the ground state [66, 67]. $\Omega_\beta^c$ is also called the critical angular velocity for symmetry breaking in the ground state.

From the discussions and numerical results in the literature [20, 4], we have

$$\Omega_0^c = \gamma_r := \gamma_x = \gamma_y, \qquad 0 \leq \Omega_\beta^c < \Omega_\beta^v \leq \gamma_r, \quad \text{for } \beta_d > 0.$$

## 10.4 Numerical methods and results for ground states

In this section, we review the continuous normalized gradient flow and its backward Euler finite difference discretization for computing the ground states of a rotating BEC.

### 10.4.1 Gradient flow with discrete normalization (GFDN)

Various algorithms, e.g., the imaginary time method [30, 4, 6], Sobolev gradient method [42, 41], finite element approximation [18], iterative method [29] etc., for finding the minimizer of the minimization problem (10.3.1) have been studied in the literature. Perhaps one of the more popular techniques for dealing with the normalization constraint (10.2.14) is through the splitting (or projection) scheme: (i). Apply the steepest descent method to an unconstrained minimization problem; (ii) project the solution back to the unit sphere $S$. This suggests that we consider gradient flow with discrete normalization (GFDN):

$$\phi_t = -\frac{\delta E_{\beta,\Omega}(\phi)}{\delta \phi^*} = \frac{1}{2}\Delta\phi - V_d(\mathbf{x})\phi - \beta_d |\phi|^2\phi + \Omega L_z\phi, \quad t_n < t < t_{n+1},$$

(10.4.1)

$$\phi(\mathbf{x}, t_{n+1}) \stackrel{\triangle}{=} \phi(\mathbf{x}, t_{n+1}^+) = \frac{\phi(\mathbf{x}, t_{n+1}^-)}{\|\phi(\cdot, t_{n+1}^-)\|}, \quad \mathbf{x} \in \mathbb{R}^d, \quad n \geq 0,$$

(10.4.2)

$$\phi(\mathbf{x}, 0) = \phi_0(\mathbf{x}), \quad \mathbf{x} \in \mathbb{R}^d \quad \text{with} \quad \|\phi_0\| = 1;$$

(10.4.3)

where $0 = t_0 < t_1 < t_2 < \cdots < t_n < \cdots$ with $\Delta t_n = t_{n+1} - t_n > 0$ and $k = \max_{n \geq 0} \Delta t_n$, and $\phi(\mathbf{x}, t_n^\pm) = \lim_{t \to t_n^\pm} \phi(\mathbf{x}, t)$. In fact, the gradient flow (10.4.1) can be viewed as applying the steepest descent method to the energy functional $E_{\beta,\Omega}(\phi)$ without constraint, and (10.4.2) then projects the solution back to the unit sphere in order to satisfy the constraint (10.2.14). From a numerical point of view, the gradient flow (10.4.1) can be solved via traditional techniques and the normalization of the gradient flow is simply achieved by a projection at the end of each time step.

Let

$$\tilde{\phi}(\cdot, t) = \frac{\phi(\cdot, t)}{\|\phi(\cdot, t)\|}, \quad t_n \leq t \leq t_{n+1}, \quad n \geq 0.$$

(10.4.4)

For the gradient flow (10.4.1), it is easy to establish the following basic facts [20].

**Lemma 3.** *Suppose $V_d(\mathbf{x}) \geq 0$ for all $\mathbf{x} \in \mathbb{R}^d$, $\beta_d \geq 0$ and $\|\phi_0\| = 1$, then*
(i) $\|\phi(\cdot, t)\| \leq \|\phi(\cdot, t_n)\| = 1$ *for $t_n \leq t < t_{n+1}$, $n \geq 0$.*
(ii) *For any $\beta_d \geq 0$, and all*
$t'$, $t$ *with $t_n \leq t' < t < t_{n+1}$:*

$$E_{\beta,\Omega}(\phi(\cdot, t)) \leq E_{\beta,\Omega}(\phi(\cdot, t')), \qquad n \geq 0. \qquad (10.4.5)$$

(iii) *For $\beta_d = 0$,*

$$E_{0,\Omega}(\tilde{\phi}(\cdot, t)) \leq E_{0,\Omega}(\tilde{\phi}(\cdot, t_n)), \qquad t_n \leq t \leq t_{n+1}, \qquad n \geq 0. \qquad (10.4.6)$$

From Lemma 3, we immediately get the following [20].

**Theorem 2.** *Suppose $V_d(\mathbf{x}) \geq 0$ for all $\mathbf{x} \in \mathbb{R}^d$ and $\|\phi_0\| = 1$. For $\beta_d = 0$, GFDN (10.4.1)–(10.4.3) is energy diminishing for any time step $k$ and initial data $\phi_0$, i.e.,*

$$E_{0,\Omega}(\phi(\cdot, t_{n+1})) \leq E_{0,\Omega}(\phi(\cdot, t_n)) \leq \cdots \leq E_{0,\Omega}(\phi(\cdot, 0)) = E_{0,\Omega}(\phi_0), \ n \geq 0. \qquad (10.4.7)$$

### 10.4.2 Continuous normalized gradient flow (CNGF)

In fact, the normalized step (10.4.2) is equivalent to solving the following ODE exactly:

$$\phi_t(\mathbf{x}, t) = \mu_\phi(t, k)\phi(\mathbf{x}, t), \quad \mathbf{x} \in \mathbb{R}^d, t_n < t < t_{n+1}, n \geq 0, \quad (10.4.8)$$
$$\phi(\mathbf{x}, t_n^+) = \phi(\mathbf{x}, t_{n+1}^-), \quad \mathbf{x} \in \mathbb{R}^d; \qquad (10.4.9)$$

where

$$\mu_\phi(t, k) \equiv \mu_\phi(t_{n+1}, \Delta t_n) = -\frac{1}{2\,\Delta t_n} \ln \|\phi(\cdot, t_{n+1}^-)\|^2, \qquad t_n \leq t \leq t_{n+1}. \qquad (10.4.10)$$

Thus the GFDN (10.4.1)–(10.4.3) can be viewed as a first-order splitting method for the gradient flow with discontinuous coefficients:

$$\phi_t = \frac{1}{2}\Delta\phi - V_d(\mathbf{x})\phi - \beta\,|\phi|^2\phi + \Omega L_z\phi + \mu_\phi(t, k)\phi, \mathbf{x} \in \mathbb{R}^d, t \geq 0, \quad (10.4.11)$$

$$\phi(\mathbf{x}, 0) = \phi_0(\mathbf{x}), \qquad \mathbf{x} \in \mathbb{R}^d \qquad \text{with} \quad \|\phi_0\| = 1. \qquad (10.4.12)$$

Letting $k \to 0$ and noticing that $\phi(\mathbf{x}, t_{n+1}^-)$ on the right-hand side of (10.4.9) is the solution of (10.4.1) at $t_{n+1} = t + \Delta t_n$, we obtain

$$\mu_\phi(t) := \lim_{k \to 0^+} \mu_\phi(t, k) = \lim_{\Delta t_n \to 0^+} \frac{1}{-2\,\Delta t_n} \ln \|\phi(\cdot, t_{n+1}^-)\|^2$$

$$= \lim_{\Delta t_n \to 0^+} \frac{1}{-2\,\Delta t_n} \ln \|\phi(\cdot, (t + \Delta t_n)^-)\|^2$$

$$= \lim_{\Delta t_n \to 0^+} \frac{\frac{d}{d\tau}\|\phi(\cdot, t + \tau)\|^2\big|_{\tau = \Delta t_n}}{-2\|\phi(\cdot, t + \Delta t_n)\|^2}$$

$$= \lim_{\Delta t_n \to 0^+} \frac{\mu_{\beta,\Omega}(\phi(\cdot, t + \Delta t_n))}{\|\phi(\cdot, t + \Delta t_n)\|^2} = \frac{\mu_{\beta,\Omega}(\phi(\cdot, t))}{\|\phi(\cdot, t)\|^2}. \qquad (10.4.13)$$

This leads us to consider the following CNGF:

$$\phi_t = \frac{1}{2}\Delta\phi - V_d(\mathbf{x})\phi - \beta_d\,|\phi|^2\phi + \Omega L_z\phi + \mu_\phi(t)\phi, \quad \mathbf{x} \in \mathbb{R}^d, \ t \geq 0, \quad (10.4.14)$$

$$\phi(\mathbf{x}, 0) = \phi_0(\mathbf{x}), \qquad \mathbf{x} \in \mathbb{R}^d \qquad \text{with} \quad \|\phi_0\| = 1. \qquad (10.4.15)$$

In fact, the right-hand side of (10.4.14) is the same as (10.2.13) if we view $\mu_\phi(t)$ as a Lagrange multiplier for the constraint (10.2.14). Furthermore, for the above CNGF, as observed in [9] for a nonrotating BEC, the solution of (10.4.14) also satisfies the following theorem [20].

**Theorem 3.** *Suppose $V_d(\mathbf{x}) \geq 0$ for all $\mathbf{x} \in \mathbb{R}^d$, $\beta_d \geq 0$ and $\|\phi_0\| = 1$. Then the CNGF (10.4.14), (10.4.15) is normalization conserving and energy diminishing, i.e.,*

$$\|\phi(\cdot, t)\|^2 = \int_{\mathbb{R}^d} |\phi(\mathbf{x}, t)|^2 \, d\mathbf{x} = \|\phi_0\|^2 = 1, \qquad t \geq 0, \qquad (10.4.16)$$

$$\frac{d}{dt} E_{\beta,\Omega}(\phi) = -2\,\|\phi_t(\cdot, t)\|^2 \leq 0, \qquad t \geq 0, \qquad (10.4.17)$$

*which in turn implies*

$$E_{\beta,\Omega}(\phi(\cdot, t_1)) \geq E_{\beta,\Omega}(\phi(\cdot, t_2)), \qquad 0 \leq t_1 \leq t_2 < \infty.$$

### 10.4.3 Fully numerical discretization

We now present a numerical method to discretize the GFDN (10.4.1)–(10.4.3) (or perform a full discretization of CNGF (10.4.14), (10.4.15)). For simplicity of notation we introduce the method for the case of 2D over a rectangle $\Omega_{\mathbf{x}} = [a, b] \times [c, d]$ with homogeneous Dirichlet boundary conditions. Generalizations to 3D are straightforward for tensor product grids and the results remain valid without modifications.

We choose the spatial mesh sizes $h_x = \Delta x > 0$, $h_y = \Delta y > 0$ with $h_x = (b - a)/M$, $h_y = (d - c)/N$ and $M, N$ even positive integers. The time step is given by $k = \Delta t > 0$ and we define grid points and time steps by

$$x_j := a + j\,h_x, \qquad j = 0, 1, \dots, M, \qquad y_l = c + l\,h_y, \qquad l = 0, 1, \dots, N,$$
$$t_n := n\,k, \qquad n = 0, 1, 2, \dots$$

Let $\phi_{j,l}^n$ be the numerical approximation of $\phi(x_j, y_l, t_n)$ and $\phi^n$ the solution vector at time $t = t_n = nk$ with components $\phi_{j,l}^n$.

We use backward Euler for time discretization and second-order centered finite difference for spatial derivatives. The detailed scheme is

$$\frac{\tilde{\phi}_{j,l} - \phi_{j,l}^n}{k} = \frac{1}{2h_x^2}\left[\tilde{\phi}_{j+1,l} - 2\tilde{\phi}_{j,l} + \tilde{\phi}_{j-1,l}\right] + \frac{1}{2h_y^2}\left[\tilde{\phi}_{j,l+1} - 2\tilde{\phi}_{j,l} + \tilde{\phi}_{j,l-1}\right]$$

$$-V_2(x_j, y_l)\tilde{\phi}_{j,l} - \beta_2 \left|\phi_{j,l}^n\right|^2 \tilde{\phi}_{j,l} + i\Omega y_l \frac{\tilde{\phi}_{j+1,l} - \tilde{\phi}_{j-1,l}}{2h_x}$$

$$-i\Omega x_j \frac{\tilde{\phi}_{j,l+1} - \tilde{\phi}_{j,l-1}}{2h_y}, \qquad j = 1, \ldots, M-1, \quad l = 1, \ldots, N-1,$$

$$\tilde{\phi}_{0,l} = \tilde{\phi}_{M,l} = \tilde{\phi}_{j,0} = \tilde{\phi}_{j,N} = 0, \qquad j = 0, \ldots, M, \quad l = 0, \ldots, N,$$

$$\phi_{j,l}^{n+1} = \frac{\tilde{\phi}_{j,l}}{\|\tilde{\phi}\|}, \quad j = 0, 1, \ldots, M, \ l = 0, \ldots, N, \quad n = 0, 1, \ldots, \qquad (10.4.18)$$

$$\phi_{j,l}^0 = \phi_0(x_j, y_l), \qquad j = 0, 1, \ldots, M; \quad l = 0, \ldots, N,$$

where the norm is defined as $\|\tilde{\phi}\|^2 = h_x h_y \sum_{j=1}^{M-1} \sum_{l=1}^{N-1} |\tilde{\phi}_{j,l}|^2$.

### 10.4.4 Numerical results

Many numerical results were reported in [20] for ground and central vortex states of a rotating BEC in 2D and 3D. Here we only present some ground state solutions in 2D of a rotating BEC for completeness. We take $d = 2$ and $\gamma_x = \gamma_y = 1$ in (10.2.8). Figures 10.1 and 10.2 plot the surface and contour, respectively, of the ground state $\phi^g(x, y) := \phi_{\beta,\Omega}^g(x, y)$ with $\beta_2 = 100$ for different $\Omega$.

## 10.5 Dynamics of a rotating BEC

In this section, we provide some analytical results on the conservation of the angular momentum expectation in a symmetric trap, i.e., $\gamma_x = \gamma_y$ in (10.2.8), derive a second-order ODE for the time evolution of the condensate width, and then present some dynamic laws of a stationary state with a shifted center in a rotating BEC.

### 10.5.1 Dynamics of angular momentum expectation and condensate width

As a measure of the vortex flux, we define the angular momentum expectation:

$$\langle L_z\rangle(t) := \int_{\mathbb{R}^d} \psi^*(\mathbf{x}, t) L_z \psi(\mathbf{x}, t) \, d\mathbf{x} = i \int_{\mathbb{R}^d} \psi^*(\mathbf{x}, t)(y\partial_x - x\partial_y)\psi(\mathbf{x}, t)d\mathbf{x},$$

$$(10.5.1)$$

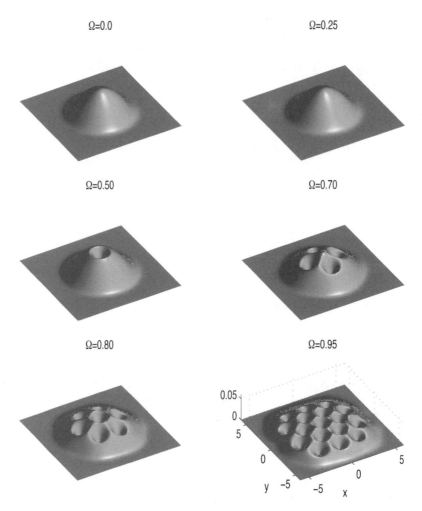

**Fig. 10.1.** Surface plots of ground state density function $|\phi^g(x,y)|^2$ in 2D with $\gamma_x = \gamma_y = 1$ and $\beta_2 = 100$ for different $\Omega$.

for any $t \geq 0$. For the dynamics of angular momentum expectation in a rotating BEC, we have the following lemma [10].

**Lemma 4.** *Suppose $\psi(\mathbf{x}, t)$ is the solution of the problem (10.2.8), (10.2.9), then we have*

$$\frac{d\langle L_z\rangle(t)}{dt} = \left(\gamma_x^2 - \gamma_y^2\right)\delta_{xy}(t), \ where \ \delta_{xy}(t) = \int_{\mathbb{R}^d} xy|\psi(\mathbf{x},t)|^2 d\mathbf{x}, \ t \geq 0 \ .$$

$$(10.5.2)$$

*Consequently, the angular momentum expectation and energy for a nonrotating part are conserved, that is, for any given initial data $\psi_0(\mathbf{x})$ in (10.2.9),*

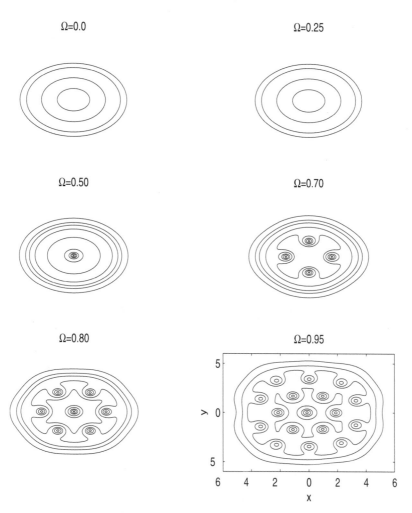

**Fig. 10.2.** Contour plots of ground state density function $|\phi^g(x,y)|^2$ in 2D with $\gamma_x = \gamma_y = 1$ and $\beta_2 = 100$ for different $\Omega$.

$$\langle L_z \rangle(t) \equiv \langle L_z \rangle(0), \quad E_{\beta,0}(\psi) \equiv E_{\beta,0}(\psi_0), \quad t \geq 0 \quad (10.5.3)$$

*at least for a radially symmetric trap in 2D or a cylindrically symmetric trap in 3D, i.e., $\gamma_x = \gamma_y$.*

Another quantity characterizing the dynamics of a rotating BEC is the condensate width defined as

$$\sigma_\alpha(t) = \sqrt{\delta_\alpha(t)}, \quad \text{where } \delta_\alpha(t) = \langle \alpha^2 \rangle(t) = \int_{\mathbb{R}^d} \alpha^2 |\psi(\mathbf{x},t)|^2 d\mathbf{x}, \quad (10.5.4)$$

for $t \geq 0$ and $\alpha$ being $x, y$ or $z$. For the dynamics of condensate widths, we have the following lemmas [10].

**Lemma 5.** *Suppose $\psi(\mathbf{x}, t)$ is the solution of problem (10.2.8), (10.2.9), then we have*

$$\frac{d^2 \delta_\alpha(t)}{dt^2} = \int_{\mathbb{R}^d} \left[ (\partial_y \alpha - \partial_x \alpha) \left( 4i\Omega \psi^*(x\partial_y + y\partial_x)\psi + 2\Omega^2(x^2 - y^2)|\psi|^2 \right) \right.$$

$$\left. + 2|\partial_\alpha \psi|^2 + \beta_d |\psi|^4 - 2\alpha |\psi|^2 \partial_\alpha (V_d(\mathbf{x})) \right] d\mathbf{x}, \quad t \geq 0,$$

$$\delta_\alpha(0) = \delta_\alpha^{(0)} = \int_{\mathbb{R}^d} \alpha^2 |\psi_0(\mathbf{x})|^2 d\mathbf{x}, \qquad \alpha = x, y, z, \tag{10.5.5}$$

$$\dot{\delta}_\alpha(0) = \delta_\alpha^{(1)} = 2 \int_{\mathbb{R}^d} \alpha \left[ -\Omega |\psi_0|^2 (x\partial_y - y\partial_x)\alpha + \mathrm{Im}\left(\psi_0^* \partial_\alpha \psi_0\right) \right] d\mathbf{x}, \tag{10.5.6}$$

*where* $\mathrm{Im}(f)$ *denotes the imaginary part of* $f$.

From Lemma 5, we then have [10] Lemma 6.

**Lemma 6.** (i) *In 2D with a radial symmetric trap, i.e., $d = 2$ and $\gamma_x = \gamma_y := \gamma_r$ in (10.2.8), for any initial data $\psi_0 = \psi_0(x, y)$, we have for any $t \geq 0$,*

$$\delta_r(t) = \frac{E_{\beta,\Omega}(\psi_0) + \Omega \langle L_z \rangle(0)}{\gamma_r^2} [1 - \cos(2\gamma_r t)] + \delta_r^{(0)} \cos(2\gamma_r t) + \frac{\delta_r^{(1)}}{2\gamma_r} \sin(2\gamma_r t), \tag{10.5.7}$$

*where* $\delta_r(t) = \delta_x(t) + \delta_y(t)$, $\delta_r^{(0)} := \delta_x(0) + \delta_y(0)$, *and* $\delta_r^{(1)} := \dot{\delta}_x(0) + \dot{\delta}_y(0)$. *Furthermore, when the initial condition $\psi_0(x, y)$ in (10.2.9) satisfies*

$$\psi_0(x, y) = f(r)e^{im\vartheta} \quad \text{with} \quad m \in \mathbb{Z} \quad \text{and} \quad f(0) = 0 \quad \text{when} \quad m \neq 0, \tag{10.5.8}$$

*we have, for any $t \geq 0$,*

$$\delta_x(t) = \delta_y(t) = \frac{1}{2}\delta_r(t) \tag{10.5.9}$$

$$= \frac{E_{\beta,\Omega}(\psi_0) + m\Omega}{2\gamma_x^2} [1 - \cos(2\gamma_x t)] + \delta_x^{(0)} \cos(2\gamma_x t) + \frac{\delta_x^{(1)}}{2\gamma_x} \sin(2\gamma_x t).$$

*This and (10.5.4) imply that*

$$\sigma_x = \sigma_y = \sqrt{\frac{E_{\beta,\Omega}(\psi_0) + m\Omega}{2\gamma_x^2} [1 - \cos(2\gamma_x t)] + \delta_x^{(0)} \cos(2\gamma_x t) + \frac{\delta_x^{(1)}}{2\gamma_x} \sin(2\gamma_x t)}. \tag{10.5.10}$$

*Thus in this case, the condensate widths $\sigma_x(t)$ and $\sigma_y(t)$ are periodic functions with frequency doubling the trapping frequency.*

(ii) *For all other cases, we have, for any $t \geq 0$*

$$\delta_\alpha(t) = \frac{E_{\beta,\Omega}(\psi_0)}{\gamma_\alpha^2} + \left(\delta_\alpha^{(0)} - \frac{E_{\beta,\Omega}(\psi_0)}{\gamma_\alpha^2}\right) \cos(2\gamma_\alpha t) + \frac{\delta_\alpha^{(1)}}{2\gamma_\alpha} \sin(2\gamma_\alpha t) + f_\alpha(t),$$

$$(10.5.11)$$

*where $f_\alpha(t)$ is the solution of the following second-order ODE:*

$$\frac{d^2 f_\alpha(t)}{dt^2} + 4\gamma_\alpha^2 \, f_\alpha(t) = F_\alpha(t), \qquad f_\alpha(0) = \frac{df_\alpha(0)}{dt} = 0, \qquad (10.5.12)$$

*with*

$$F_\alpha(t) = \int_{\mathbb{R}^d} \left[ 2|\partial_\alpha \psi|^2 - 2|\nabla \psi|^2 - \beta_d |\psi|^4 + \left(2\gamma_\alpha^2 \alpha^2 - 4V_d(\mathbf{x})\right) |\psi|^2 + 4\Omega \psi^* L_z \psi \right.$$

$$\left. + \left(\partial_y \alpha - \partial_x \alpha\right) \left(4i\Omega \psi^* \left(x\partial_y + y\partial_x\right) \psi + 2\Omega^2 (x^2 - y^2)|\psi|^2\right) \right] d\mathbf{x}.$$

## 10.5.2 Dynamics of a stationary state with its center shifted

Let $\phi_e(\mathbf{x})$ be a stationary state of the GPE (10.2.8) with a chemical potential $\mu_e$ [20], i.e., $(\mu_e, \phi_e)$ satisfying

$$\mu_e \phi_e(\mathbf{x}) = -\frac{1}{2}\Delta \phi_e + V_d(\mathbf{x})\phi_e + \beta_d |\phi_e|^2 \phi_e - \Omega L_z \phi_e, \qquad \|\phi_e\|^2 = 1.$$

$$(10.5.13)$$

If the initial data $\psi_0(\mathbf{x})$ in (10.2.9) is chosen as a stationary state with a shift in its center, one can construct an exact solution of the GPE (10.2.8) with a harmonic oscillator potential. This kind of analytical construction can be used, in particular, in the benchmark and validation of numerical algorithms for the GPE. In [44], a similar kind of solution was constructed for GPEs and a second-order ODE system was derived for the dynamics of the center, but the results there were valid only for nonrotating BECs, i.e., $\Omega = 0$. Modifications must be made for the rotating BEC, i.e., $\Omega \neq 0$. Later, in [22], similar results were extended to the case of a general Hamiltonian but without specifying the initial data for the ODE system. Here we present the dynamic laws for the rotating BEC [10].

**Lemma 7.** *If the initial data $\psi_0(\mathbf{x})$ in (10.2.9) is chosen as*

$$\psi_0(\mathbf{x}) = \phi_e(\mathbf{x} - \mathbf{x}_0), \qquad \mathbf{x} \in \mathbb{R}^d, \qquad (10.5.14)$$

*where $\mathbf{x}_0$ is a given point in $\mathbb{R}^d$, then the exact solution of (10.2.8), (10.2.9) satisfies*

$$\psi(\mathbf{x}, t) = \phi_e(\mathbf{x} - \mathbf{x}(t)) \, e^{-i\mu_e t} \, e^{iw(\mathbf{x},t)}, \qquad \mathbf{x} \in \mathbb{R}^d, \quad t \geq 0, \qquad (10.5.15)$$

*where for any time $t \geq 0$, $w(\mathbf{x}, t)$ is linear for $\mathbf{x}$, i.e.,*

$$w(\mathbf{x}, t) = \mathbf{c}(t) \cdot \mathbf{x} + g(t), \quad \mathbf{c}(t) = (c_1(t), \dots, c_d(t))^T, \quad \mathbf{x} \in \mathbb{R}^d, \ t \geq 0, \tag{10.5.16}$$

*and $\mathbf{x}(t)$ satisfies the following second-order ODE system:*

$$\ddot{x}(t) - 2\Omega \dot{y}(t) + \left(\gamma_x^2 - \Omega^2\right) x(t) = 0, \tag{10.5.17}$$

$$\ddot{y}(t) + 2\Omega \dot{x}(t) + \left(\gamma_y^2 - \Omega^2\right) y(t) = 0, \quad t \geq 0, \tag{10.5.18}$$

$$x(0) = x_0, \quad y(0) = y_0, \quad \dot{x}(0) = \Omega y_0, \quad \dot{y}(0) = -\Omega x_0. \tag{10.5.19}$$

*Moreover, if in 3D, another ODE needs to be added:*

$$\ddot{z}(t) + \gamma_z^2 z(t) = 0, \quad z(0) = z_0, \quad \dot{z}(0) = 0. \tag{10.5.20}$$

### 10.5.3 Analytical solutions for the center of mass

Without loss of generality, in this subsection, we assume $\gamma_x = 1$ and $\gamma_x \leq \gamma_y$ in (10.5.17)–(10.5.20). From (10.5.13) and (10.5.15), changing variables, we get

$$\langle \mathbf{x} \rangle (t) := \int_{\mathbb{R}^d} \mathbf{x} |\psi(\mathbf{x}, t)|^2 \, d\mathbf{x} = \int_{\mathbb{R}^d} \mathbf{x} |\phi_e(\mathbf{x} - \mathbf{x}(t))|^2 \, d\mathbf{x}$$

$$= \int_{\mathbb{R}^d} (\mathbf{x} + \mathbf{x}(t)) |\phi_e(\mathbf{x})|^2 \, d\mathbf{x} = \mathbf{x}(t), \quad t \geq 0. \tag{10.5.21}$$

This immediately implies that the dynamics of the center of mass is the same as that of $\mathbf{x}(t)$, i.e., it satisfies the ODE system (10.5.17)–(10.5.20). It is easy to see that the solution of (10.5.20) is

$$z(t) = z_0 \cos(\gamma_z t), \quad t \geq 0, \tag{10.5.22}$$

thus, $z(t)$ is a periodic function with period $T_z = 2\pi/\gamma_z$. Furthermore, when $\Omega \neq 0$, dividing both sides of (10.5.17) by $2\Omega$, we get

$$\dot{y}(t) = \frac{1}{2\Omega} \left( \ddot{x}(t) + \left(\gamma_x^2 - \Omega^2\right) x(t) \right), \quad t \geq 0. \tag{10.5.23}$$

Differentiating (10.5.18) with respect to $t$, we obtain

$$y^{(3)}(t) + 2\Omega \ddot{x}(t) + \left(\gamma_y^2 - \Omega^2\right) \dot{y}(t) = 0, \quad t \geq 0. \tag{10.5.24}$$

Inserting (10.5.23) into (10.5.24), we get the following fourth-order ODE for $x(t)$:

$$x^{(4)}(t) + \left(\gamma_x^2 + \gamma_y^2 + 2\Omega^2\right) \ddot{x}(t) + \left(\gamma_x^2 - \Omega^2\right) \left(\gamma_y^2 - \Omega^2\right) x(t) = 0, \ t \geq 0. \tag{10.5.25}$$

The characteristic equation of (10.5.25) is

$$\lambda^4 + \left(\gamma_x^2 + \gamma_y^2 + 2\Omega^2\right)\lambda^2 + \left(\gamma_x^2 - \Omega^2\right)\left(\gamma_y^2 - \Omega^2\right) = 0. \qquad (10.5.26)$$

In the following, we will discuss the solutions of the ODE system (10.5.17)–(10.5.19) in different parameter regimes of trapping frequencies and angular rotation speed $\Omega$.

For a nonrotating BEC, i.e., $\Omega \equiv 0$ in GPE (10.2.8), the second-order ODE system (10.5.17)–(10.5.19) collapses to

$$\ddot{x}(t) + \gamma_x^2 x(t) = 0, \qquad \ddot{y}(t) + \gamma_y^2 y(t) = 0, \qquad t \geq 0, \qquad (10.5.27)$$
$$x(0) = x_0, \quad y(0) = y_0, \quad \dot{x}(0) = \dot{y}(0) = 0. \qquad (10.5.28)$$

It is straightforward to see that the solution of (10.5.27), (10.5.28) is

$$x(t) = x_0 \cos(\gamma_x t), \qquad y(t) = y_0 \cos(\gamma_y t), \qquad t \geq 0, \qquad (10.5.29)$$

which implies that both $x(t)$ and $y(t)$ are periodic functions with periods $T_x = 2\pi/\gamma_x$ and $T_y = 2\pi/\gamma_y$, respectively.

For a rotating BEC with a symmetric trap, i.e., $\Omega \neq 0$ in (10.2.8) and $\gamma_x \equiv \gamma_y$, we have the following solution for the second-order ODE system (10.5.17)–(10.5.19) [76]:

**Lemma 8.** *When $\Omega \neq 0$ and $\gamma_x \equiv \gamma_y$ in (10.5.17)–(10.5.19), the solutions of $x(t)$ and $y(t)$ for the motion of the center are*

$$x(t) = \frac{x_0}{2}\left[\cos(at) + \cos(bt)\right] + \frac{|\Omega|y_0}{2\Omega}\left[\sin(at) - \sin(bt)\right], \qquad (10.5.30)$$

$$y(t) = \frac{y_0}{2}\left[\cos(at) + \cos(bt)\right] + \frac{|\Omega|x_0}{2\Omega}\left[-\sin(at) + \sin(bt)\right], t \geq 0, \qquad (10.5.31)$$

*where*

$$a = \gamma_x + |\Omega|, \qquad b = \gamma_x - |\Omega|.$$

*Furthermore, we can obtain the distance between the center of mass and the trap center as a periodic function with period $T = \pi/\gamma_x$, i.e.,*

$$|\mathbf{x}(t)| := \sqrt{x^2(t) + y^2(t)} = \sqrt{x_0^2 + y_0^2}\,|\cos(\gamma_x t)|, \qquad t \geq 0. \qquad (10.5.32)$$

For a rotating BEC with an anisotropic trap, i.e., $\Omega \neq 0$ in (10.2.8) and $\gamma_x < \gamma_y$, we will present the analytical solutions in four different cases: (a) $|\Omega| = \gamma_x$; (b) $|\Omega| = \gamma_y$; (c) $0 < |\Omega| < \gamma_x$ or $|\Omega| > \gamma_y$; and (d) $\gamma_x < \Omega < \gamma_y$. For $|\Omega| = \gamma_x$, we have [76] the following.

**Lemma 9.** *When $|\Omega| = \gamma_x < \gamma_y$ in (10.5.17)–(10.5.19), the solutions of $x(t)$ and $y(t)$ for the motion of the center are*

$$x(t) = \frac{x_0}{a^2}\left[(\gamma_y^2 + \Omega^2) + 2\Omega^2 \cos(at)\right]$$
$$+ \frac{\Omega y_0}{a^2}\left[-(\gamma_y^2 - \Omega^2)t + \frac{2(\gamma_y^2 + \Omega^2)}{a}\sin(at)\right],$$

$$(10.5.33)$$

$$y(t) = \frac{y_0}{a^2}\left[2\Omega^2 + (\gamma_y^2 + \Omega^2)\cos(at)\right] - \frac{\Omega x_0}{a}\sin(at), \qquad t \geq 0; \quad (10.5.34)$$

*where $a = \sqrt{\gamma_y^2 + 3\Omega^2}$. This implies that the center moves on an ellipse when $y_0 = 0$, and moves to infinity when $y_0 \neq 0$.*

Similarly for $\gamma_x < |\Omega| = \gamma_y$, we have [76] the following.

**Lemma 10.** *When $\gamma_x < \gamma_y = |\Omega|$ in (10.5.17)–(10.5.19), the solutions of $x(t)$ and $y(t)$ for the motion of the center are*

$$x(t) = \frac{x_0}{a^2}\left[2\Omega^2 + (\gamma_x^2 + \Omega^2)\cos(at)\right] + \frac{\Omega y_0}{a}\sin(at), \qquad t \geq 0, \quad (10.5.35)$$

$$y(t) = \frac{y_0}{a^2}\left[(\gamma_x^2 + \Omega^2) + 2\Omega^2 \cos(at)\right]$$
$$+ \frac{\Omega x_0}{a^2}\left[(\gamma_x^2 - \Omega^2)t - \frac{2(\gamma_x^2 + \Omega^2)}{a}\sin(at)\right],$$

$$(10.5.36)$$

*where $a = \sqrt{\gamma_x^2 + 3\Omega^2}$. Again this implies that the center moves on an ellipse when $x_0 = 0$, and moves to infinity when $x_0 \neq 0$.*

If $\Omega \neq 0$, $\gamma_x$ or $\gamma_y$, let

$$\delta_1 = (\gamma_x^2 + \gamma_y^2 + 2\Omega^2)/2,$$
$$\delta_2 = \sqrt{\delta_1^2 - (\gamma_x^2 - \Omega^2)(\gamma_y^2 - \Omega^2)},$$

$a = \sqrt{|\delta_1 - \delta_2|}$ and $b = \sqrt{\delta_1 + \delta_2}$. When $0 < |\Omega| < \gamma_x$ or $|\Omega| > \gamma_y$, we have $0 < \delta_2 < \delta_1$. Thus we get the four roots for the characteristic equation (10.5.26)

$$\lambda_{1,2} = \pm i\sqrt{\delta_1 - \delta_2} = \pm a\, i, \qquad \lambda_{3,4} = \pm i\sqrt{\delta_1 + \delta_2} = \pm b\, i. \quad (10.5.37)$$

Thus, in this case, we get the solution of the ODE system (10.5.17)–(10.5.19) [76].

**Lemma 11.** *When $\gamma_x < \gamma_y$, and $0 < |\Omega| < \gamma_x$ or $|\Omega| > \gamma_y$, we have the solution $x(t)$ and $y(t)$ of the ODE system (10.5.17)–(10.5.19)*

$$x(t) = c_1 \cos(at) + c_2 \sin(at) + c_3 \cos(bt) + c_4 \sin(bt), \tag{10.5.38}$$

$$y(t) = c_5 \cos(at) + c_6 \sin(at) + c_7 \cos(bt) + c_8 \sin(bt), \quad t \geq 0, \tag{10.5.39}$$

*where*

$$
c_1 = \frac{\left(\gamma_x^2 + \Omega^2 - b^2\right) x_0}{a^2 - b^2}, \qquad c_2 = \frac{a\Omega\left(\gamma_x^2 - \Omega^2 + b^2\right) y_0}{\left(\gamma_x^2 - \Omega^2\right)\left(a^2 - b^2\right)},
$$

$$
c_3 = -\frac{\left(\gamma_x^2 + \Omega^2 - a^2\right) x_0}{a^2 - b^2}, \qquad c_4 = -\frac{b\Omega\left(\gamma_x^2 - \Omega^2 + a^2\right) y_0}{\left(\gamma_x^2 - \Omega^2\right)\left(a^2 - b^2\right)},
$$

$$
c_5 = -\frac{\left(\gamma_x^2 - \Omega^2 - a^2\right)\left(\gamma_x^2 - \Omega^2 + b^2\right) y_0}{2\left(\gamma_x^2 - \Omega^2\right)\left(a^2 - b^2\right)},
$$

$$
c_6 = \frac{\left(\gamma_x^2 - \Omega^2 - a^2\right)\left(\gamma_x^2 + \Omega^2 - b^2\right) x_0}{2a\Omega\left(a^2 - b^2\right)},
$$

$$
c_7 = \frac{\left(\gamma_x^2 - \Omega^2 + a^2\right)\left(\gamma_x^2 - \Omega^2 - b^2\right) y_0}{2\left(\gamma_x^2 - \Omega^2\right)\left(a^2 - b^2\right)},
$$

$$
c_8 = -\frac{\left(\gamma_x^2 - \Omega^2 - b^2\right)\left(\gamma_x^2 + \Omega^2 - a^2\right) x_0}{2b\Omega\left(a^2 - b^2\right)}.
$$

*This implies that the graph of the trajectory is a bounded set.*

Similarly, when $\gamma_x < |\Omega| < \gamma_y$, we have $\delta_2 > \delta_1$. Thus we get the four roots for the characteristic equation (10.5.26)

$$\lambda_{1,2} = \pm\sqrt{\delta_2 - \delta_1} = \pm a, \qquad \lambda_{3,4} = \pm i\sqrt{\delta_1 + \delta_2} = \pm b\, i. \tag{10.5.40}$$

Thus, in this case, we get the solution of the ODE system (10.5.17)–(10.5.19) [76].

**Lemma 12.** *When $\gamma_x < |\Omega| < \gamma_y$, we have the solution $x(t)$ and $y(t)$ of the ODE system (10.5.17)–(10.5.19)*

$$x(t) = d_1 e^{at} + d_2 e^{-at} + d_3 \cos(bt) + d_4 \sin(bt), \tag{10.5.41}$$
$$y(t) = d_5 e^{at} + d_6 e^{-at} + d_7 \cos(bt) + d_8 \sin(bt), \quad t \geq 0, \tag{10.5.42}$$

*where*

$$
d_1 = \frac{1}{2}(c_1 - c_2), \qquad d_2 = -\frac{1}{2}(c_1 + c_2), \qquad d_3 = c_3,
$$

$$
d_4 = c_4, \qquad d_7 = c_7, \qquad d_8 = c_8,
$$

$$
d_5 = \frac{\left(\gamma_x^2 - \Omega^2 + a^2\right)}{4a\Omega}(c_1 - c_2), \qquad d_6 = \frac{\left(\gamma_x^2 - \Omega^2 + a^2\right)}{4a\Omega}(c_1 + c_2),
$$

with $c_1, \ldots, c_8$ constants defined in Lemma 11. From the above solution, we can see that if $c_1 = c_2$, i.e., $y_0 = \frac{(\gamma_x^2 - \Omega^2)(\gamma_x^2 + \Omega^2 - b^2)x_0}{a\Omega(\gamma_x^2 - \Omega^2 + b^2)}$, the graph of the trajectory is a bounded set; otherwise, the center will move to infinity exponentially fast and satisfies

$$\lim_{t \to \infty} \frac{y(t)}{x(t)} = \frac{c_5}{c_1} = \frac{(\gamma_x^2 - \Omega^2 + a^2)}{2a\Omega}. \tag{10.5.43}$$

### 10.5.4 Dynamics of the total density in the presence of dissipation

Consider a more general GPE of the form

$$(i - \lambda)\partial_t \psi(\mathbf{x}, t) = -\frac{1}{2}\Delta\psi + V(\mathbf{x}, t)\psi + \beta_d|\psi|^2\psi - \Omega L_z\psi, \quad \mathbf{x} \in \mathbb{R}^d, \tag{10.5.44}$$

$$\psi(\mathbf{x}, 0) = \psi_0(\mathbf{x}), \quad \mathbf{x} \in \mathbb{R}^d; \tag{10.5.45}$$

where $\lambda \geq 0$ is a real parameter that models a dissipation mechanism [5, 10] and $V(\mathbf{x}, t) = V_d(\mathbf{x}) + W(\mathbf{x}, t)$ with $W(\mathbf{x}, t)$ an external driven field [24, 25, 50]. Typical external driven fields used in the physics literature include a delta-kicked potential [50]

$$W(x, t) = K_s \cos(k_s x) \sum_{n=-\infty}^{\infty} \delta(t - n\tau), \tag{10.5.46}$$

with $K_s$ being the kick strength, $k_s$ the wavenumber, $\tau$ the time interval between kicks and $\delta(\tau)$ the Dirac delta function; or a far-blue detuned Gaussian laser beam stirrer [10, 24, 25]

$$W(\mathbf{x}, t) = W_s(t) \exp\left[-\left(\frac{|\mathbf{x} - \mathbf{x}_s(t)|^2}{w_s/2}\right)\right], \tag{10.5.47}$$

with $W_s(t)$ being the height, $w_s$ the width and $\mathbf{x}_s(t)$ the position of the stirrer. In addition, we note that to study the onset of energy dissipation in a BEC stirred by a laser field, another possibility is to view the beam as a translating *obstacle* [5] instead of introducing the Gaussian potential.

While the total density remains constant with $\lambda = 0$, in the more general case, we have the following lemma for the dynamics of the total density [10].

**Lemma 13.** Let $\psi(\mathbf{x}, t)$ be the solution of (10.5.44), (10.5.45), then the total density satisfies

$$\dot{N}(\psi)(t) = \frac{d}{dt} \int_{\mathbb{R}^d} |\psi(\mathbf{x}, t)|^2 \, d\mathbf{x} = -\frac{2\lambda}{1 + \lambda^2} \mu_{\beta, \Omega}(\psi), \qquad t \geq 0, \tag{10.5.48}$$

*where*

$$\mu_{\beta,\Omega}(\psi) = \int_{\mathbb{R}^d} \left[ \frac{1}{2}|\nabla\psi|^2 + V(\mathbf{x},t)|\psi|^2 + \beta_d|\psi|^4 - \Omega\mathrm{Re}(\psi^* L_z\psi) \right] d\mathbf{x}.$$

*Consequently, the total density decreases when $\lambda > 0$ and $|\Omega| \leq \gamma_{xy} :=$*
*$\min\{\gamma_x, \gamma_y\}$.*

## 10.6 Numerical methods for computing dynamics in a rotating BEC

In this section, we review the efficient and accurate numerical methods proposed recently to solve the following GPE for the dynamics of a rotating BEC.

Due to the trapping potential $V_d(\mathbf{x})$, the solution $\psi(\mathbf{x},t)$ of (10.5.44), (10.5.45) decays to zero exponentially fast when $|\mathbf{x}| \rightarrow \infty$. Thus in practical computation, we truncate the problem (10.5.44), (10.5.45) into a bounded computational domain with the homogeneous Dirichlet boundary condition:

$$(i - \lambda)\partial_t\psi(\mathbf{x},t) = -\frac{1}{2}\Delta\psi + V(\mathbf{x},t)\psi + \beta_d|\psi|^2\psi - \Omega L_z\psi, \ \mathbf{x} \in \Omega_{\mathbf{x}},$$
$$(10.6.1)$$

$$\psi(\mathbf{x},t) = 0, \qquad \mathbf{x} \in \Gamma = \partial\Omega_{\mathbf{x}}, \qquad t \geq 0, \qquad (10.6.2)$$

$$\psi(\mathbf{x},0) = \psi_0(\mathbf{x}), \quad \mathbf{x} \in \bar{\Omega}_{\mathbf{x}}; \qquad (10.6.3)$$

where $\Omega_{\mathbf{x}}$ is a bounded computational domain to be specified later. The use of more sophisticated radiation boundary conditions is an interesting topic that remains to be examined in the future.

### 10.6.1 Time splitting

We choose a time step size $\Delta t > 0$. For $n = 0, 1, 2, \ldots$, from time $t = t_n = n\Delta t$ to $t = t_{n+1} = t_n + \Delta t$, the GPE (10.6.1) is solved in two splitting steps. One solves first

$$(i - \lambda) \, \partial_t\psi(\mathbf{x},t) = -\frac{1}{2}\Delta\psi - \Omega L_z\psi \qquad (10.6.4)$$

for the time step of length $\Delta t$, and then solves

$$(i - \lambda) \, \partial_t\psi(\mathbf{x},t) = V(\mathbf{x},t)\psi + \beta_d|\psi|^2\psi, \qquad (10.6.5)$$

for the same time step. Equation (10.6.4) will be discretized in detail in the next two subsections. For $t \in [t_n, t_{n+1}]$, after dividing (10.6.5) by $(i - \lambda)$, multiplying it by $\psi^*$ and adding with its complex conjugate, we obtain the following ODE for $\rho(\mathbf{x}, t) = |\psi(\mathbf{x}, t)|^2$:

$$\partial_t \rho(\mathbf{x}, t) = -\frac{2\lambda}{1 + \lambda^2} \left[ V(\mathbf{x}, t)\rho(\mathbf{x}, t) + \beta_d \rho^2(\mathbf{x}, t) \right], \quad \mathbf{x} \in \Omega_{\mathbf{x}}, \ t_n \leq t \leq t_{n+1}.$$
(10.6.6)

The ODE for the phase angle $\phi(\mathbf{x}, t)$ (determined as $\psi = \sqrt{\rho} e^{i\phi}$) is given by

$$\phi_t = -\frac{1}{1 + \lambda^2} \left[ V(\mathbf{x}, t) + \beta_d \rho(\mathbf{x}, t) \right], \quad \mathbf{x} \in \Omega_{\mathbf{x}}, \ t_n \leq t \leq t_{n+1}.$$
(10.6.7)

For $\lambda \neq 0$, by (10.6.6), this is equivalent to

$$\phi_t = \frac{1}{2\lambda} \partial_t \ln \rho, \quad \mathbf{x} \in \Omega_{\mathbf{x}}, \ t_n \leq t \leq t_{n+1}.$$
(10.6.8)

Denote $V_n(\mathbf{x}, t) = \int_{t_n}^t V(\mathbf{x}, \tau) d\tau$, we can solve (10.6.6) to get

$$\rho(\mathbf{x}, t) = \frac{\rho(\mathbf{x}, t_n) \ \exp[\frac{-2\lambda V_n(\mathbf{x},t)}{1+\lambda^2}]}{1 + \rho(\mathbf{x}, t_n)\frac{2\lambda\beta_d}{1+\lambda^2} \int_{t_n}^t \exp[\frac{-2\lambda V_n(\mathbf{x},\tau)}{1+\lambda^2}] \, d\tau}.$$
(10.6.9)

Consequently, in the special case $V(\mathbf{x}, t) = V(\mathbf{x})$, we have some exact analytical solutions given by

$$\rho(\mathbf{x}, t) = \begin{cases} \rho(\mathbf{x}, t_n), & \lambda = 0, \\[2ex] \dfrac{(1 + \lambda^2)\rho(\mathbf{x}, t_n)}{(1 + \lambda^2) + 2\lambda\beta_d(t - t_n)\rho(\mathbf{x}, t_n)}, & V(\mathbf{x}) = 0, \\[3ex] \dfrac{V(\mathbf{x})\rho(\mathbf{x}, t_n) \ \exp[\frac{-2\lambda V(\mathbf{x})(t-t_n)}{1+\lambda^2}]}{V(\mathbf{x}) + \left(1 - \exp[\frac{-2\lambda V(\mathbf{x})(t-t_n)}{1+\lambda^2}]\right)\beta_d\rho(\mathbf{x}, t_n)}, & V(\mathbf{x}) \neq 0. \end{cases}$$
(10.6.10)

Inserting (10.6.9) into (10.6.7), we get for $t \in [t_n, t_{n+1}]$,

$$\psi(\mathbf{x}, t) = \psi(\mathbf{x}, t_n)\sqrt{U_n(\mathbf{x}, t)} \exp\left[-\frac{i}{1 + \lambda^2}\left(V_n(\mathbf{x}, t) + \beta_d \int_{t_n}^t \rho(\mathbf{x}, \tau)d\tau\right)\right],$$
(10.6.11)

where

$$U_n(\mathbf{x}, t) = \frac{\exp[\frac{-2\lambda V_n(\mathbf{x},t)}{1+\lambda^2}]}{1 + |\psi(\mathbf{x}, t_n)|^2 \frac{2\lambda\beta_d}{1+\lambda^2} \int_{t_n}^t \exp[\frac{-2\lambda V_n(\mathbf{x},\tau)}{1+\lambda^2}] \, d\tau}.$$
(10.6.12)

Again, with $V(\mathbf{x}, t) = V(\mathbf{x})$, we can integrate exactly to get

$$\psi(\mathbf{x}, t) = \psi(\mathbf{x}, t_n) \begin{cases} \exp\left[-i(\beta_d|\psi(\mathbf{x}, t_n)|^2 + V(\mathbf{x}))(t - t_n)\right], & \lambda = 0, \\ \sqrt{\hat{U}_n(\mathbf{x}, t)} \exp[\frac{i}{2\lambda} \ln \hat{U}_n(\mathbf{x}, t)], & \lambda \neq 0; \end{cases}$$

$$(10.6.13)$$

where

$$\hat{U}_n(\mathbf{x}, t) = \begin{cases} \dfrac{1 + \lambda^2}{1 + \lambda^2 + 2\lambda\beta_d(t - t_n)|\psi(\mathbf{x}, t_n)|^2}, & V(\mathbf{x}) = 0, \\[3mm] \dfrac{V(\mathbf{x}) \exp[-\frac{2\lambda(t-t_n)V(\mathbf{x})}{1+\lambda^2}]}{V(\mathbf{x}) + \left(1 - \exp[-\frac{2\lambda(t-t_n)V(\mathbf{x})}{1+\lambda^2}]\right) \beta_d|\psi(\mathbf{x}, t_n)|^2}, & V(\mathbf{x}) \neq 0. \end{cases}$$

*Remark 2.* If the function $V_n(\mathbf{x}, t)$ as well as other integrals in (10.6.9), (10.6.11) and (10.6.12) cannot be evaluated analytically, numerical quadrature can be used, e.g.,

$$V_n(\mathbf{x}, t_{n+1}) = \int_{t_n}^{t_{n+1}} V(\mathbf{x}, \tau)\, d\tau$$

$$\approx \frac{\Delta t}{6} \left[V(\mathbf{x}, t_n) + 4V(\mathbf{x}, t_n + \Delta t/2) + V(\mathbf{x}, t_{n+1})\right].$$

## 10.6.2 Discretization by using polar/cylindrical coordinates

To solve (10.6.4), we choose $\Omega_{\mathbf{x}} = \{(x, y),\ r = \sqrt{x^2 + y^2} < R\}$ in 2D, and respectively $\Omega_{\mathbf{x}} = \{(x, y, z),\ r = \sqrt{x^2 + y^2} < R,\ a < z < b\}$ in 3D, with $R$, $|a|$ and $b$ sufficiently large, and try to formulate the equation in a variable separable form. When $d = 2$, we use the polar coordinates $(r, \vartheta)$, and discretize in the $\vartheta$-direction by a Fourier pseudo-spectral method, in the $r$-direction by a finite element method (FEM) and in time by a Crank-Nicolson (C-N) scheme. Assume

$$\psi(r, \vartheta, t) = \sum_{l=-L/2}^{L/2-1} \hat{\psi}_l(r, t)\, e^{il\vartheta}, \qquad (10.6.14)$$

where $L$ is an even positive integer and $\hat{\psi}_l(r, t)$ is the Fourier coefficient for the $l$th mode. Inserting (10.6.14) into (10.6.4), noticing the orthogonality of the Fourier functions, we obtain for $-\frac{L}{2} \leq l \leq \frac{L}{2} - 1$ and $0 < r < R$:

$$(i - \lambda) \partial_t \hat{\psi}_l(r, t) = -\frac{1}{2r} \frac{\partial}{\partial r}\left(r \frac{\partial \hat{\psi}_l(r, t)}{\partial r}\right) + \left(\frac{l^2}{2r^2} - l\Omega\right) \hat{\psi}_l(r, t), \quad (10.6.15)$$

$$\hat{\psi}_l(R, t) = 0 \quad \text{(for all } l\text{)}, \qquad \hat{\psi}_l(0, t) = 0 \quad \text{(for } l \neq 0\text{)}. \qquad (10.6.16)$$

Let $P^k$ denote all polynomials with degree at most $k$, $M > 0$ be a chosen integer and $0 = r_0 < r_1 < r_2 < \cdots < r_M = R$ be a partition for the interval $[0, R]$ with a mesh size $h = \max_{0 \leq m < M} \{r_{m+1} - r_m\}$. Define an FEM subspace by

$$U^h = \left\{ u^h \in C[0, R] \mid u^h \big|_{[r_m, r_{m+1}]} \in P^k, \ 0 \leq m < M, \ u^h(R) = 0 \right\}$$

for $l = 0$, and for $l \neq 0$,

$$U^h = \left\{ u^h \in C[0, R] \mid u^h \big|_{[r_m, r_{m+1}]} \in P^k, \ 0 \leq m < M, \ u^h(0) = u^h(R) = 0 \right\}.$$

Then we obtain the FEM approximation for (10.6.15), (10.6.16): Find $\hat{\psi}_l^h = \hat{\psi}_l^h(\cdot, t) \in U^h$ such that for all $\phi^h \in U^h$ and $t_n \leq t \leq t_{n+1}$,

$$(i - \lambda) \frac{d}{dt} A(\hat{\psi}_l^h(\cdot, t), \phi^h) = B(\hat{\psi}_l^h(\cdot, t), \phi^h) + l^2 C(\hat{\psi}_l^h, \phi^h) - l\Omega A(\hat{\psi}_l^h, \phi^h),$$

$$\text{(10.6.17)}$$

where

$$A(u^h, v^h) = \int_0^R r \, u^h(r) \, v^h(r) \, dr, \qquad B(u^h, v^h) = \int_0^R \frac{r}{2} \frac{du^h(r)}{dr} \frac{dv^h(r)}{dr} \, dr,$$

$$C(u^h, v^h) = \int_0^R \frac{1}{2r} u^h(r) \, v^h(r) \, dr, \qquad u^h, \ v^h \in U^h.$$

The ODE system (10.6.17) is then discretized by the standard Crank-Nicolson scheme in time. Although an implicit time discretization is applied for (10.6.17), the 1D nature of the problem makes the coefficient matrix for the linear system band-limited. For example, if the piecewise linear polynomial is used, i.e., $k = 1$ in $U^h$, the matrix is tridiagonal. Fast algorithms can be applied to solve the resulting linear systems.

In practice, we always use the second-order Strang splitting [72], i.e., from time $t = t_n$ to $t = t_{n+1}$: i) first evolve (10.6.5) for half time step $\Delta t/2$ with initial data given at $t = t_n$; ii) then evolve (10.6.4) for one time step $\Delta t$ starting with the new data; iii) and evolve (10.6.5) for half time step $\Delta t/2$ with the newer data. Other ways to discretize (10.6.15) were also proposed in [10]. This method was demonstrated to be of spectral accuracy in the transverse direction, second- or fourth-order accuracy in the radial direction and second-order accuracy in time [10].

### 10.6.3 Discretization by using ADI technique

To solve (10.6.4) in another way, we choose $\Omega_{\mathbf{x}} = [a, b] \times [c, d]$ in 2D, and resp., $\Omega_{\mathbf{x}} = [a, b] \times [c, d] \times [e, f]$ in 3D, with $|a|$, $b$, $|c|$, $d$, $|e|$ and $f$ sufficiently large. For simplicity of notation, here we assume $\lambda = 0$ and $V(\mathbf{x}, t) = V(\mathbf{x})$ in (10.6.1).

When $d = 2$ in (10.6.4), we choose mesh sizes $\Delta x > 0$ and $\Delta y > 0$ with $\Delta x = (b - a)/M$ and $\Delta y = (d - c)/N$ for $M$ and $N$ even positive integers, and let the grid points be

$$x_j = a + j\Delta x, \quad j = 0, 1, 2, \dots, M; \quad y_k = c + k\Delta y, \quad k = 0, 1, 2, \dots, N.$$

Let $\psi_{jk}^n$ be the approximation of $\psi(x_j, y_k, t_n)$ and $\psi^n$ be the solution vector with component $\psi_{jk}^n$.

From time $t = t_n$ to $t = t_{n+1}$, we solve (10.6.4) first

$$i\, \partial_t \psi(\mathbf{x}, t) = -\frac{1}{2} \partial_{xx} \psi(\mathbf{x}, t) - i\Omega y \partial_x \psi(\mathbf{x}, t), \tag{10.6.18}$$

for the time step of length $\Delta t$, and then solve

$$i\, \partial_t \psi(\mathbf{x}, t) = -\frac{1}{2} \partial_{yy} \psi(\mathbf{x}, t) + i\Omega x \partial_y \psi(\mathbf{x}, t), \tag{10.6.19}$$

for the same time step.

For each fixed $y$, the operator in the equation (10.6.18) is in the $x$-direction with constant coefficients and thus we can discretize it in the $x$-direction by a Fourier pseudo-spectral method. Assume

$$\psi(x, y, t) = \sum_{p=-M/2}^{M/2-1} \widehat{\psi}_p(y, t)\, \exp[i\mu_p(x - a)], \tag{10.6.20}$$

where $\mu_p = \frac{2p\pi}{b-a}$ and $\widehat{\psi}_p(y, t)$ is the Fourier coefficient for the $p$th mode in the $x$-direction. Inserting (10.6.20) into (10.6.18), noticing the orthogonality of the Fourier functions, we obtain for $-\frac{M}{2} \le p \le \frac{M}{2} - 1$ and $c \le y \le d$:

$$i\, \partial_t \widehat{\psi}_p(y, t) = \left( \frac{1}{2}\mu_p^2 + \Omega y \mu_p \right) \widehat{\psi}_p(y, t), \qquad t_n \le t \le t_{n+1}. \tag{10.6.21}$$

This linear ODE can be integrated in time *exactly* and we obtain

$$\widehat{\psi}_p(y, t) = \exp\left[ -i\left( \frac{1}{2}\mu_p^2 + \Omega y \mu_p \right)(t - t_n) \right] \widehat{\psi}_p(y, t_n), \qquad t_n \le t \le t_{n+1}. \tag{10.6.22}$$

Similarly, for each fixed $x$, the operator in the equation (10.6.19) is in the $y$-direction with constant coefficients and thus we can discretize it in the $y$-direction by a Fourier pseudo-spectral method. Assume

$$\psi(x, y, t) = \sum_{q=-N/2}^{N/2-1} \widehat{\psi}_q(x, t)\, \exp[i\lambda_q(y - c)], \tag{10.6.23}$$

where $\lambda_q = \frac{2q\pi}{d-c}$ and $\widehat{\psi}_q(x,t)$ is the Fourier coefficient for the $q$th mode in the $y$-direction. Inserting (10.6.23) into (10.6.19), noticing the orthogonality of the Fourier functions, we obtain for $-\frac{N}{2} \leq q \leq \frac{N}{2} - 1$ and $a \leq x \leq b$:

$$i\,\partial_t\widehat{\psi}_q(x,t) = \left(\frac{1}{2}\lambda_q^2 - \Omega x\lambda_q\right)\widehat{\psi}_q(x,t), \qquad t_n \leq t \leq t_{n+1}. \tag{10.6.24}$$

Again this linear ODE can be integrated in time *exactly* and we obtain

$$\widehat{\psi}_q(x,t) = \exp\left[-i\left(\frac{1}{2}\lambda_q^2 - \Omega x\lambda_q\right)(t - t_n)\right]\widehat{\psi}_q(x,t_n), \qquad t_n \leq t \leq t_{n+1}. \tag{10.6.25}$$

From time $t = t_n$ to $t = t_{n+1}$, we combine the splitting steps via the standard second-order Strang splitting [72, 19]:

$$\psi_{jk}^{(1)} = \sum_{p=-M/2}^{M/2-1} e^{-i\Delta t(\mu_p^2 + 2\Omega y_k\mu_p)/4}\,\widehat{(\psi_k^n)}_p\,e^{i\mu_p(x_j-a)}, \quad 0 \leq j \leq M, 0 \leq k \leq N,$$

$$\psi_{jk}^{(2)} = \sum_{q=-N/2}^{N/2-1} e^{-i\Delta t(\lambda_q^2 - 2\Omega x_j\lambda_q)/4}\,\widehat{(\psi_j^{(1)})}_q\,e^{i\lambda_q(y_k-c)}, \quad 0 \leq k \leq N, 0 \leq j \leq M,$$

$$\psi_{jk}^{(3)} = e^{-i\Delta t[V(x_j,y_k)+\beta_2|\psi_{jk}^{(2)}|^2]}\,\psi_{jk}^{(2)},$$

$$\psi_{jk}^{(4)} = \sum_{q=-N/2}^{N/2-1} e^{-i\Delta t(\lambda_q^2 - 2\Omega x_j\lambda_q)/4}\,\widehat{(\psi_j^{(3)})}_q\,e^{i\lambda_q(y_k-c)}, \quad 0 \leq k \leq N, 0 \leq j \leq M,$$

$$\psi_{jk}^{n+1} = \sum_{p=-M/2}^{M/2-1} e^{-i\Delta t(\mu_p^2 + 2\Omega y_k\mu_p)/4}\,\widehat{(\psi_k^{(4)})}_p\,e^{i\mu_p(x_j-a)}, \tag{10.6.26}$$

where for each fixed $k$, $\widehat{(\psi_k^\alpha)}_p$ $(p = -M/2, \ldots, M/2 - 1)$ with $\alpha$ an index, the Fourier coefficients of the vector $\psi_k^\alpha = (\psi_{0k}^\alpha, \psi_{1k}^\alpha, \ldots, \psi_{(M-1)k}^\alpha)^T$, are defined as

$$\widehat{(\psi_k^\alpha)}_p = \frac{1}{M}\sum_{j=0}^{M-1} \psi_{jk}^\alpha\,e^{-i\mu_p(x_j-a)}, \qquad p = -\frac{M}{2}, \ldots, \frac{M}{2} - 1; \tag{10.6.27}$$

similarly, for each fixed $j$, $\widehat{(\psi_j^\alpha)}_q$ $(q = -N/1, \ldots, N/2 - 1)$, the Fourier coefficients of the vector $\psi_j^\alpha = (\psi_{j0}^\alpha, \psi_{j1}^\alpha, \ldots, \psi_{j(N-1)}^\alpha)^T$ are defined as

$$\widehat{(\psi_j^\alpha)}_q = \frac{1}{N}\sum_{k=0}^{N-1} \psi_{jk}^\alpha\,e^{-i\lambda_q(y_k-c)}, \qquad q = -\frac{N}{2}, \ldots, \frac{N}{2} - 1. \tag{10.6.28}$$

For algorithm (10.6.26), the total memory requirement is $O(MN)$ and the total computational cost per time step is $O(MN\ln(MN))$. The scheme is time reversible when $W(\mathbf{x}) \equiv 0$, just as it holds for the GPE (10.2.8), i.e., the scheme is unchanged if we interchange $n \leftrightarrow n+1$ and $\Delta t \leftrightarrow -\Delta t$ in (10.6.26). Also, a main advantage of the numerical method is its time-transverse invariance when $W(\mathbf{x}) \equiv 0$, just as it holds for the GPE (10.2.8) itself. If a constant $\alpha$ is added to the external potential $V$, then the discrete wave functions $\psi_{jk}^{n+1}$ obtained from (10.6.26) will be multiplied by the phase factor $e^{-i\alpha(n+1)\Delta t}$, which leaves the discrete quadratic observable $|\psi_{jk}^{n+1}|^2$ unchanged. This method was demonstrated to be of spectral accuracy in space and second accuracy in time [19].

### 10.6.4 The leap-frog spectral method

Another way to discretize (10.6.1) is the leap-frog spectral method. We choose $\Omega_{\mathbf{x}} = [a,b] \times [c,d]$ in 2D, and resp., $\Omega_{\mathbf{x}} = [a,b] \times [c,d] \times [e,f]$ in 3D, with $|a|$, $b$, $|c|$, $d$, $|e|$ and $f$ sufficiently large. Again, for simplicity of notation, here we assume $\lambda = 0$ and $V(\mathbf{x},t) = V(\mathbf{x})$ in (10.6.1). When $d = 2$, choose spatial mesh sizes $\Delta x = (b-a)/J$ and $\Delta y = (d-c)/K$ with $J$, $K$ and $L$ even integers, denote the grid points as

$$x_j = a + j\Delta x, \quad j = 0,1,\ldots,J, \qquad y_k = c + k\Delta y, \quad k = 0,1,\ldots,K,$$

and let $\psi_{j,k}^n$ be the approximation of $\psi(x_j, y_k, t_n)$. For $n = 1,2,\ldots$, from time $t = t_{n-1} = (n-1)\Delta t$ to $t = t_{n+1} = t_n + \Delta t$, the GPE (10.5.44) is discretized in space by the Fourier pseudo-spectral method and in time by the leap-frog scheme, i.e., for $j = 0,1,\ldots,J$ and $k = 0,1,\ldots,K$

$$
\begin{aligned}
i\frac{\psi_{j,k}^{n+1} - \psi_{j,k}^{n-1}}{2\Delta t} = &-\frac{1}{2}\left(\nabla_h^2 \psi^n\right)\big|_{j,k} + V_2(x_j, y_k)\psi_{j,k}^n \\
&+ \beta_2 |\psi_{j,k}^n|^2 \psi_{j,k}^n - \Omega\left(L_h \psi^n\right)\big|_{j,k},
\end{aligned}
\tag{10.6.29}
$$

where $\nabla_h^2$ and $L_h$, the pseudo-spectral differential operators approximating the operators $\nabla^2$ and $L_z$ respectively, are defined as

$$\left(\nabla_h^2 \psi^n\right)\big|_{j,k} = -\sum_{p=-J/2}^{J/2-1} \sum_{q=-K/2}^{K/2-1} \left(\mu_p^2 + \lambda_q^2\right) \widehat{(\psi^n)}_{p,q}\, e^{i\mu_p(x_j-a)}\, e^{i\lambda_q(y_k-c)},$$

$$\left(L_h \psi^n\right)\big|_{j,k} = x_j \left(D_y^h \psi^n\right)\big|_{j,k} - y_k \left(D_x^h \psi^n\right)\big|_{j,k}, \quad 0 \le j \le J,\ 0 \le k \le K,$$

$$\left(D_x^h \psi^n\right)\big|_{j,k} = \sum_{p=-J/2}^{J/2-1} \sum_{q=-K/2}^{K/2-1} \mu_p\, \widehat{(\psi^n)}_{p,q}\, e^{i\mu_p(x_j-a)}\, e^{i\lambda_q(y_k-c)},$$

$$\left(D_y^h \psi^n\right)\big|_{j,k} = \sum_{p=-J/2}^{J/2-1} \sum_{q=-K/2}^{K/2-1} \lambda_q\, \widehat{(\psi^n)}_{p,q}\, e^{i\mu_p(x_j-a)}\, e^{i\lambda_q(y_k-c)},$$

with

$$\mu_p = \frac{2p\pi}{b-a}, \quad p = -\frac{J}{2}, \dots, \frac{J}{2} - 1; \qquad \lambda_q = \frac{2q\pi}{d-c}, \quad q = -\frac{K}{2}, \dots, \frac{K}{2} - 1,$$

$$\widehat{(\psi^n)}_{p,q} = \frac{1}{JK} \sum_{j=0}^{J-1} \sum_{k=0}^{K-1} \psi_{j,k}^n \, e^{-i\mu_p(x_j - a)} \, e^{-i\lambda_q(y_k - c)}.$$

As stated in the introduction, here we use the leap-frog scheme for time discretization since we want to have an explicit and time-reversible time integrator. In order to compute $\psi_{j,k}^1$, we apply the modified trapezoidal rule in time on the interval $[t_0, t_1]$:

$$i\frac{\psi_{j,k}^{(1)} - \psi_{j,k}^0}{\Delta t} = -\frac{1}{2} \left. \left( \nabla_h^2 \psi^0 \right) \right|_{j,k} + V_2(x_j, y_k)\psi_{j,k}^0$$
$$+ \beta_2 |\psi_{j,k}^0|^2 \psi_{j,k}^0 - \Omega \left. \left( L_h \psi^0 \right) \right|_{j,k},$$

$$i\frac{\psi_{j,k}^{(2)} - \psi_{j,k}^{(1)}}{\Delta t} = -\frac{1}{2} \left. \left( \nabla_h^2 \psi^{(1)} \right) \right|_{j,k} + V_2(x_j, y_k)\psi_{j,k}^{(1)}$$
$$+ \beta_2 |\psi_{j,k}^{(1)}|^2 \psi_{j,k}^{(1)} - \Omega \left. \left( L_h \psi^{(1)} \right) \right|_{j,k},$$

$$\psi_{j,k}^1 = \frac{1}{2} \left( \psi_{j,k}^{(1)} + \psi_{j,k}^{(2)} \right), \qquad j = 0, 1, \dots, J, \ k = 0, 1, \dots, K.$$
$$(10.6.30)$$

The initial data (10.6.3) is discretized as

$$\psi_{j,k}^0 = \psi_0(x_j, y_k), \qquad j = 0, 1, \dots, J, \quad k = 0, 1, \dots, K. \qquad (10.6.31)$$

The leap-frog Fourier pseudo-spectral discretization (10.6.29) is explicit and time reversible. The total memory requirement is $O(JK)$ and the total computational cost per time step is $O(JK \ln(JK))$. Following the standard von Neumann analysis and frozen coefficient technique, the stability condition for (10.6.29) is

$$\Delta t < \frac{2(\Delta x)^2}{\pi^2 \left[1 + \left(\frac{\Delta x}{\Delta y}\right)^2\right] + \max_{\mathbf{x} \in \Omega_{\mathbf{x}}} \left[\pi \left(|x|\Delta x + |y|\frac{(\Delta x)^2}{\Delta y}\right) + V_2(\mathbf{x}) + \beta_2 |\psi(\mathbf{x}, t)|^2\right]}.$$

This method was demonstrated to be of spectral accuracy in space and second-order accuracy in time [76]

### 10.6.5 Numerical results

Many numerical results were reported in [10, 19, 76] to demonstrate the efficiency and accuracy of the above numerical methods. Here we only report the dynamics of a quantized vortex lattice with 81 vortices in a rotating BEC. We take $d = 2$, $\beta_2 = 2000$, $\Omega = 0.9$. The initial condition in (10.6.3) is taken as the ground state [20, 10, 3] of the GPE computed numerically with the

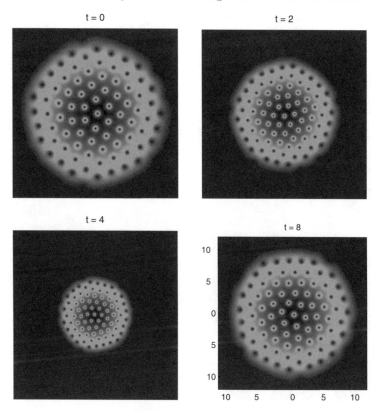

**Fig. 10.3.** The contour plots of the density function $|\psi(\mathbf{x}, t)|$ of the vortex lattices at different times for changing from $\gamma_x = \gamma_y = 1$ to $\gamma_x = \gamma_y = 1.5$.

same parameter values and $\gamma_x = \gamma_y = 1$. Then at $t = 0$, we change the trap frequency by setting $\gamma_x = \gamma_y = 1.5$, or $\gamma_x = 1.2$ and $\gamma_y = 1.5$ respectively. We take $\Omega_\mathbf{x} = [-24, 24] \times [-24, 24]$ and choose mesh size $\Delta x = \Delta y = 3/64$ and time step $\Delta = 0.0001$. Figures 10.3 and 10.4 show contour plots of the density function $|\psi(\mathbf{x}, t)|^2$ at different times.

From these figures, at $t = 0$, there are 81 quantized vortices in the ground state. During the time evolution, the lattice is rotated due to the angular momentum term (see Figure 10.4), and shrunk or expanded due to the changing of the trapping frequencies (see Figure 10.3). This clearly demonstrates the high resolution of the leap-frog Fourier pseudo-spectral method for a rotating BEC.

## 10.7 Conclusion

We have reviewed our recent works for the ground state and dynamics of the Gross–Pitaevskii equation (GPE) with an angular momentum rotation

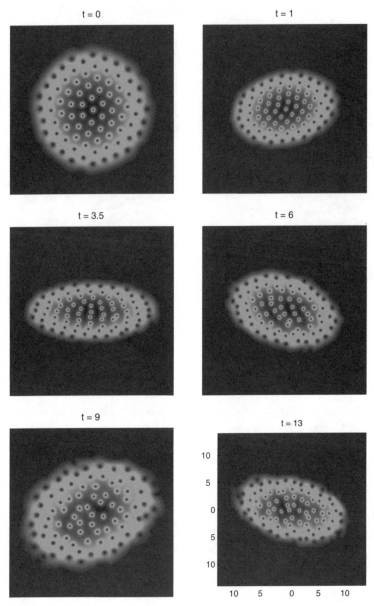

**Fig. 10.4.** The contour plots of the density function $|\psi(\mathbf{x}, t)|$ of the vortex lattices at different times for changing from $\gamma_x = \gamma_y = 1$ to $\gamma_x = 1.2$ and $\gamma_y = 1.5$.

term for rotating BECs. Along the analytical front, we provided asymptotics of the energy and chemical potential of the ground state in the semiclassical regime, and showed that the ground state is a global minimizer of the energy functional over the unit sphere and all excited states are saddle points in the linear case. We proved the conservation of the angular momentum expectation when the external trapping potential is radially symmetric in 2D, and respectively cylindrically symmetric in 3D. A second-order ODE was also derived to describe the time evolution of the condensate width as a periodic function with/without a perturbation, and where the frequency of the periodic function doubles the trapping frequency. We also presented an ODE system with complete initial data that governs the dynamics of a stationary state with a shifted center and we illustrated the decrease in the total density when a damping term is applied in the GPE. On the numerical side, we reviewed the continuous normalized gradient flow with backward Euler finite difference discretization for computing the ground states in a rotating BEC and three efficient and accurate numerical methods for computing the dynamics of rotating BEC. Finally the dynamics of a quantized vortex lattice with 81 vortices is reported to demonstrate the spectral resolution of our numerical methods.

## Acknowledgments

The author thanks his collaborators Peter A. Markowich, Qiang Du, Yanzhi Zhang and Hanquan Wang for their very fruitful collaboration on this subject and acknowledges support by the National University of Singapore grant No. R-146-000-081-112.

# References

[1]    J. R. Abo-Shaeer, C. Raman, J. M. Vogels and W. Ketterle, Observation of vortex lattices in Bose–Einstein condensates, *Science*, 292 (2001), p. 476.

[2]    S. K. Adhikari and P. Muruganandam, Effect of an implusive force on vortices in a rotating Bose–Einstein condensate, *Phys. Lett. A*, 301 (2002), pp. 333–339.

[3]    A. Aftalion and I. Danaila, Three-dimensional vortex configurations in a rotating Bose–Einstein condensate, *Phys. Rev. A*, 68 (2003), article 023603.

[4]    A. Aftalion and Q. Du, Vortices in a rotating Bose–Einstein condensate: critical angular velocities and energy diagrams in the Thomas–Fermi regime, *Phys. Rev. A*, 64 (2001), article 063603.

[5]    A. Aftalion, Q. Du and Y. Pomeau, Dissipative flow and vortex shedding in the Painlevé boundary layer of a Bose Einstein condensate, *Phy. Rev. Lett.*, 91 (2003), article 090407.

[6]    A. Aftalion and T. Riviere, Vortex energy and vortex bending for a rotating Bose–Einstein condensate, *Phys. Rev. A*, 64 (2001), article 043611.

[7]    M. H. Anderson, J. R. Ensher, M. R. Matthews, C. E. Wieman, and E. A. Cornell, Observation of Bose–Einstein condensation in a dilute atomic vapor, *Science*, 269 (1995), p. 198.

[8]    W. Bao, Ground states and dynamics of multi-component Bose–Einstein condensates, *Multiscale Modeling and Simulation*, 2 (2004), pp. 210–236.

[9] W. Bao and Q. Du, Computing the ground state solution of Bose–Einstein condensates by a normalized gradient flow, *SIAM J. Sci. Comput.*, 25 (2004), pp. 1674–1697.

[10] W. Bao, Q. Du and Y. Zhang, Dynamics of rotating Bose–Einstein condensates and their efficient and accurate numerical computation, *SIAM J. Appl. Math.*, 66 (2006), pp. 758–786.

[11] W. Bao and D. Jaksch, An explicit unconditionally stable numerical method for solving damped nonlinear Schrödinger equations with a focusing nonlinearity, *SIAM J. Numer. Anal.*, 41 (2003), pp. 1406–1426.

[12] W. Bao, D. Jaksch and P. A. Markowich, Numerical solution of the Gross-Pitaevskii Equation for Bose–Einstein condensation, *J. Comput. Phys.*, 187 (2003), pp. 318–342.

[13] W. Bao, D. Jaksch and P. A. Markowich, Three dimensional simulation of jet formation in collapsing condensates, *J. Phys. B: At. Mol. Opt. Phys.*, 37 (2004), pp. 329–343.

[14] W. Bao, S. Jin and P. A. Markowich, On time-splitting spectral approximation for the Schrödinger equation in the semiclassical regime, *J. Comput. Phys.*, 175 (2002), pp. 487–524.

[15] W. Bao, S. Jin and P. A. Markowich, Numerical study of time-splitting spectral discretizations of nonlinear Schrödinger equations in the semi-classical regimes, *SIAM J. Sci. Comput.*, 25 (2003), pp. 27–64.

[16] W. Bao, F.Y. Lim and Y. Zhang, Energy and chemical potential asymptotics for the ground state of Bose–Einstein condensates in the semiclassical regime, *Trans. Theory Stat. Phys.*, to appear.

[17] W. Bao and J. Shen, A fourth-order time-splitting Laguerre–Hermite pseudo-spectral method for Bose–Einstein condensates, *SIAM J. Sci. Comput.*, 26 (2005), pp. 2010–2028.

[18] W. Bao and W. Tang, Ground state solution of trapped interacting Bose–Einstein condensate by directly minimizing the energy functional, *J. Comput. Phys.*, 187 (2003), pp. 230–254.

[19] W. Bao and H. Wang, An efficient and spectrally accurate numerical method for computing dynamics of rotating Bose–Einstein condensates, *J. Comput. Phys.*, 217 (2006), pp. 612-626

[20] W. Bao, H. Wang and P. A. Markowich, Ground, symmetric and central vortex states in rotating Bose–Einstein condensates, *Comm. Math. Sci.*, 3 (2005), pp. 57–88.

[21] W. Bao and Y. Zhang, Dynamics of the ground state and central vortex states in Bose–Einstein condensation, *Math. Mod. Meth. Appl. Sci.*, 15 (2005), pp. 1863–1896.

[22] I. Bialynicki-Birula and Z. Bialynicki-Birula, Center-of-mass motion in the many-body theory of Bose–Einstein condensates, *Phys. Rev. A*, 65 (2002), article 063606.

[23] C. C. Bradley, C. A. Sackett and R. G. Hulet, Evidence of Bose–Einstein condensation in an atomic gas with attractive interactions, *Phys. Rev. Lett.*, 75 (1995), p. 1687.

[24] B. M. Caradoc-Davis, R. J. Ballagh and P. B. Blakie, Three-dimensional vortex dynamics in Bose–Einstein condensates, *Phys. Rev. A*, 62 (2000), article 011602.

[25]  B. M. Caradoc-Davis, R. J. Ballagh and P. B. Blakie, Coherent dynamics of vortex formation in trapped Bose–Einstein condensates, *Phys. Rev. Lett.*, 83 (1999), p. 895.

[26]  Y. Castin and R. Dum, Bose–Einstein condensates with vortices in rotating traps, *Eur. Phys. J. D*, 7 (1999), pp. 399–412.

[27]  M. M. Cerimele, M. L. Chiofalo, F. Pistella, S. Succi and M. P. Tosi, Numerical solution of the Gross–Pitaevskii equation using an explicit finite-difference scheme: An application to trapped Bose–Einstein condensates, *Phys. Rev. E*, 62 (2000), pp. 1382–1389.

[28]  M. M. Cerimele, F. Pistella and S. Succi, Particle-inspired scheme for the Gross–Pitaevskii equation: An application to Bose–Einstein condensation, *Comput. Phys. Comm.*, 129 (2000), pp. 82–90.

[29]  S.-M. Chang, W.-W. Lin and S.-F. Shieh, Gauss–Seidel-type methods for energy states of a multi-component Bose–Einstein condensate, *J. Comput. Phys.*, 64 (2001), article 053611.

[30]  M. L. Chiofalo, S. Succi and M. P. Tosi, Ground state of trapped interacting Bose–Einstein condensates by an explicit imaginary-time algorithm, *Phys. Rev. E*, 62 (2000), p. 7438.

[31]  F. Dalfovo and S. Giorgini, Theory of Bose–Einstein condensation in trapped gases, *Rev. Mod. Phys.*, 71 (1999), p. 463.

[32]  K. B. Davis, M. O. Mewes, M. R. Andrews, N. J. van Druten, D. S. Durfee, D. M. Kurn and W. Ketterle, Bose–Einstein condensation in a gas of sodium atoms, *Phys. Rev. Lett.*, 75 (1995), p. 3969.

[33]  R. J. Donnelly, *Quantizied Vortices in Helium II*, Cambridge University Press, London, 1991.

[34]  P. Engels, I. Coddington, P. Haijan and E. Cornell, Nonequilibrium effects of anisotropic compression applied to vortex lattices in Bose–Einstein condensates, *Phy. Rev. Lett.*, 89 (2002), article 100403.

[35]  J. R. Ensher, D. S. Jin, M. R. Matthews, C. E. Wieman and E. A. Cornell, Bose–Einstein condensation in a dilute gas: Measurement of energy and ground-state occupation, *Phys. Rev. Lett.*, 77 (1996), p. 4984.

[36]  D. L. Feder, C. W. Clark and B. I. Schneider, Nucleation of vortex arrays in rotating anisotropic Bose–Einstein condensates, *Phys. Rev. A*, 61 (1999), article 011601.

[37]  D. L. Feder, C. W. Clark and B. I. Schneider, Vortex stability of interacting Bose–Einstein condensates confined in anisotropic harmonic traps, *Phys. Rev. Lett.*, 82 (1999), p. 4956.

[38]  D. L. Feder, A. A. Svidzinsky, A. L. Fetter and C. W. Clark, Anomalous modes drive vortex dynamics in confined Bose–Einstein condensates, *Phys. Rev. Lett.*, 86 (2001), pp. 564–567.

[39]  A. L. Fetter and A. A. Svidzinsky, Vortices in a trapped dilute Bose–Einstein condensate, *J. Phys. Condens. Matter*, 13 (2001), R135–194.

[40]  J. J. Garcia-Ripoll and V. M. Perez-Garcia, Vortex bending and tightly packed vortex latices in Bose–Einstein condensates, *Phys. Rev. A*, 64 (2001), article 053611.

[41]  J. J. Garcia-Ripoll and V. M. Perez-Garcia, Stability of vortices in inhomogeneous Bose–Einstein condensates subject to rotation: A three dimensional analysis, *Phys. Rev. A*, 60 (1999), pp. 4864–4874.

[42]  J. J. Garcia-Ripoll and V. M. Perez-Garcia, Optimizing Schrödinger function-
      als using Sobolev gradients: application to quantum mechanics and nonlinear
      optics, *SIAM J. Sci. Comput.*, 23 (2001), pp. 1316–1334.

[43]  J. J. Garcia-Ripoll and V. M. Perez-Garcia, Vortex nucleation and hysteresis
      phenomena in rotating Bose–Einstein condensate, *Phys. Rev. A*, 63 (2001),
      article 041603.

[44]  J. J. Garcia-Ripoll, V. M. Perez-Garcia and V. Vekslerchik, Construction of
      exact solutions by spatial translations in inhomogeneous nonlinear Schrödinger
      equations, *Phys. Rev. E*, 64 (2001), article 056602.

[45]  R. Glowinski and P. LeTallec, *Augmented Lagrangians and Operator Splitting
      Methods in Nonlinear Mechanics*, SIAM, Philadelphia, 1989.

[46]  E. P. Gross, Structure of a quantized vortex in boson systems, *Nuovo. Ci-
      mento.*, 20 (1961), p. 454.

[47]  D. S. Hall, M. R. Mattthews, J. R. Ensher, C. E. Wieman and E. A. Cor-
      nell, Dynamics of component separation in a binary mixture of Bose–Einstein
      condensates, *Phys. Rev. Lett.*, 81 (1998), pp. 1539–1542.

[48]  B. Jackson, J. F. McCann and C. S. Adams, Vortex formation in dilute inho-
      mogeneous Bose–Einstein condensates, *Phys. Rev. Lett.*, 80 (1998), p. 3903.

[49]  B. Jackson, J. F. McCann and C. S. Adams, Vortex line and ring dynamics in
      trapped Bose–Einstein condensates, *Phys. Rev. A*, 61 (1999), article 013604.

[50]  D. Jaksch, C. Bruder, J. I. Cirac, C. W. Gardiner and P. Zoller, Cold bosonic
      atoms in optical lattices, *Phys. Rev. Lett.*, 81 (1998), pp. 3108–3111.

[51]  K. Kasamatsu, M. Tsubota and M. Ueda, Nonlinear dynamics of vortex lattice
      formation in a rotating Bose–Einstein condensate, *Phys. Rev. A*, 67 (2003),
      article 033610.

[52]  L. Laudau and E. Lifschitz, *Quantum Mechanics: Non-Relativistic Theory*,
      Pergamon Press, New York, 1977.

[53]  E. H. Lieb, R. Seiringer and J. Yngvason, Bosons in a trap: A rigorous deriva-
      tion of the Gross–Pitaevskii energy functional, *Phys. Rev. A*, 61 (2000), p.
      3602.

[54]  E. M. Lifshitz and L. P. Pitaevskii, *Statistical Physics*, 3rd edition, Pergamon,
      Oxford, 1980.

[55]  E. Lundh, C. J. Pethick and H. Smith, Vortices in Bose–Einstein-condensated
      atomic clouds, *Phys. Rev. A*, 58 (1998), pp. 4816–4823.

[56]  K. W. Madison, F. Chevy, W. Wohlleben and J. Dalibard, Vortex formation
      in a stirred Bose–Einstein condensate, *Phys. Rev. Lett.*, 84 (2000), p. 806.

[57]  M. R. Matthews, B. P. Anderson, P. C. Haljan, D. S. Hall, C. E. Wieman and
      E. A. Cornell, Vortices in a Bose–Einstein condensate, *Phys. Rev. Lett.*, 83
      (1999), p. 2498.

[58]  A. Minguzzi, S. Succi, F. Toschi, M. P. Tosi and P. Vignolo, Numerical methods
      for atomic quantum gases with applications to Bose–Einstein condensates and
      to ultracold fermions, *Phys. Rep.*, 395 (2004), pp. 223–355.

[59]  M. Modugno, L. Pricoupenko and Y. Castin, Bose–Einstein condensates with
      a bent vortex in rotating traps, *Eur. Phys. J. D*, 22 (2003), pp. 235–257.

[60]  A. A. Penckwitt and R.J. Ballagh, The nucleation, growth and stabilization of
      vortex lattices, *Phys. Rev. Lett.*, 89 (2002), article 268402.

[61]  C. J. Pethick and H. Smith, *Bose–Einstein Condensation in Dilute Gases*,
      Cambridge University Press, London, 2002.

[62]  L. P. Pitaevskii, Vortex lines in a imperfect Bose gas, *Zh. Eksp. Teor. Fiz.*, 40
      (1961), p. 646. (*Sov. Phys. JETP*, 13 (1961), p. 451).

[63] L. P. PITAEVSKII AND S. STRINGARI, *Bose–Einstein condensation*, Clarendon Press, Oxford, New York, 2003.

[64] C. Raman, J. R. Abo-Shaeer, J. M. Vogels, K. Xu and W. Ketterle, Vortex nucleation in a stirred Bose–Einstein condensate, *Phys. Rev. Lett.*, 87 (2001), article 210402.

[65] P. Rosenbuch, V. Bretin and J. Dalibard, Dynamics of a single vortex line in a Bose–Einstein condensate, *Phys. Rev. Lett.*, 89 (2002), article 200403.

[66] R. Seiringer, Gross–Pitaevskii theory of the rotating Bose gas, *Comm. Math. Phys.*, 229 (2002), pp. 491–509.

[67] R. Seiringer, Ground state asymptotics of a dilute rotating gas, *J. Phys. A: Math. Gen.*, 36 (2003), pp. 9755–9778.

[68] L. Simon, Asymptotics for a class of nonlinear evolution equations, with applications to geometric problems, *Ann. Math.*, 118 (1983), pp. 525–571.

[69] T. P. Simula, P. Engels, I. Coddington, V. Schweikhard, E. A. Cornell and R. J. Ballagh, Observations on sound propagation in rapidly rotating Bose–Einstein condensates, *Phys. Rev. Lett.*, 94 (2005), article 080404.

[70] T. P. Simula, A. A. Penckwitt and R. J. Ballagh, Giant vortex lattice deformations in rapidly rotating Bose–Einstein condensates, *Phys. Rev. Lett.*, 92 (2004), article 060401.

[71] S. Sinha and Y. Castin, Dynamic instability of a rotating Bose–Einstein condensate, *Phys. Rev. Lett.*, 87 (2001), article 190402.

[72] G. Strang, On the construction and comparison of difference schemes, *SIAM J. Numer. Anal.*, 5 (1968), pp. 505–517.

[73] A. A. Svidzinsky and A. L. Fetter, Dynamics of a vortex in a trapped Bose–Einstein condensate, *Phys. Rev. A*, 62 (2000), article 063617.

[74] M. Tsubota, K. Kasamatsu and M. Ueda, Vortex lattice formation in a rotating Bose–Einstein condensate, *Phys. Rev. A*, 65 (2000), article 023603.

[75] J. E. Williams and M. J. Hooand, Preparing topological states of a Bose–Einstein condensate, *Nature*, 401 (1999), p. 568.

[76] Y. Zhang and W. Bao, Dynamics of the center of mass in rotating Bose–Einstein condensates, Appl. Numer. Math., to appear.

# Two inverse problems in photon transport theory: evaluation of a time-dependent source and of a time-dependent cross section

Aldo Belleni-Morante

Dipartimento di Ingegneria Civile, Università degli Studi di Firenze, via di S. Marta 3, 50139 Firenze, Italia, `abelleni@dicea.unifi.it`

## 11.1 Introduction

In photon transport theory, two types of *inverse* problems are considered:

(a) identification of some physical or geometrical quantity (such as a cross section, or a photon source, or the shape of the surface that bounds the host medium), evaluating its dependence on spatial and/or angle variables, under the assumption that photon transport is time independent and starting, for instance, from the knowledge of the exiting photon flux;

(b) identification of some physical or geometrical quantity that characterizes the host medium, evaluating its dependence on spatial and/or angle variables *and* on time, under the assumption that photon transport is time dependent and starting, for instance, from the knowledge of the time behaviour of the exiting photon flux.

The literature on time-independent inverse problems for photon transport is rather abundant, see the references listed in [1]. On the other hand, only a few papers deal with time-dependent inverse problems for photon transport, see [2] through [8].

We remark that in most (a) and (b) inverse problems, it is assumed that for instance the *whole* exiting photon flux is known (or measured), due to a given photon flux entering the host medium under consideration. However, if the host medium is an interstellar cloud occupying the region $V \subset \mathbb{R}^3$ (see Remark 1), as it is in this chapter, one can measure only *one value* of the photon density at a location $\hat{\mathbf{x}}$ far from the cloud in the stationary case, or record the time dependence of such a single density in the time-dependent case. This, in principle, allows us to determinate only one parameter, see [9, 10].

*Remark 1.* Interstellar clouds are astronomical objects that occupy large regions of the galactic space: the diameter of an average cloud may range from

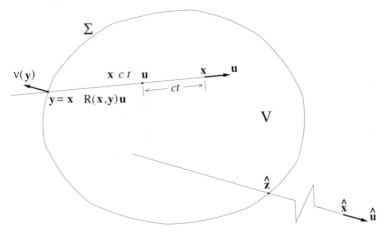

**Fig. 11.1.** The interstellar cloud occupying the region $V \subset \mathbb{R}^3$, bounded by the regular surface $\Sigma = \partial V$.

$10^3$ to $10^5$ times the diameter of our solar system. Clouds are composed of a low density mixture of gases and dust grains (mainly hydrogen molecules with some 1–2% of silicon grains); typical particle densities may be of the order of $10^4$ particles/cm³, i.e., $10^{-15}$ times the density of the earth's atmosphere at sea level, see [11, 12].

In this chapter, we shall first consider photon transport in an interstellar cloud with a space- and time-dependent source $q = q(\mathbf{x}, t)$ of UV-photons. We shall assume that the cloud occupies the strictly convex region $V \subset \mathbb{R}^3$, bounded by the closed "regular" surface $\Sigma = \partial V$, and that the values of the photon number density $N$ are recorded at a location $\widehat{\mathbf{x}} \in \mathbb{R}^3 \setminus V$ "far" from the cloud, during a suitable time interval. See Figure 11.1.

Then, our first inverse problem can be stated as follows: is it possible to determine $q(\mathbf{x}, t)$ within a suitable family of functions, starting from the knowledge of $N$ at $\widehat{\mathbf{x}}$ as a function of time?

We shall prove, rather surprisingly, that only a time average of $q$ can be evaluated. In fact, in [7] we considered a similar inverse problem and, by using a time-discretized model of photon transport, we proved that the set $\{q(\mathbf{x}, t_0), q(\mathbf{x}, t_1), \ldots, q(\mathbf{x}, t_n)\}$ could be indentified within a suitable family of functions of $\mathbf{x} \in V$. However, time discretization is reasonable if the measurements of the photon density at $\mathbf{x}$ are made only at times $t_0$,

$$t_1 = t_0 + \tau, \ldots, t_n = t_0 + n\tau.$$

On the other hand, if the measurements at $\widehat{\mathbf{x}}$ are taken continuously in time, one should be in a position to obtain more complete results on $q(\mathbf{x}, t)$. The reasons why this does not seem possible are perhaps due to the difficulties arising when $\tau \to 0^+$.

Finally, our second inverse problem deals with the evolution of a time-dependent (and unknown) cross section.

## 11.2 The mathematical model

Let $N(\mathbf{x}, \mathbf{u}, t)$ be the photon number density at any $(\mathbf{x}, \mathbf{u}) \in V \times S$ and $t \geq 0$, where $V$ is a bounded and strictly convex subset of $\mathbb{R}^3$ and $S$ is the surface of the unit sphere. Then, if $\sigma_s$, $\sigma_c$ and $\sigma = \sigma_s + \sigma_c$ are the scattering, the capture and the total cross sections of the interstellar cloud under consideration, the photon density $N$ satisfies the following system [13]:

$$\frac{\partial}{\partial t} N(\mathbf{x}, \mathbf{u}, t) = -\, c\mathbf{u} \cdot \nabla_{\mathbf{x}} N(\mathbf{x}, \mathbf{u}, t) - c\sigma N(\mathbf{x}, \mathbf{u}, t)$$

$$+ c\frac{\sigma}{4\pi} \int_S N(\mathbf{x}, \mathbf{u}', t)\, d\mathbf{u}' + q(\mathbf{x}, t), \quad (\mathbf{x}, \mathbf{u}) \in V_i \times S, \quad t > 0,$$

$$\tag{11.2.1}$$

$$N(\mathbf{y}, \mathbf{u}, t) = 0 \qquad \mathbf{y} \in \partial V, \quad \mathbf{u} \cdot \nu(\mathbf{y}) < 0, \quad t > 0, \tag{11.2.2}$$

$$N(\mathbf{x}, \mathbf{u}, 0) = N_0(\mathbf{x}, \mathbf{u}), \qquad (\mathbf{x}, \mathbf{u}) \in V_i \times S. \tag{11.2.3}$$

In (11.2.1), $c$ is the speed of light and $V_i$ is the interior of $V = V_i \cup \partial V$. Relation (11.2.2) is a non re-entry boundary condition at any $(\mathbf{y}, \mathbf{u})$ with $\mathbf{y} \in \partial V$ and $\mathbf{u} \cdot \nu(\mathbf{y}) < 0$, where $\nu(\mathbf{y})$ is the outward directed normal at $\mathbf{y}$. Finally, (11.2.3) is a given initial condition.

Note that condition (11.2.2) is certainly satisfied if the scattering cross section $\sigma_s$ and the given source term $q(\mathbf{x}, t)$ are equal to zero if $\mathbf{x} \notin V$.

Since

$$\frac{\partial}{\partial r} N\left(\mathbf{x} - r\mathbf{u}, \mathbf{u}, t - \frac{r}{c}\right) = -\frac{1}{c}\frac{\partial}{\partial t} N\left(\mathbf{x} - r\mathbf{u}, \mathbf{u}, t - \frac{r}{c}\right)$$

$$-\, \mathbf{u} \cdot \nabla_{\mathbf{x}} N\left(\mathbf{x} - r\mathbf{u}, \mathbf{u}, t - \frac{r}{c}\right),$$

equation (11.2.1) (with $\mathbf{x}$ substituted by $\mathbf{x} - r\mathbf{u}$ and $t$ by $t - r/c$) becomes

$$0 = \frac{\partial}{\partial r}\left\{\exp(-\sigma r)N\left(\mathbf{x} - r\mathbf{u}, \mathbf{u}, t - \frac{r}{c}\right)\right\}$$

$$+ \frac{\sigma_s}{4\pi}\exp(-\sigma r)\int_S N\left(\mathbf{x} - r\mathbf{u}, \mathbf{u}', t - \frac{r}{c}\right) d\mathbf{u}' + \exp(-\sigma r)q(\mathbf{x} - r\mathbf{u}, t - r/c).$$

Hence, we obtain

$$0 = -N(\mathbf{x}, \mathbf{u}, t)$$

$$+ \int_0^{R(\mathbf{x},\mathbf{u})} dr \, \exp(-\sigma \, r)$$

$$\times \left\{ \frac{\sigma_s}{4\pi} \int_S N(\mathbf{x} - r\mathbf{u}, \mathbf{u}', t - r/c) \, d\mathbf{u}' + q(\mathbf{x} - r\mathbf{u}, t - r/c) \right\}$$

if $t > R(\mathbf{x}, \mathbf{u})/c$, and

$$0 = \exp(\sigma ct) N_0(\mathbf{x} - ct\mathbf{u}, \mathbf{u}) - N(\mathbf{x}, \mathbf{u}, t)$$

$$+ \int_0^{ct} dr \, \exp(-\sigma \, r)$$

$$\times \left\{ \frac{\sigma_s}{4\pi} \int_S N(\mathbf{x} - r\mathbf{u}, \mathbf{u}', t - r/c) \, d\mathbf{u}' + q(\mathbf{x} - r\mathbf{u}, t - r/c) \right\}$$

if $t < R(\mathbf{x}, \mathbf{u})/c$, where we used conditions (11.2.2) and (11.2.3) and where $R(\mathbf{x}, \mathbf{u})$ is such that $\mathbf{x} - R(\mathbf{x}, \mathbf{u})\mathbf{u} \in \partial V$; see Figure 11.1.

Thus, the integral form of system (11.2.1–11.2.3) can be written as follows:

$$N(\mathbf{x}, \mathbf{u}, t) = \exp(-\sigma ct) \, N_0(\mathbf{x} - ct\mathbf{u}, \mathbf{u}) + \int_0^{R^*} dr \, \exp(-\sigma \, r) q(\mathbf{x} - r\mathbf{u}, t - r/c)$$

$$+ \frac{\sigma_s}{4\pi} \int_0^{R^*} dr \, \exp(\sigma r) \int_S N(\mathbf{x} - r\mathbf{u}, \mathbf{u}', t - r/c) \, d\mathbf{u}',$$

$$(11.2.4)$$

where

$$\begin{aligned}
R^* &= R^*(\mathbf{x}, \mathbf{u}, t) = R(\mathbf{x}, \mathbf{u}), & \text{if } t > R(\mathbf{x}, \mathbf{u})/c \\
R^* &= R^*(\mathbf{x}, \mathbf{u}, t) = ct, & \text{if } t < R(\mathbf{x}, \mathbf{u})/c
\end{aligned} \right\} \qquad (11.2.5)$$

and $N_0(\mathbf{x} - ct\mathbf{u}, \mathbf{u})$ is understood to be zero if $ct > R(\mathbf{x}, \mathbf{u})$, i.e., if $\mathbf{x} - ct\mathbf{u} \notin V$.

The integral equation (11.2.4) is the "mathematical model" that will be used to investigate our inverse problems. Such an equation will be studied in the Banach space $X = C(V \times S \times [0, \bar{t}])$ with the usual norm $\|\varphi\| = \max\{|\varphi(\mathbf{x}, \mathbf{u}, t)|, \ (\mathbf{x}, \mathbf{u}, t) \in V \times S \times [0, \bar{t}]\}$ (where $\bar{t} > 0$ is given) and under the following assumptions:

(i) $V$ is a bounded and strictly convex region of $\mathbb{R}^3$ (thus, given any $\mathbf{x}, \mathbf{y} \in V$, $\lambda \mathbf{x} + (1 - \lambda)\mathbf{y} \in V_i \ \forall \lambda \in (0, 1)$);

(ii) $\partial V$ is a "regular" surface, i.e., $R = R(\mathbf{x}, \mathbf{u})$ is a continuous function of $(\mathbf{x}, \mathbf{u}) \in V \times S$;

(iii) $N_0(\mathbf{x}, \mathbf{u})$ is a continuous function of $(\mathbf{x}, \mathbf{u}) \in V \times S$.

We now introduce the operators

$$(F\varphi)(\mathbf{x}, \mathbf{u}, t) = \begin{cases} \exp(-\sigma c t)\, \varphi(\mathbf{x} - ct\mathbf{u}, \mathbf{u}, t) & \text{if } \mathbf{x} - ct\mathbf{u} \in V \\ 0 & \text{if } \mathbf{x} - ct\mathbf{u} \notin V \end{cases} \quad (11.2.6)$$

$$(B\varphi)(\mathbf{x}, \mathbf{u}, t) = \int_0^{R^*} dr\, \exp(-\sigma r)\, \varphi(\mathbf{x} - r\mathbf{u}, \mathbf{u}, t - r/c) \quad (11.2.7)$$

$$(K\varphi)(\mathbf{x}, t) = \frac{\sigma_s}{4\pi} \int_S \varphi(\mathbf{x}, \mathbf{u}', t)\, d\mathbf{u}', \quad (11.2.8)$$

with $D(F) = D(B) = D(K) = X$ where for instance $D(F)$ is the domain of $F$. We remark that $B\varphi$ maps $X$ into itself because $V$ is strictly convex.

By using (11.2.6), (11.2.7) and (11.2.8), equation (11.2.4) becomes

$$N = FN_0 + Bq + BKN. \quad (11.2.9)$$

Note that $FN_0$ takes care of first-flight photons due to the initial photon density $N_0$ (i.e., of photons arriving directly $(\mathbf{x}, \mathbf{u})$ without interacting with the host medium contained in a cloud), whereas $Bq$ is the contribution of first-flight photons from the source $q$.

We have from definitions (11.2.6), (11.2.7) and (11.2.8):

$$\|F\varphi\| \le \|\varphi\|, \qquad \|B\varphi\| \le \|\varphi\| \int_0^{R^*} \exp(-\sigma r)\, dr \le \frac{\|\varphi\|}{\sigma}, \qquad \|K\varphi\| \le \sigma_s \|\varphi\|$$

and so

$$\|F\| \le 1, \qquad \|B\| \le \frac{1}{\sigma}, \qquad \|K\| \le \sigma_s, \qquad \|BK\| \le \frac{\sigma_s}{\sigma} < 1, \quad (11.2.10)$$

where we recall that $\sigma = \sigma_s + \sigma_c$.

Since $\|BK\| < 1$, the unique solution of equation (11.2.9) has the form

$$N(\mathbf{x}, \mathbf{u}, t) = \left((I - BK)^{-1} FN_0\right)(\mathbf{x}, \mathbf{u}, t) + \left((I - BK)^{-1} Bq\right)(\mathbf{x}, \mathbf{u}, t),$$

$$\forall (\mathbf{x}, \mathbf{u}, t) \in V \times S \times [0, \bar{t}],$$

$$(11.2.11)$$

where

$$(I - BK)^{-1} = I + BK + (BK)^2 + \cdots, \qquad \|(I - BK)^{-1}\| \le \frac{\sigma}{\sigma - \sigma_s}. \quad (11.2.12)$$

*Remark 2.* If we assume that $\sigma_s$, $q$ and *also* $\sigma_c$ are zero if $\mathbf{x} \notin V$ (which is reasonable from a physical viewpoint), then we have that

$$0 = \frac{\partial}{\partial r} N\left(\mathbf{x} - r\mathbf{u}, \mathbf{u}, t - \frac{r}{c}\right).$$

Hence,

$$N\left(\widehat{\mathbf{x}}, \widehat{\mathbf{u}}, t + \frac{|\widehat{\mathbf{x}} - \widehat{\mathbf{u}}|}{c}\right) = N(\widehat{\mathbf{z}}, \widehat{\mathbf{u}}, t),$$

because $\widehat{\mathbf{x}} = \widehat{\mathbf{z}} + |\widehat{\mathbf{x}} - \widehat{\mathbf{z}}|\widehat{\mathbf{u}}$; see Figure 11.1.

## 11.3 The first inverse problem: evaluation of $q(\mathbf{x}, t)$

Relation (11.2.11) gives $N(\mathbf{x}, \mathbf{u}, t)$ at any $(\mathbf{x}, \mathbf{u}, t) \in V \times S \times [0, \overline{t}]$ in terms of the initial photon density $N_0$ and of the source term $q$. However, in our first inverse problem, $q(\mathbf{x}, t)$ is unknown (it is "the unknown"), whereas $N_0$ is given and $N(\widehat{\mathbf{z}}, \widehat{\mathbf{u}}, t)$ is measured at $(\widehat{\mathbf{z}}, \widehat{\mathbf{u}})$, $\forall t \in [0, \overline{t}]$ (in fact, $N(\widehat{\mathbf{x}}, \widehat{\mathbf{u}}, t')$ is measured at $(\widehat{\mathbf{x}}, \widehat{\mathbf{u}})$, i.e., "far" from the cloud, $\forall t' \in \left[\frac{|\widehat{\mathbf{x}} - \widehat{\mathbf{z}}|}{c}, \frac{|\widehat{\mathbf{x}} - \widehat{\mathbf{z}}|}{c} + \overline{t}\right]$; see Remark 1.

In order to evaluate $q(\mathbf{x}, t)$, we need to prove some "monotone" properties of the operators appearing on the right-hand side (r.h.s.). of (11.2.11). Let $\varphi$, $\varphi_1 \in X$ and assume that $\varphi(\mathbf{x}, \mathbf{u}, t) < \varphi_1(\mathbf{x}, \mathbf{u}, t)$ $\forall(\mathbf{x}, \mathbf{u}, t) \in V \times S \times [0, \overline{t}]$. Then, we obviously have from (11.2.7) and (11.2.8)

$$\begin{aligned}
&(B\varphi)(\mathbf{x}, \mathbf{u}, t) < (B\varphi_1)(\mathbf{x}, \mathbf{u}, t) \\
&(K\varphi)(\mathbf{x}, \mathbf{u}, t) < (K\varphi_1)(\mathbf{x}, \mathbf{u}, t) \qquad \forall(\mathbf{x}, \mathbf{u}, t) \in V \times S \times [0, \overline{t}]. \quad (11.3.1) \\
&(BK\varphi)(\mathbf{x}, \mathbf{u}, t) < (BK\varphi_1)(\mathbf{x}, \mathbf{u}, t)
\end{aligned}$$

Inequalities (11.3.1) and the first of (11.2.12) imply that

$$\begin{aligned}
&\left((I - BK)^{-1}\varphi\right)(\mathbf{x}, \mathbf{u}, t) \\
&\qquad < \left((I - BK)^{-1}\varphi_1\right)(\mathbf{x}, \mathbf{u}, t) \quad \forall(\mathbf{x}, \mathbf{u}, t) \in V \times S \times [0, \overline{t}]. \quad (11.3.2)
\end{aligned}$$

The physical meaning of (11.3.2) is clear: the photon density $N$, given by (11.2.11), is an "increasing function" of the source term.

Consider now (11.2.11) with $\mathbf{x} = \widehat{\mathbf{z}}$ and $\mathbf{u} = \widehat{\mathbf{u}}$:

$$N(\widehat{\mathbf{z}}, \widehat{\mathbf{u}}, t) = \left((I - BK)^{-1}FN_0\right)(\widehat{\mathbf{z}}, \widehat{\mathbf{u}}, t) + \left((I - BK)^{-1}Bq\right)(\widehat{\mathbf{z}}, \widehat{\mathbf{u}}, t). \quad (11.3.3)$$

If we set

$$\widehat{N}(t) = N(\widehat{\mathbf{z}}, \widehat{\mathbf{x}}, t), \qquad \widehat{M_0}(t) = \left((I - BK)^{-1}FN_0\right)(\widehat{\mathbf{z}}, \widehat{\mathbf{u}}, t)$$

$$\widehat{Q}(q, t) = \left((I - BK)^{-1}Bq\right)(\widehat{\mathbf{z}}, \widehat{\mathbf{u}}, t),$$

then we obtain from (11.3.3)

$$\widehat{N}(t) = \widehat{M_0}(t) + \widehat{Q}(q, t), \qquad t \in [0, \overline{t}], \quad (11.3.4)$$

where we recall that $\widehat{N}(t)$ is measured and $\widehat{M_0}(t)$ is known. Note also that $\widehat{Q}(q, t)$ depends on $q$ linearly and "monotonically"; see (11.3.1) and (11.3.2).

As in [10], we introduce a family of "physically reasonable" sources $\mathcal{P}_\lambda(\mathbf{x}, t)$ defined by

$$\mathcal{F} = \{\mathcal{P}_\lambda : \mathcal{P}_\lambda = (1 - \lambda)\mathcal{P}_{\min} + \lambda\mathcal{P}_{\max}, \ \lambda \in [0, 1]\}, \tag{11.3.5}$$

where $\mathcal{P}_{\min}(\mathbf{x}, t)$ and $\mathcal{P}_{\max}(\mathbf{x}, t)$ are respectively a "minimal" and a "maximal" source.

The sources $\mathcal{P}_{\min}$ and $\mathcal{P}_{\max}$ are chosen accordingly to their physical plausibility and must be such that

$$\mathcal{P}_{\min}, \mathcal{P}_{\max} \in C(V \times [0, \bar{t}]), \tag{11.3.6}$$

$$\mathcal{P}_{\min}(\mathbf{x}, t) < \mathcal{P}_{\max}(\mathbf{x}, t) \qquad \forall(\mathbf{x}, t) \in C(V \times [0, \bar{t}]), \tag{11.3.7}$$

$$\widehat{N}_{\min}(t) = \widehat{M_0}(t) + Q(\mathcal{P}_{\min}, t) \leq \widehat{N}(t)$$
$$\leq \widehat{M_0}(t) + Q(\mathcal{P}_{\max}, t) \leq \widehat{N}_{\max}(t) \qquad \forall t \in [0, \bar{t}]. \tag{11.3.8}$$

*Remark 3.* (11.3.5) is perhaps the simplest way to define a family of possible sources. One could give more elaborate definitions of $\mathcal{F}$, but this would unnecessarily complicate the procedures. On the other hand, since we gave just one mesurement, it is reasonable to consider a one-parameter family of sources.

If $\widehat{N}_\lambda(t)$ is defined by

$$\widehat{N}_\lambda(t) = \widehat{M_0}(t) + \widehat{Q}(\mathcal{P}_{\min}, t), \tag{11.3.9}$$

we have from (11.3.5) and (11.3.8)

$$\widehat{N}_\lambda(t) = (1 - \lambda)\left[\widehat{M_0}(t) + \widehat{Q}(\mathcal{P}_{\min}, t)\right] + \lambda\left[\widehat{M_0}(t) + \widehat{Q}(\mathcal{P}_{\max}, t)\right]$$
$$= (1 - \lambda)\widehat{N}_{\min}(t) + \lambda\widehat{N}_{\max}(t)$$

and so

$$\widehat{N}_{\min}(t) \leq \widehat{N}_\lambda(t) \leq \lambda\widehat{N}_{\max}(t), \qquad \forall t \in [0, \bar{t}], \quad \forall \lambda \in [0, 1]. \tag{11.3.10}$$

Note that (11.3.8) indicates that the measured $\widehat{N}(t)$ is such that

$$\widehat{N}_{\min}(t) \leq \widehat{N}(t) \leq \lambda\widehat{N}_{\max}(t), \qquad \forall t \in [0, \bar{t}]. \tag{11.3.11}$$

However, (11.3.10) and (11.3.11) do *not* imply that a (*time-independent*) $\widehat{\lambda} \in [0.1]$ exists such that (see Figure 11.2)

$$\widehat{N}_{\widehat{\lambda}}(t) = \widehat{N}(t), \qquad \forall t \in [0, \bar{t}],$$

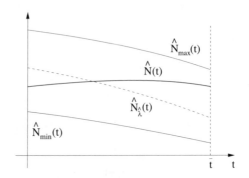

**Fig. 11.2.** Photon densities at $(\widehat{\mathbf{z}}, \widehat{\mathbf{u}})$.

i.e., such that

$$\widehat{N}(t) = \widehat{M_0}(t) + \widehat{Q}(\mathcal{P}_{\min}, t), \tag{11.3.12}$$

because of (11.3.9) with $\lambda = \widehat{\lambda}$.

Since (11.3.12) is not necessarily true, it may happen that no $\mathcal{P}_{\widehat{\lambda}} \in \mathcal{F}$ exists that can be taken as the "best approximation" within $\mathcal{F}$ to the physical source $q(\mathbf{x}, t)$.

However, we can still use the family $\mathcal{F}$ as follows. Let $t_1 \in (0, \overline{t}]$ be given and assume that $\lambda_1$ is defined by

$$\lambda_1 = \frac{\displaystyle\int_0^{t_1} \widehat{N}(t)\, dt - \int_0^{t_1} \widehat{N}_{\min}(t)\, dt}{\displaystyle\int_0^{t_1} \widehat{N}_{\max}(t)\, dt - \int_0^{t_1} \widehat{N}_{\min}(t)\, dt}. \tag{11.3.13}$$

Then, we have

$$(1 - \lambda_1) \int_0^{t_1} \widehat{N}_{\min}(t)\, dt + \lambda_1 \int_0^{t_1} \widehat{N}_{\max}(t)\, dt = \int_0^{t_1} \widehat{N}(t)\, dt$$

and so

$$\int_0^{t_1} \widehat{N}_{\lambda_1}(t)\, dt = \int_0^{t_1} \widehat{N}(t)\, dt, \tag{11.3.14}$$

where

$$\widehat{N}_{\lambda_1}(t)\, dt = (1 - \lambda_1)\widehat{N}_{\min}(t) + \lambda_1 \widehat{N}_{\max}(t) = \widehat{M_0}(t) + \widehat{\varphi}(\mathcal{P}_{\lambda_1}, t), \tag{11.3.15}$$

i.e., $\widehat{N}_{\lambda_1}(t)$ is the photon density produced by the source $\mathcal{P}_{\lambda_1}(\mathbf{x}, t) = (1 - \lambda_1) \mathcal{P}_{\min}(\mathbf{x}, t)(\mathbf{x}, t) + \lambda_1\, \mathcal{P}_{\max}(\mathbf{x}, t)\ \forall \mathbf{x} \in V,\ \forall t \in [0, t_1]$. Relation (11.3.14) shows that the physical source $q$ and the "approximate" source $\mathcal{P}_{\lambda_1}$ produce the same *total number* of photons arriving at $(\widehat{\mathbf{z}}, \widehat{\mathbf{u}})$ *during the time interval* $[0, t_1]$.

In an analogous way, given $[t_{j-1}, t_j] \subset [0, \overline{t}]$, if $\lambda_j$ is defined by

$$\lambda_j = \frac{\displaystyle\int_{t_{j-1}}^{t_j} \widehat{N}(t)\, dt - \int_{t_{j-1}}^{t_j} \widehat{N}_{\min}(t)\, dt}{\displaystyle\int_{t_{j-1}}^{t_j} \widehat{N}_{\max}(t)\, dt - \int_{t_{j-1}}^{t_j} \widehat{N}_{\min}(t)\, dt}, \tag{11.3.16}$$

we have

$$\int_{t_{j-1}}^{t_j} \widehat{N}_{\lambda_1}(t)\, dt = \int_{t_{j-1}}^{t_j} \widehat{N}(t)\, dt, \tag{11.3.17}$$

where

$$\widehat{N}_{\lambda_j}(t)\, dt = \widehat{M}_0(t) + \widehat{\varphi}(\mathcal{P}_{\lambda_j}, t). \tag{11.3.18}$$

Note that, for a fixed $t_{j-1}$, $\lambda_j$ is a continuous function of $t_j$ and

$$\lambda_j \rightarrow \frac{\widehat{N}(t_{j-1}) - \widehat{N}_{\min}(t_{j-1})}{\widehat{N}_{\max}(t_{j-1}) - \widehat{N}_{\min}(t_{j-1})}, \qquad \text{as } t \rightarrow t_{j-1}.$$

We stress that $\mathcal{P}_{\lambda_j}(\mathbf{x}, t) = (1 - \lambda_j)\mathcal{P}_{\min}(\mathbf{x}, t)(\mathbf{x}, t) + \lambda_j \mathcal{P}_{\max}(\mathbf{x}, t)$ may be considered a reasonable approximation of $q(\mathbf{x}, t)\ \forall \mathbf{x} \in V,\ t \in [0, t_j]$ if the quantity of interest is $\int_{t_{j-1}}^{t_j} \widehat{N}(t)\, dt$, i.e., the *total number* of photons arriving at $(\widehat{\mathbf{z}}, \widehat{\mathbf{u}})$ *during the time interval* $[t_{j-1}, t_j]$.

*Remark 4.* The above procedures suggest that we should look for a time-dependent $\lambda = \lambda(t)$ and consider the equation

$$\widehat{N}_{\lambda(t)}(t) = \widehat{M}_0(t) + \widehat{Q}(\mathcal{P}_{\lambda(t)}, t), \tag{11.3.19}$$

see (11.3.9). Correspondingly, if we want $\widehat{N}_{\lambda(t)}(t) = \widehat{N}(t)\ \forall t \in [0, \overline{t}]$, we obtain from (11.3.19)

$$\widehat{Q}(\mathcal{P}_{\lambda(t)}, t) = \widehat{N}_{\lambda(t)}(t) - \widehat{M}_0(t),$$

i.e.,

$$\widehat{Q}((\mathcal{P}_{\max} - \mathcal{P}_{\min})\lambda(t), t) = \widehat{N}_{\lambda(t)}(t) - \widehat{M}_0(t) - \widehat{Q}(\mathcal{P}_{\min}, t), \quad t \in [0, \overline{t}]. \tag{11.3.20}$$

Since (11.3.20) is a linear abstract equation of the *first kind* for the unknown $\lambda(t)$, as such, it is not easy to deal with. However, assume that $t_j = t_{j-1} + \tau$, $j = 1, 2, \ldots, n$, with $t_n = \overline{t}$ and with $\tau \ll \tau_0$ a characteristic time of the interstellar cloud under consideration. Then, we have from (11.3.3), (11.3.4), (11.3.17) and (11.3.18)

$$\tau \widehat{N}_{\lambda_j}(t) \simeq \tau \widehat{N}(t),$$

i.e.,

$$\widehat{Q}(\mathcal{P}_{\lambda_j}, t) \simeq \widehat{Q}(q, t), \qquad t \in [t_{j-1}, t_j]. \tag{11.3.21}$$

Relation (11.3.21) suggests that

$$q(\mathbf{x}, t) \simeq (1 - \lambda_j)\mathcal{P}_{\min}(\mathbf{x}, t) + \lambda_j \mathcal{P}_{\max}(\mathbf{x}, t), \quad \mathbf{x} \in V, \ t \in [t_{j-1}, t_j].$$

## 11.4 The second inverse problem: evaluation of $\sigma(t)$

In this section, we assume that $\sigma = \sigma(t) = \sigma_c(t) + \sigma$, i.e., that the capture cross section in equation (11.2.1) is time dependent (in fact, it is "the unknown"). The knowledge of the behaviour of $\sigma_c(t)$ may give some indications of the (time-dependent) composition of the interstellar cloud under consideration. By a procedure similar to that of Section 11.2, the transport equation with $\sigma = \sigma(t)$ can be written as follows:

$$0 = \frac{\partial}{\partial r} \left\{ \exp\left[c \int_0^{t-r/c} \sigma(s)\, ds\right] N(\mathbf{x} - r\mathbf{u}, \mathbf{u}, t - r/c) \right\} + \exp\left[c \int_0^{t-r/c} \sigma(s)\, ds\right]$$

$$\cdot \left\{ \frac{\sigma}{4\pi} \int_S N(\mathbf{x} - r\mathbf{u}, \mathbf{u}', t - r/c)\, d\mathbf{u}' + q(\mathbf{x} - r\mathbf{u}, t - r/c) \right\}.$$

Hence, we obtain

$$N(\mathbf{x}, \mathbf{u}, t) = \exp\left[-c \int_0^t \sigma(s)\, ds\right] N_0(\mathbf{x} - ct\mathbf{u}, \mathbf{u})$$

$$+ \frac{\sigma}{4\pi} \int_0^{R^*} dr \exp\left[-c \int_{t-r/c}^t \sigma(s)\, ds\right] \int_S N(\mathbf{x} - r\mathbf{u}, \mathbf{u}', t - r/c)\, d\mathbf{u}'$$

$$+ \int_0^{R^*} dr \exp\left[-c \int_{t-r/c}^t \sigma(s)\, ds\right] q(\mathbf{x} - ct\mathbf{u}, t - r/c),$$

$$\forall(\mathbf{x}, \mathbf{u}, t) \in V \times S \times [0, \bar{t}], \tag{11.4.1}$$

where $R^*$ is defined by (11.2.5), $N_0(\mathbf{x} - ct\mathbf{u}, \mathbf{u})$ is still understood to be zero if $ct < R(\mathbf{x}, \mathbf{u})$) and where we assume that $\bar{t} \gg \max\{R(\mathbf{x}, \mathbf{u}), (\mathbf{x}, \mathbf{u}) \in V \times S\}/c$.

In order to write the abstract version of (11.4.1) in the Banach space $X = C(V \times S \times [0, \bar{t}])$, we introduce the following operators (see (11.2.6) and (11.2.7)):

$$(F_\sigma \varphi)(\mathbf{x}, \mathbf{u}, t) = \begin{cases} \exp\left[-c \int_0^t \sigma(s)\, ds\right] \varphi(\mathbf{x} - c\, t\mathbf{u}, \mathbf{u}, t) & \text{if } \mathbf{x} - c\, t\mathbf{u} \in V \\ 0 & \text{if } \mathbf{x} - c\, t\mathbf{u} \notin V \end{cases}$$

$$(11.4.2)$$

$$(B_\sigma \varphi)(\mathbf{x}, \mathbf{u}, t) = \int_0^{R^*} dr \, \exp\left[-c \int_{t-r/c}^t \sigma(s)\, ds\right] \varphi(\mathbf{x} - r\mathbf{u}, \mathbf{u}, t - r/c)$$

$$(11.4.3)$$

with $D(F_\sigma) = D(B_\sigma) = X$.

From definitions (11.4.2) and (11.4.3) we have that

$$\|F_\sigma \varphi\| \le \|\varphi\|, \qquad \|B_\sigma \varphi\| \le \|\varphi\| \int_0^{R^*} \exp(-\sigma_0 r)\, dr \le \frac{\|\varphi\|}{\sigma_0},$$

where $0 < \sigma_0 \le \sigma(t), \ \forall t \in [0, \bar{t}]$. Hence,

$$\|F_\sigma\| \le 1, \qquad \|B_\sigma\| \le \frac{1}{\sigma_0}, \qquad \|B_\sigma K\| \le \frac{\sigma_s}{\sigma_0}, \tag{11.4.4}$$

where $\sigma_s/\sigma_0 < 1$ because $\sigma_0 = \sigma_s + \min\{\sigma_c(t), \ t \in [0, \bar{t}]\}$, with $\min\{\sigma_c(t), \ t \in [0, \bar{t}]\} > 0$.

By using (11.2.8), (11.4.2) and (11.4.3), equation (11.4.1) becomes

$$N = F_\sigma N_0 + B_\sigma q + B_\sigma K N. \tag{11.4.5}$$

Since $\|B_\sigma K\| \le \sigma_s/\sigma_0 < 1$, the unique solution of equation (11.4.5) has the form

$$N(\mathbf{x}, \mathbf{u}, t) = ((I - B_\sigma K)^{-1} F_\sigma N_0)(\mathbf{x}, \mathbf{u}, t)$$

$$+ ((I - B_\sigma)^{-1} B_\sigma q)(\mathbf{x}, \mathbf{u}, t), \qquad (\mathbf{x}, \mathbf{u}, t) \in V \times S \times [0, \bar{t}], \tag{11.4.6}$$

where

$$(I - B_\sigma)^{-1} = I + B_\sigma K + (B_\sigma K)^2 + \cdots. \tag{11.4.7}$$

As already remarked, the unknown in our (second) inverse problem is the function $\sigma(t)$, which is contained in the definition of the operators $F_\sigma$ and $B_\sigma$, whereas $N_0$ and $q$ are known and $N(\widehat{\mathbf{z}}, \widehat{\mathbf{u}}, t) = N(\mathbf{x}, \mathbf{u}, t + |\widehat{\mathbf{z}} - \widehat{\mathbf{x}}|/c)$ is registered by some suitable instrument. Consider then (11.4.6) at $\mathbf{x} = \widehat{\mathbf{z}}$, $\mathbf{u} = \mathbf{u}'$:

$$\widehat{N}_\sigma = \widehat{H}(\sigma)(t), \tag{11.4.8}$$

where

$$\widehat{N}_\sigma = N(\widehat{\mathbf{z}}, \widehat{\mathbf{u}}, t), \quad \widehat{H}(\sigma)(t) = ((I - B_\sigma K)^{-1})(F_\sigma N_0 + B_\sigma q)(\widehat{\mathbf{z}}, \widehat{\mathbf{u}}, t). \quad (11.4.9)$$

We note that (11.4.8) is a *nonlinear* abstract equation of the *first kind* for the unknown $\sigma(t)$. This implies that it is rather difficult to find even an approximate form of $\sigma(t)$, and suggests that we use some suitable time-discretization method.

Let $t_j = j\tau$, $j = 0, 1, \ldots, n$, with $t_n = \bar{t}$ and $\tau \ll \tau_0$; further, assume that $\sigma(t)$ may be approximated by $\widetilde{\sigma}(t) = \sigma_0 + k_1 t \ \forall t \in [0, t_1]$, where $\sigma_0 = \sigma(0)$ is known and $k_1$ must be suitably chosen.

*Remark 5.* That $\sigma(t) \simeq \widetilde{\sigma}(t) = \sigma_0 + k_1 t, \ \forall t \in [0, t_1] = [0, \tau]$ for some $k_1$ is reasonable because $\tau \ll \tau_0 = $ a characteristic time of the interstellar cloud under consideration.

Substitution of $\widetilde{\sigma}(t)$ into the r.h.s. of (11.4.8) with $t = t_1$ gives

$$\widehat{N}_\sigma(t_1) \simeq \widehat{H}(\sigma_0 + k_1 t)(t_1). \quad (11.4.10)$$

Since

$$(B_{\widetilde{\sigma}}\varphi)(\mathbf{x}, \mathbf{u}, t)$$

$$= \int_0^{R^*} dr \, \exp\left\{-\sigma_0 r - \frac{ck_1}{2}\left[t^2 - (t - r/c)^2\right]\right\} \varphi(\mathbf{x} - r\mathbf{u}, \mathbf{u}, t - r/c),$$

it is easy to see that $(B_{\widetilde{\sigma}}\varphi$ is a decreasing function of $k_1 \ \forall t \in [0, t_1]$ for any given $\varphi \in X_+ = $ the positive cone of $X$. This implies that $\widehat{H}(\sigma_0 + k_1 t)(t_1)$ is a decreasing function of $k_1$ and so a unique $\widetilde{k}_1$ exists such that (11.4.10) is satisfied with $k_1 = \widetilde{k}_1$.

Correspondingly, $\widetilde{\sigma}(t) = \sigma_0 + \widetilde{k}_1 t \ \forall t \in [0, t_1]$.

In general, let $t \in [t_{j-1}, t_j] = [t_{j-1}, t_{j-1} + \tau]$ and $\widetilde{\sigma}(t) = \widetilde{\sigma}(t_{j-1}) + k_j(t - t_{j-1})$. Since

$$(B_{\widetilde{\sigma}}\varphi)(\mathbf{x}, \mathbf{u}, t)$$

$$= \int_0^{R^*} dr \, \exp\left\{\int_{t-r/c}^t \left[\widetilde{\sigma}(t_{j-1}) + k_j(s - t_{j-1})\right] ds\right\} \varphi(\mathbf{x} - r\mathbf{u}, \mathbf{u}, t - r/c)$$

$$= \int_0^{R^*} \exp\left\{-\widetilde{\sigma}(t_{j-1})r - \frac{ck_j}{2}\left[(t - t_{j-1})^2 - (t = r/c - t_{j-1})^2\right]\right\}$$

$$\cdot \varphi(\mathbf{x} - r\mathbf{u}, \mathbf{u}, t - r/c)$$

if $t - r/c > t_{j-1}$, and

$(B_{\widetilde{\sigma}}\varphi)(\mathbf{x}, \mathbf{u}, t)$

$$= \int_0^{R^*} dr \, \exp\left\{ -c\widetilde{\sigma}(t_{j-1})(t_{j-1} - t + r/c) - \frac{ck_1}{2}(t - t_j)^2 \right\}$$

$$\cdot \varphi(\mathbf{x} - r\mathbf{u}, \mathbf{u}, t - r/c)$$

if $t - r/c < t_{j-1}$, $B_{\widetilde{\sigma}}\varphi$ is a decreasing function of $k_j \, \forall t \in [t_{j-1}, t_j]$, for any given $\varphi \in X_+$. This implies that $\widehat{H}(\widetilde{\sigma}(t_{j-1}) + \widetilde{k}_j(t - t_{j-1}))(t_j)$ is a decreasing function of $k_j$ and so a unique $\widetilde{k}_j$ exists such that

$$\widehat{N}_\sigma(t_j) = \widehat{H}(\widetilde{\sigma}(t_{j-1}) + \widetilde{k}_j(t - t_{j-1}))(t_j).$$

Correspondingly, we have

$$\widetilde{\sigma}(t) = \widetilde{\sigma}(t_{j-1}) + \widetilde{k}_j(t - t_{j-1})$$

$$= \widetilde{\sigma}(t_{j-2}) + \widetilde{k}_{j-1}(t_{j-1} - t_{j-2}) + \widetilde{k}_j(t - t_{j-1})$$

$$= \sigma_0 + \widetilde{k}_1 t_1 + \widetilde{k}_2(t_2 - t_1) + \cdots + \widetilde{k}_{j-1}(t - t_j), \quad t \in [t_{j-1}, t_j].$$
$$(11.4.11)$$

Relation (11.4.11) is a reasonable approximation of the physical cross section $\sigma(t)$. We conclude this section by considering briefly an alternate time-discretization procedure, starting directly from the evolution equation

$$\frac{d}{dt}N(t) = TN(t) - c\sigma(t)N(t) + cKN(t) + q(t), \qquad t > 0; \; N(0) = N_0,$$
$$(11.4.12)$$

where

$$(T_\gamma)(\mathbf{x}, \mathbf{u}) = -c\mathbf{u} \cdot \nabla \gamma(\mathbf{x}, \mathbf{u}),$$

$$D(T) = \{\gamma \colon \gamma \in X_0, \, c\mathbf{u} \cdot \nabla \gamma \in X_0\}, \quad X_0 = C(V \times S)$$

and where $N(t) = N(\cdot, \cdot, t)$ and $q(t) = q(\cdot, t)$ are to be considered as maps from $[0, t]$ into $X_0$.

If $u_j(\mathbf{x}, \mathbf{u})$ approximates $N(\mathbf{x}, \mathbf{u}, t_j)$, we discretize (11.4.12) as follows:

$$\frac{1}{\tau}[u_{j+1} - u_j] = Tu_{j+1} - c\sigma(t_j)u_j + cKu_j + q(t_j), \quad j = 0, 1, \ldots, n$$

with $u_0 = N_0$ [10, 7]. Hence, we obtain

$$u_{j+1}(\mathbf{x}, \mathbf{u}) = ((I - \tau T)^{-1}u_j)(\mathbf{x}, \mathbf{u}) - c\tau\sigma(t_j)((I - \tau T)^{-1}u_j)(\mathbf{x}, \mathbf{u})$$

$$+ (\tau(I - \tau T)^{-1}[cKu_j + q(t_j)])(\mathbf{x}, \mathbf{u}), \qquad j = 0, 1, \ldots, n,$$
$$(11.4.13)$$

where the explicit expression of the operator $(I - \tau T)^{-1}$ is given by

$$((I - \tau T)^{-1}\gamma)(\mathbf{x}, \mathbf{u}) = \frac{1}{c\tau} \int_0^{R(\mathbf{x}, \mathbf{u})} \exp(-r/c\tau)\,\gamma(\mathbf{x} - r\mathbf{u}, \mathbf{u})\, dr.$$

Consider now (11.4.13) with $j = 0$, $\mathbf{x} = \hat{\mathbf{z}}$ and $\mathbf{u} = \hat{\mathbf{u}}$:

$$u_{j+1}(\hat{\mathbf{z}}, \hat{\mathbf{u}}) = ((I - \tau T)^{-1}u_j)(\hat{\mathbf{z}}, \hat{\mathbf{u}}) - c\tau\sigma(t_j)((I - \tau T)^{-1}u_j)(\hat{\mathbf{z}}, \hat{\mathbf{u}})$$

$$+ (\tau(I - \tau T)^{-1}[cKu_j + q(t_j)])(\hat{\mathbf{z}}, \hat{\mathbf{u}}), \quad j = 0, 1, \ldots, n.$$
$$(11.4.14)$$

Since $u_0 = N_0$ and $u_1(\hat{\mathbf{z}}, \hat{\mathbf{u}})$ is measured, (11.4.14) gives $\sigma(t_0) = \sigma(0)$ explicitly. Further, by using (11.4.13) with $j = 0$ and with the value $\sigma(t_0)$ just found, we obtain $u_1(\mathbf{x}, \mathbf{u})$ $\forall (\mathbf{x}, \mathbf{u}) \in V \times S$.

Then, (11.4.13) with $j = 1$, $\mathbf{x} = \hat{\mathbf{z}}$ gives $\sigma(t_1)$ because $u_1(\mathbf{x}, \mathbf{u})$ is known and $u_2(\hat{\mathbf{z}}, \hat{\mathbf{u}})$ is measured. Again (11.4.13) with $j = 1$ and with the value $\sigma(t_1)$ just found also gives $u_2(\mathbf{x}, \mathbf{u})$, $\forall (\mathbf{x}, \mathbf{u}) \in V \times S$. And so on.

By this procedure, we obtain the set $(\sigma(t_0)), \sigma(t_1), \ldots, \sigma(t_n))$; correspondingly, the approximate expression $\tilde{\sigma}(t)$ of the total cross section $\sigma(t)$ may be taken as follows:

$$\tilde{\sigma}(t) = \sigma(t_{j-1}) + \frac{\sigma(t_j) - \sigma(t_{j-1})}{t_j - t_{j-1}}, \quad t \in [t_{j-1}, t_j], \quad j = 1, 2, \ldots, n.$$
$$(11.4.15)$$

Note that (11.4.11) and (11.4.15) are rather similar. However, (11.4.15) is easier to derive because the values $\sigma(t_0), \sigma(t_1), \ldots, \sigma(t_n)$ are found explicitly step by step.

## 11.5 Concluding remarks

As remarked in Section 11.1, the peculiarity of inverse problems for photon transport in interstellar clouds is that one has to start with the knowledge of a single value of the photon density (in the time-independent case) or of a time behaviour of such a density (in the time-dependent case). This implies that only *one* parameter $\lambda$, constant or time-dependent, can be evaluated. If, for instance, the unknown is the photon source $q(\mathbf{x})$, we have first to choose some one-parameter family $\mathcal{F} = \{\mathcal{P}_\lambda : \mathcal{P}_\lambda(\mathbf{x}) = (1 - \lambda)\mathcal{P}_{\min}(\mathbf{x}) + \lambda\mathcal{P}_{\max}(\mathbf{x}), \lambda \in [0, 1]\}$. Correspondingly, it is possible to identify a value $\hat{\lambda}$ of the parameter such that $\mathcal{P}_{\hat{\lambda}}(\mathbf{x})$ produces the known value of the photon density. Of course, this does not mean that $q(\mathbf{x}) = \mathcal{P}_{\hat{\lambda}}(\mathbf{x})$ but only that $\mathcal{P}_{\hat{\lambda}}(\mathbf{x})$ is a reasonable approximation of $q(\mathbf{x})$, compatible with the single datum available.

The situation is "worse" in the time-dependent case, i.e., if $q = q(\mathbf{x}, t)$. In fact, only a suitable constant $\hat{\lambda}$ can be found such that $\mathcal{P}_{\hat{\lambda}} \in \mathcal{F} =$

$\{\mathcal{P}_\lambda : \mathcal{P}_\lambda(\mathbf{x}, t) = (1 - \lambda)\mathcal{P}_{\min}(\mathbf{x}, t) + \lambda\mathcal{P}_{\max}(\mathbf{x}, t), \ \lambda \in [0, 1]\}$ produces the same *total* number of photons arriving at $\widehat{\mathbf{x}}$ (where the measurements are made) as the physical source $q$ during a given time interval. On the other hand, looking for a $\widehat{\lambda} = \widehat{\lambda}(t) \in [0, 1]$ such that $\mathcal{P}_{\widehat{\lambda}(t)}$ approximates $q$ (in the sense specified above) leads to an abstract equation of the first kind, which is not easy to deal with.

# References

[1]   F. Mugelli and A. Belleni-Morante. Identification of the boundary surface of an interstellar cloud from a measurement of the photon far field. *Math. Meth. Appl. Sci.* (27):627–642, 2004.

[2]   A.I. Prilepko and N.P. Volkov. Inverse problem of finding parameters of a nonstationary transport equation from integral overdeterminations. *Differ. Equations*, (23):91–101, 1987.

[3]   G.M. Sydykov and A.D. Sariev. On inverse problems for a time-dependent transport equation in plane-parallel geometry. *Differ. Uravn.*, (27):1617–1625, 1991.

[4]   A.I. Prilepko and I.V. Tikhonov. Reconstruction of the inhomogeneous term in an abstract evolution equation. *Russian Acad. Sci. Izv. Math.*, (44):373–394, 1995.

[5]   J. Ying, S. He, S. Ström, and W. Sun. A two-dimensional inverse problem for the time-dependent transport equation in a stratified half-space. *Math. Engrg. Indust.*, (5):337–347, 1996.

[6]   A.I. Prilepko, D.G. Orlovsky, and I.A. Vasin. *Methods for Solving Inverse Problems in Mathematical Physics.* Marcel Dekker, New York, 2000.

[7]   A. Belleni-Morante. A time-dependent inverse problem in photon transport, *Proceedings of the Internat. Meeting "New Trends in Mathematical Physics,"* 1–11, World Scientific, Singapore, 2004.

[8]   S. Pieraccini, R. Riganti, and A. Belleni-Morante. Numerical treatment of a time-dependent inverse problem in photon transport, *Boll. U.M.I.* (8)*8-B*: 773–779, 2005.

[9]   A. Belleni-Morante. An inverse problem for photon transport in interstellar clouds. *Transp. Theory and Statistic. Phys.*, 32:73–91, 2003.

[10]  F. Mugelli and A. Belleni-Morante. Identification of a time-dependent source in an interstellar cloud. *Ann. Univ. Ferrara, Sez. VII, Sc. Mat.*, in print.

[11]  J.E. Dyson and D.A. Williams. *The physics of interstellar medium.* Inst. of Phys. Publ., Bristol, 1997.

[12]  M.A. Dopite and R.S. Sutherland, *Astrophysics of the Diffuse Universe,* Springer, Berlin, 2003.

[13]  G.C. Pomraning, *Radiation Hydrodynamics,* Pergamon Press, Oxford, 2003.